This book is
the property of:
P.R. Thackray

HYDRAULIC MACHINERY AND CAVITATION

VOLUME I

HYDRAULIC MACHINERY AND CAVITATION

Proceedings of the XVIII IAHR Symposium on
Hydraulic Machinery and Cavitation

VOLUME I

Edited by

E. CABRERA, V. ESPERT and F. MARTÍNEZ
Polytechnic University of Valencia, Spain

Polytechnic University of Valencia
Hydraulic and Environmental Eng. Dept.
FLUID MECHANICS GROUP

Co-sponsors:

Centro de estudios y experimentación
de obras públicas

IBERDROLA

RED ELECTRICA

KLUWER ACADEMIC PUBLISHERS
DORDRECHT / BOSTON / LONDON

A C.I.P. Catalogue record for this book is available from the Library of Congress.

ISBN 0-7923-4208-9 (Volume I)
ISBN 0-7923-4209-7 (Volume II)
ISBN 0-7923-4210-0 (Set of 2 Volumes)

Published by Kluwer Academic Publishers,
P.O. Box 17, 3300 AA Dordrecht, The Netherlands.

Kluwer Academic Publishers incorporates
the publishing programmes of
D. Reidel, Martinus Nijhoff, Dr W. Junk and MTP Press.

Sold and distributed in the U.S.A. and Canada
by Kluwer Academic Publishers,
101 Philip Drive, Norwell, MA 02061, U.S.A.

In all other countries, sold and distributed
by Kluwer Academic Publishers Group,
P.O. Box 322, 3300 AH Dordrecht, The Netherlands.

Printed on acid-free paper

Printed in the Netherlands

TABLE OF CONTENTS

Acknowledgements   xvii
Preface   xix
Foreword   xxi

VOLUME I

INVITED LECTURES

The hydroelectricity in the world-Present and future.
    Brekke, H.   3

Analysis of transients caused by hydraulic machinery.
    Chaudhry, M.H.   17

Some present trends in hydraulic machinery research.
    Fanelli, M.   23

Rapid prototyping of hydraulic machinery.
    Schilling R., Riedel, N., Bader, R., Ascherbrenner, T.,
    Weber, Ch., Fernandez, A.   40

Fluid transients in flexible piping systems.
A perspective on recent developments
    Wiggert, D.C.   58

SYMPOSIUM CONTRIBUTIONS

GROUP 1   **HYDRAULIC TURBINES. ANALYSIS AND DESIGN**

1. 1 NUMERICAL ANALYSIS OF COMPONENTS

A decision aid system for hydraulic power station refurbishment
procedure.
    Bellet, L., Parkinson, E. Avellan, F., Cousot, T., Laperrousaz, E. 71

A three dimensional spiral casing Navier-Stokes flow simulation.
    Bruttin, C.H., Kueny, J.L., Boyer, B., Héon, K,. Vu, T.C.
    and Parkinson, E.   81

Tip clearance flow in turbomachines- experimental flow analysis.
    Ciocan, G.D. and Kueny, J.L.   91

Numerical prediction of hydraulic losses in the spiral casing of a
Francis turbine.
    Drtina, P. and Sebestyen, A.   101

Improvements of a graphical method for calculation of flow on a Pelton
bucket.
    Hana, M.   111
Flow behavior and performance of draft tubes for bulb turbines.
    Kanemoto, T.,Uno, M., Nemoto, M., and Kashiwabara, T.   120

Performance analysis of draft tube for GAMM francis turbine.
    Kubota, T., Han, F., and Avellan, F.   130

Modelling complex draft-tube flows using near-wall turbulence closures.
    Ventikos, Y., Sotiropoulos, F. and Patel, V.C.   140

1.2 NUMERICAL ANALYSIS AND DESIGN OF HYDRAULIC MACHINERY

Fluid flow interactions in hydraulic machinery.
    Aschenbrenner, T., Riedel, N. and Schilling, R.   150

Numerical optimization of high head pump-turbines
    Buchmaicher, H,. Quaschnowitz, B., Moser, W. and Klemm, D.   160

Numerical hill chart prediction by means of CFD stage simulation
for a complete Francis turbine.
    Keck, H. Drtina, P. and Sick, M.   170

Development of a new generation of high head pump-turbines,
Guangzhou II.
    Klemm, D., Jaeger, E.U., and Hauff, C.   180

Development of integrated CAE tools for design assessment and
analysis of hydraulic turbines.
    Massé, B,. Pastorel, H. and Magnan, R.   190

Design and analysis of a two stage pump turbine
    *Mazzouji, F. , Francois, M. Hebrard, F., Houdeline, B. and
Bazin, D.*   200

Study of high speed and high head reversible pump-turbine
    *Nakamura, T., Nishizawa, H., Yasuda, M., Suzuki, T. and
Tanaka, H.*   210

Analysis of the performance of a bulb turbine using 3-D viscous
numerical techniques
    Qian, Y., Suzuki, R. and Arakawa, C.   220

Analysis of the inlet reverse flows in a pump turbine using 3-D
viscous numerical techniques
    Qian, Y., Suzuki, R. and Arakawa, C.   230

Importance of interaction between turbine components in flow field
simulation
    Riedelbauch, S,. Klemm, D. and Hauff, C.   238

From components to complete turbine numerical simulation
    Sabourin, M., Labrecque, Y. and de Henau, V.   248

Validation of a stage calculation in a Francis turbine
    Sick M., Casey, M.V. and Galpin, P.F.   257

Simulation of flow through Francis turbine by LES method
    *Song, C.C.S., Xiangying, Ch., Ikohagi,T., Sato, J., Shinmei, K.
and Tani, K.*   267

Simulation of flow through pump-turbine
    *Song, C.C.S., Changsi, C., Ikohagi,T., Sato, J., Shinmei, K. and
Tani, K.*   277

1.3   LOSS ANALYSIS AND SCALE EFFECTS

The scale effect in Kaplan turbines. New relationships for the
calculation of scalable and non-scalable hydraulic losses and of the
coefficient V.
    Anton, I.M.   284

Analysis of losses in hydraulic turbines.
    Brekke, H.   294

Scaling-up head discharge characteristics from model to prototype.
    Couston, M. and Philibert, R.   304

Recent development of studies on scale effect
    Ida, T., Kubota,T. , Kurokawa, J. and Tanaka, H.   313

Prediction of scalable loss in Francis runners of different specific speed
    Kitahora, T., Jurokawa, J., Matumoto, M. and Suzuki, R.   323

Scale effect of jet interference in multinozzle Pelton turbines
    Nakanishi, Y. and Kubota, T.   333

Further development of step-up formula considering surface roughness
    Nichtawitz, A.   342

Numerical simulations of jet in a Pelton turbine.
    Nonoshita, T. , Matsumoto, Y., Kubota,T. and Ohashi, H.   352

An assessment of the loss distribution in Francis turbines
    Suzuki, R., Qian, Y., Kitahora, T. and Kurokawa, J.   361

GROUP 2.   HYDRAULIC PUMPS

Unsteady flow calculation in a centrifugal pump using a finite element method.
    Bert, P.F., Combes, J.F. and Kueny, J.L.   371

Steady and unsteady flow pattern between stay and guide vanes in a pump-turbine
    Ciocan, G., Kueny, J. L. and Mesquita, A.L.A.   381

Self-sustained oscillation of gas-liquid flow in a centrifugal pump with semi-open impeller
    Kurokawa, J., Matsui, J., Takada, H. and Hirayama, T.   391

Measurements in the dynamic pressure field of the volute of a centrifugal pump
    Parrondo, J.L., Fernández, J., Santolaria, C. and González, J.   401

Functional modelling of pump volute geometry
    Thackray, P.R. and James, R.D.   411

Analysis of flow measurements in the impeller and vaned diffuser of a centrifugal pump operating at part load.
    Toussaint, M. and Hureau, F.   419

Influence of the blade roughness on the hydraulic performance of a mixed-flow pump. A viscous analysis
    Undreiner, S. and Dueymes, E.   428

Improvement of performance of centrifugal pumps based on computational and theoretical methods and experimental design
    Vinokurov, A.F., Volkov, A.V., Morgunov, G.M. and Pankratov, S.N.   438

Liquid-particulate two-phase flow in centrifugal impeller by turbulent
simulation
     Wu, Y., Dai, J., Mei, Z., Oba, S. and Ikoagi, T.   445

GROUP 3.   **HYDRAULIC ELEMENTS. DYNAMIC CHARACTERIZATION
AND HYDRAULIC BEHAVIOUR.**

Application of the method of kinetic balance for flow passages forming.
     Benisek, M., Cantrak, S., Ignatovic, B. and Pokrajac, D.   455

Dynamics of large hydrogenerators.
     Brito, G.C., Weber, H.I. and Fuerst, A.G.A.   464

Instabilities in a flow-control valve.
     Cigada, A., Guadagnini, A. and Orsi, E.   474

Study of stayvane vibration by hydroelastic model.
     Deniau, J.L.   484

Study of dynamic behaviour of non-return valves.
     François, P.   494

Optimum hydraulic design of two-way inlet conduit of Wangyuhe
pumping station
     Linguang, L., Jiren, Z. and Rentian, Z.   504

Flow analysis for the intake of low-head hydro power plants
     Ruprecht, A., Maihofer, M. and Gode, E.   514

GROUP 4.   **CAVITATION AND SAND EROSION**

Efficiency alteration of Francis turbines by travelling bubble cavitation.
     Arn, Ch., Dupont, Ph. and Avellan, F.   524

Cavitation erosion prediction on Francis turbines - Part 1. Measurements
on the prototype.
     *Bourdon, P., Pfarhat, M., Simoneau, R., Pereira, F.,
Dupont, P., Avellan, F. and Dorey, J.M.*   534

Determination of critical cavitation limit in the pressure control devices.
     Castorani, A., De Martino, G. and Fratino, U.   544

Stability of air cavities in tip vortices.
Crespo, A., Castro, F., Manuel, F. and Fruman, D.H.   554

Cavitation erosion prediction on Francis turbines - Part 3 Methodologies
of prediction.
Dorey, J.M., Laperrousaz, E., Avellan, F., Dupont, P.,
Simoneau, R. and Bourdon, P.   564

Cavitation erosion prediction on Francis turbines - Part 2. Model test and
flow analysis.
Dupont, P.H., Caron, J.F., Avellan, F., Bourdon, P.,
Lavigne, M., Farhat, M., Simoneau, R., Dorey, J.M.,
Archer, A., Laperrousaz, E. and Couston, M.   574

Impact of vapour production and cavity dynamics on the estimation of
thermal effects in cavitation.
Fruman, D.H., Reboud, J.L. and Stutz, B.   584

Aireation versus cavitation in dam spillways: self-aeration and artificial
aeration (aerators).
Gutiérrez , R.   594

Leading edge cavitation in a centrifugal pump: Numerical predictions
compared with model tests
Hirschi, R., Dupont, P.H., Avellan, F., Favre, J.N.,
Guelich, F., and Handloser, W.   604

The relation between erosion ripples on the wetted surface of hydraulic
turbine and instability waves in the turbulent boundary layer
Huang, S. and Cheng, L.   614

Acoustic method and its applications on measuring and judging
cavitation of hydraulic turbine.
Kehuang, L. and Chun, Y.   622

Numerical simulation for dilute sandy water flow in plane cascade
Liu, X.B., Zeng, Q.C. and Zhang, L.D.   632

Review of research on abrasion and cavitation of silt-laden flows
through hydraulic turbines in China
Mei, Z. and Wu, Y.   641

Effect of the leading edge design on sheet cavitation around a blade
section
Reboud, J.L., Rebattet, C. and Morel, P.   651

VOLUME II

GROUP 5.   HYDRAULIC TRANSIENTS AND CONTROL SYSTEMS
RELATED WITH HYDRAULIC MACHINERY AND PLANTS

Optimal closure of a valve for minimizing waterhammer.
Abreu, J., Cabrera, E., García-Serra, J. and López, P.A.   661

Qualitative flow visualizations during fast start-up of centrifugal pumps.
Barrand, J.P. and Picavet, A.   671

Analysis of a numerical model for the oscillatory properties of a Francis
turbine group.
Cattanei, A., Capozza, A., and Molinaro, P.   681

Transients analysis and dynamic criteria for HPP exploatation
Gajic, A., Pejovic, S., Krsmanovic, L.J. and Stojanovic, Z.   691

Modelling a protection device in a low pressure lifting system.
Giustolisi, O. and Mastrorilli, M.   701

Dynamic compression of entrapped air pockets by elastic water
columns.
Guarga, R., Acosta, A. and Lorenzo, E.   710

Generalization of pump station boundary condition in hydraulic
transient simulation.
Izquierdo, J., Iglesias, P., Espert, V. and Fuertes, V.   720

Analysis of unsteady characteristics of flows through a
centrifugal-pump impeller by an advanced vortex method.
Kamemoto, K., Kurasawa, H., Matsumoto, H. and Yokoi, Y.   729

Expert system for analysis of pumped storage schemes.
Koelle, E., Andrade, J.G.P. and Luvizotto Jr., E.   739

Model-based analysis of active PID-control of transient flow in
hydraulic networks.
Lauria, J.C. and Koelle, E.   749

Simulation of transients in pressurized hydraulic systems with visual
tools.
Martínez, F., Izquierdo, J., Pérez, R. and Vela, A.   759

xii

Dynamic behaviour of governing turbines sharing the same electrical grid
  Nielsen, T.K.   769

Prediction of natural frequencies in a hydro power plant supplying an electric network by itself having a known load type
  Raabe, J.   779

Modelling and practical analysis of the transient overspeed effect of small Francis turbines.
  Ramos, H. and Betamio, A.   789

Simulation of turbine governing in time domain.
  Stuksrud, D.B.   799

Parametrical modelling of power characteristics of the Francis and Kaplan hydraulic turbines.
  Tolea, M.F. and Kueny, J.L.   809

Unsteady frictions in pipelines.
  Vennatro, R.   819

GROUP 6.   OSCILLATORY   AND   VIBRATION   PROBLEMS   IN HYDRAULIC MACHINERY AND POWER STATIONS

Swirl flow in conical diffusers.
  Dahlhaug, O.G.   827

Experimental investigation of vortex core in reverse swirl flow from Francis runner.
  Furuie, Y., Mita, H. and Hosoi, Y.   835

Hydraulic oscillation analysis using the fluid-structure interaction model.
  Gajic, A., Pejovic, S. and Stojanovic, Z.   845

Francis turbine surge: discussion and data base.
  Jacob, T. and Prenat, J.E.   855

Non-stationary flow in reversible Francis turbine runner due to wakes trailing the guide vanes.
  Jernsletten, J.   865

The swirling inlet flow effects on the pressure recovery of a low head water turbine draft tube.
    Kikuyama, K., Hasegawa, Y., Augusto, G., Nishibori, K. and Nakamura, S. 875

Two kinds of whirl on fixed-blade propeller type turbine.
    Léonard, F. 885

Self-excited hydraulic oscillations dued of unstable valve behaviour. A case study.
    Mateos, C., Pérez-Andújar, T., Andreu, M. and Cabrera, E. 895

An experimental study on fins. Their role in control of the draft tube surging
    Nishi, M., Wang, X.M., Yoshida, K., Takahashi, T. and Tsukamoto, T. 905

Model for vortex rope dynamics in Francis turbine outlet.
    Pedrizzetti, G. and Angelico, G. 915

Vortices rotating in the vaneless space of a Kaplan turbine operating under off-cam high swirl flow conditions.
    Pulpitel, L., Skotak, A. and Kontnik, J. 925

Experimental investigation of frequency characteristics of draft tube pressure pulsations for Francis turbines.
    Qinghua, S. 935

On the suppression of coupled liquid/pipe vibrations.
    Tijsseling, A.S. and Vardy, A.E. 945

Unsteady hydraulic force on an impeller due to rotor-stator interaction in a diffuser pump.
    Tsukamoto, H., Uno, M., Qian, W., Teshima, T., Sakamoto, K., and Okamura, T. 955

Swirling flow with helical vortex core in a draft tube predicted by a vortex method.
    Wang, X.M. and Nishi, M. 965

GROUP 7.   **EXPERIMENTAL INVESTIGATIONS RELATED WITH HYDRAULIC MACHINARY AND ITS APPLICATIONS**

Friction loss of rough wall passage in a turbomachinery
    Akaike, S. 975

Redesign of sharp heel draft tube- Results from tests in model and prototype.
Dahlbäck, N.   985

Model and prototype draft tube pressure pulsations
Kerkan V., Bajid, M. Djelic, V. Lipej, A. and Jost, D.   994

Fish bypass system impact upon turbine runner performance at Rocky Reach dam.
Lang, A. and Christman, B.   1004

LDV measurements in an impeller-generated turbulent jet developing in a new coflow.
Peterson, P., Larson, M and Jönsson, L.   1014

Inline radial force measurement of turbine runners
Riener, J., Egger, A. and Schnur, G.   1024

Different types and locations of part-load recirculations in centrifugal pumps found from LDV measurements.
Stoffel, B. and Weiss, K.   1034

Turbulent 3D flows near the impeller of a mixed-flow pump
Wang, B. and Hellman, D.H.   1044

GROUP 8   PRACTICAL APPLICATIONS OF THE HYDRAULIC MACHINERY

Overload in Kaplan turbines of Salto Grande hydropower complex.
Baccino, M. and Bonecarrere, E.   1053

Full sized tests on a french main coolant pump under two-phase flow.
Huchard, J.C., Bore, C. and Dueymes, E.   1063

Computer simulations of dynamic performance of 400 Mw adjustable speed pumped storage units.
Kita, E., Nakagawa, H., Kuwabara, T. and Harada, M.   1073

Performance comparison of nuclear reactor recirculation pumps tested under large reynolds number difference
Saiki, K., Ikura, T., Matsumoto, K., Komita, H.,
Kobayashi, M., Saito, T. and Tanaka, H.   1083

Performance of Candu heat transport pumps under two-phase flow conditions.
Samarasekera, H. and Kumar, A.N.   1093

Interdependence of draft tube and tailwater flow in bulb turbine power plants.
Schneider, C.H., Knapp, W. and Schilling, R.   1103

Study of hydraulic transients using the bond graphs method.
Tiago Filho, G.L.   1113

GROUP 9.   **MONITORING, PREDICTIVE MAINTENANCE AND REFURBISHMENT.**

Numerical flow analysis of a Kaplan turbine.
Jost, D., Lipej, A., Oberdank, K., Jamnik, M. and Velensek, B.   1123

18-Paths acoustic flowmeter and transducer protrusion.
Lévesque, J.-M., Néron, J. and Tran, C.M.   1133

Step-up in rehabilitation: not a myth, a science
Mahé, B., de Henau, V. and Sabourin, M.   1142

Super system: a hydroelectric unit condition monitoring system in operation at Hydro-Quebec.
Mossoba, Y.   1152

Application of rotor response analysis to fault detection in hydro powerplants.
Nascimento, L.P. and Egusquiza, E.   1162

The placement exploration of diagnostic measuring points on operating unit equipment.
Liu, X.T.   1172

Later the horizontal Kaplan bulb turbines with conical arrangement of the guide vanes were developed. This type of turbines are often located in the dam for heads up to approximately 15 m. The reason for this is that a Bulb turbine requires less space and less depth than a vertical Kaplan turbine.

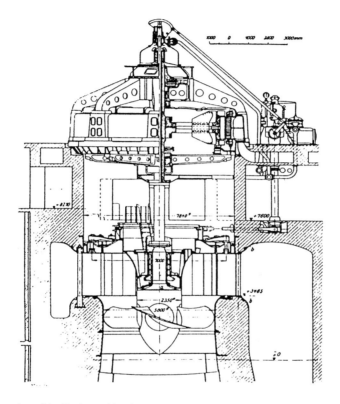

Figure 4. Cross section of the Kaplan turbine for Lilla Edet power station in Sweden.

2.2. IMPULSE TURBINES

The dominating impulse turbine used world wide is the Pelton turbine. This turbine type was patented by Lester Allen Pelton in 1880 in USA. The development of the bucket shape is illustrated in fig. 5.

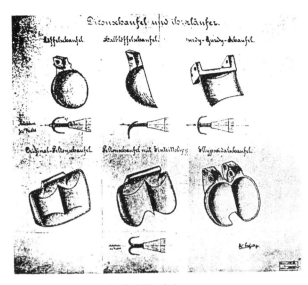

Figure 5. Schematic drawings of the Pelton bucket (Raabe).

The development of this turbine type included in the first time single and twin jet horizontal shaft turbines. Later the vertical types were developed with a number of jets up to 6 and output up to 315 MW. Within 2 years turbines with 400 MW for 1882 m net head will be in operation in Switzerland.

2.3.   REVERSIBLE PUMP TURBINES

The last turbine type developed after world war II is the reversible pump turbines. This type of turbine was developed in the fifties and has been designed for increasing head especially in the industrialized part of the world for peak load operation. A diagram of the development of commissioned and planned reversible pump turbine projects up to 1985 is shown in fig. 6. This diagram was presented during the IAHR Symposium in Tokyo 1980, by Dr. Hitoshi Muray [Ref. 2].

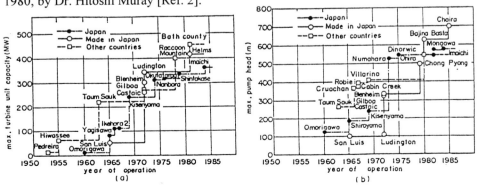

Figure 6. Trends in (a) unit capacity and (b) total head of pump turbines.

2.4.   STATE OF THE ART IN DEVELOPMENT

As the present state of the art of Francis turbines, Itaipu in Brazil/Uruguay with 740 MW for 118.4 m, could be mentioned as the largest Francis turbines. For Pelton turbines Sy Sima in Norway with 315 MW for 885 m head is the largest. For Kaplan turbines the turbine for Ligga III in Sweden with 182 MW for 39 m head could be mentioned as one of the largest turbines of this type.

3.   Development of hydraulic machines for electricity production in the future

In general all turbine types will be developed for increased output and head for a given specific speed. An increase in the efficiency by 0.5-1.0% may be expected. Simplified design of small hydro turbines for utilization of non-regulated small rivers will also be a challenge in order to give a contribution to the production of clean renewable energy for electricity production.

In the following a brief description of the design challenges for future turbines will be given. The initial data for a turbine design will be the flow (Q) and the range of operating head (H). From this a decision will be made, based on experience from the manufacturer and consultants. In the future, border lines will be broken concerning specific speed versus head and range of operation.

In order to do so it is important to establish an analytical approach to the problem besides the available CFD tool which is improving very fast. I believe that it is important to be able to make an initial design for a complete new runner by means of an analytical study of the different geometry parameters. This is because the requirements for new turbines may be outside the available data base for runner design based on the experience of the turbine manufacturer.

Such initial analytical parameter design must of course be analysed by modern CFD codes and tested in laboratory for the fine tuning of blades.

The main problem at present time is that young engineers may have been "too computerized" and have too little geometry feeling and decreased physical thinking. The engineer's education does not include creative drawing board training. This training in three-dimensional shaping without a computerized geometri shaping aid will always be valuable for a turbine runner designer.

In order to clarify and explain the meaning of an analytical preliminary design a brief description is given for the use of some parameters.

3.1. THE INFLUENCE FROM THE SPEED NUMBER $*\Omega$ ON CAVITATION

The speed number is defined as $*\Omega = *\omega \ *Q^{1/2}/(2g \ *H)^{3/4}$ based on the best efficiency point (BEP, denoted *). The relation between the defined speed number and the specific speed is:

$$n_s = {}^*\Omega \cdot K^{1/2}(2g)^{3/4}30/\pi, \qquad \text{where } K = Q_n/{}^*Q.$$

The speed number $^*\Omega$ is a dimensionless parameter which is more convenient to be used instead of the specific speed in this lecturer. The speed number may also be expressed as a function of the circumferential speed of the blade outlet and the blade angle:

$$^*\Omega = \sqrt{\pi} \ \underline{u}_2^{3/2} \sqrt{\tan(\pi - \beta_2)} \tag{1}$$

where $\underline{u}_2 = u_2/\sqrt{2gH}$ which is the dimensionless outlet velocity and β_2 = blade angle

in the stream way direction (see fig. 7).
From eq. (1) we obtain

$$\underline{u}_2 = {}^*\Omega^{3/2}/(\pi\tan(\pi - \beta_2))^{1/3}$$

From this equation we find that within the same speed number an increased diameter and circumferential speed may be compensated by a decreased outlet blade angle within a limited range of angles.

3.2. THE REACTION RATIO

Besides the specific speed the reaction ratio, which is a measure of the pressure drop through the runner, is a very important parameter for the blade loading and cavitation performance. The reaction ratio can be increased by changing the blade shape for a given speed number and thus inlet cavitation problems may be solved. The reaction ratio is defined by the pressure drop from runner inlet to outlet divided by the total available net head at best efficiency flow for $c_{u2} = 0$. By introducing the dimensionless expression for pressure $\underline{h} = h/H$ the reaction ratio yields:

$$\underline{h}_1 - \underline{h}_2 = \frac{h_1 - h_2}{H} \tag{3}$$

f friction losses are ignored we can establish the following equation by combining the Euler turbine equation, the absolute specific energy (E) and the relative specific stagnation energy (I) (rothalpy):

$$gH\eta_h = u_1 c_{u1} - u_2 c_{u2} = (E_1 - I_1) - (E_2 - I_2) \tag{4}$$

Because the relative stagnation energy is constant along a streamline through the runner if friction is ignored, we get $I_1 = I_2$ and then

$$gH\eta_h = u_1 c_{u1} - u_2 c_{u2} = E_1 - E_2 = (gh_1 + \frac{c_1^2}{2}) - (gh_2 + \frac{c_2^2}{2}) \qquad (5)$$

Here the difference in geodetic height from inlet to outlet is ignored. When introducing reduced or dimensionless variables $\underline{c} = c/\sqrt{2gH} = \underline{c}_u^2 + \underline{c}_m^2$, remembering that $c_{u2} = 0$ at best efficiency point (BEP) and assuming $\underline{c}_{m1} = \underline{c}_{m2}$ i.e. constant meridional cross section through the runner we can establish eq. (6) by substituting for η_h by eq. (4) in eq. (5):

$$\frac{h_1 - h_2}{H} = \underline{h}_1 - \underline{h}_2 = 2\underline{u}_1 \underline{c}_{u1} - \underline{c}_{u1}^2 = \underline{u}_1^2 [2\frac{\underline{c}_{u1}}{\underline{u}_1} - (\frac{\underline{c}_{u1}}{\underline{u}_1})^2] \qquad (6)$$

By studying the vector diagrams for runners with increasing speed number and constant hydraulic efficiency $\eta_h = 2\underline{u}_1\underline{c}_{u1}$ as shown in fig. 7, we find increasing pressure difference from inlet to outlet when $\underline{u}_1$ and $\underline{u}_2$ are increased.

However, it must be emphasized that an increased speed number increases $\underline{u}_2$. Thus, decreased pressure at the outlet $= \underline{h}_2$ may also lead to a reduction of the inlet pressure $\underline{h}_1$ even if $\underline{h}_1 - \underline{h}_2$ is increased, if the reduction in $\underline{h}_2$ is not compensated by an increased submergence.

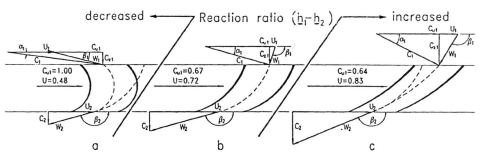

decreased ◄────── Reaction ratio $(\underline{h}_1 - \underline{h}_2)$ ──────► increased

$C_{u1} = 1.00$
$U = 0.48$

$C_{u1} = 0.67$
$U = 0.72$

$C_{u1} = 0.64$
$U = 0.83$

a   b   c

Blade cascades for different specific speed if η_h=const and uc_u =const.

Figure 7. Blade cascades for different reaction ratios for $c_{u2}=0$.

Two examples illustrate the increase in the reaction ratio for $\eta_h = 2\underline{u}_1\underline{c}_{u1}$ = const. For a high head turbine $\underline{c}_{u1} = 0.69$ and $\underline{u}_1 = 0.72$: $\underline{h}_1 - \underline{h}_2 = 0.518$. For a low head turbine $\underline{c}_{u1} = 0.6$ and $\underline{u}_1 = 0.83$: $\underline{h}_1 - \underline{h}_2 = 0.634$.

However, for a given reduced circumferential speed $\underline{u}_1$ a maximum pressure ratio is obtained for $\underline{c}_{u1}/\underline{u}_1 = 1$ i.e. $\beta_1 = 90°$. From this we find that the inlet pressure may be increased locally by bending the inlet of the blade towards $\beta_1 = 90°$. The efficiency will also increase with decreasing β_1 if no rotation occurs at the outlet of the runner, but this is impossible with the given boundary conditions. However, local bending of the blade on the shroud side may also lead to a local negative blade lean which in turn leads to a low pressure zone near the blade inlet with cavitation problems.

3.3. THE INFLUENCE FROM THE MERIDIONAL CROSS SECTION OF CROWN AND SHROUD

During the initial stage of design work on a runner the choice of inlet and outlet angles have been proven to be of great importance. The meridional cross section of the runner i.e. crown and shroud shape will also be influenced by the blade angles.

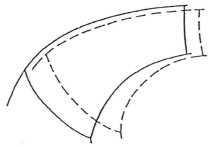

Figure 8. Different shapes of Francis runners for the same specific speed i.e. same speed, output and head.

For example, the blade outlet angle [= $(\pi - \beta_2)$, see fig. 7] may have a variation range from 13° to 18° and the inlet angle [= $(\pi - \beta_1)$] may have a variation 50° to 70° for a medium head turbine. The meridional cross section may have a variation in shape and outlet diameters as shown schematically in fig. 8, for the same speed number of a high head turbine.

At best efficiency point of operation both a large outlet diameter and a small outlet diameter may operate stably. However, at off-design, with swirl flow in the draft tube, it will be more difficult to avoid large pressure surges for runners with large outlet diameters.

The advantage of a large outlet diameter will be that the required submergence to avoid cavitation of the turbine, is smaller according to eq. (2), but the disadvantage may be a more unstable operation at off-design head or at off-design flow.

3.4. THE INFLUENCE FROM BLADE LEAN

The blade lean (also called blade raking) is defined by the blade angle with the meridional plane normal to the streamlines (Θ) and other geometry parameters as illustrated in fig. 9. The dimensionless pressure gradient $d\underline{h}/dy = d(h/H)/dy$ in a stream surface can be established by eq. (7) expressed by the dimensionless meridional velocity $\underline{c}_m = c_m/(2gH)^{0.5}$ and angular speed $\underline{\omega} = \omega/(2gH)^{0.5}$ (m^{-1}).

"cone" and its interaction with the draft tube "elbow". Some partial analytical and numerical models of the flow field generated by a *precessing helicoidal vortex filament* inside a *cylindrical boundary* have been set up (this problem can be approximated by a potential flow, inviscid incompressible fluid scheme); however, the modeling of a precessing helicoidal vortex filament inside an *elbow* is immensely more complicated and calls essentially for integration of the full Navier-Stokes equations under non-permanent conditions.

The final result of such models would ideally provide an estimate of the *intensity and frequency of the excitation term* as well as indications about its *dependence on the index n* of partial load.

At present, two main mechanisms are being considered for the excitation: the "*elbow pressure gradient sweep*" and the "*parametric head loss excitation*".

Elbow pressure gradient sweep : it is considered that the cavitated vortex rope, in its portion near the elbow, sweeps periodically zones of higher pressure (near the outside of the elbow) and zones of lower pressure (near the inside of the elbow). By setting up the equation for the volume variations of the cavity under the joint action of the synchronous pressure variations and of the elbow gradient sweep, together with the equation for dynamic equilibrium of the draft tube water column under the synchronous pressure variations, the excitation is made to depend explicitly on the ratio between the radius of the cone and the radius of curvature of the elbow axis:

$$
h_e \cong \frac{n^2.Q_0^2}{2g.S_0^2} \cdot \left[1 - \frac{1}{\left(1 - \varepsilon.\kappa.\sin\omega_T.t\right)^2} \right] \, ,
\tag{3.3.1}
$$

where S_0 = cone exit section, ε = excentricity of helicoidal vortex rope at cone exit,

$\kappa = \dfrac{r}{R}$ = ratio between radius of cone exit section and radius of curvature of elbow

axis (ε = function of n).

It is seen, in contrast with (3.1.6), that under this hypothesis *the excitation would not be exactly sinusoidal* in time (it would be nearly so only for $\varepsilon.\kappa \ll 1$); it is also seen that the excitation would be all the more intense as the radius of curvature R grows smaller, which is in accordance with known facts.

Parametric head loss excitation : it is assumed that the head losses in the draft tube, and particularly those concentrated immediately downstream of the elbow, are dependent on the position of the helicoidal vortex rope. Indeed, the zone of low pressure surrounding the helicoidal vortex filament interacts more or less with the zone of low pressure near the inside of the elbow during the precessionary motion of the vortex rope; thus more or less pronounced flow separation will occur, with quasi-periodic variations in the head loss factor. This makes for a *quasi-periodic blocking effect*; the conjecture is in accordance with experience, in the sense that numerical models based on a periodic variation of the head loss factor in the draft tube produce (by suitable *calibration* of said variations) effects in   general agreement with experimental data (see DOERFLER, 1982).

The two above-mentioned effects could well coexist in the physical reality. More detailed experimental data (including *phase* informations, see § 3.1) will be necessary to confirm or disprove either model, as well as to "identify" the values to be assigned to the relevant parameters.

A non-linear numerical model based on the above considerations and in which the finite value of celerity in the draft tube is introduced *("distributed parameters model")* is under development by the Author; in it, besides the "control parameters" already seen for the elementary non-linear model of § 3.2, two new control parameters appear, related to propagation times along the draft tube axis. More accurate identification tools could thus be derived.

4. Conclusions

From the foregoing synthetic overview it is easy to see what will be the likely development of *future research* in the field.

As concerns the *analytical-numerical models*, an effort of *synthesis* is necessary to fuse together the partial models so far developed: dynamics of the cavitated vortex rope, its interactions with the machine on one hand, with the draft tube elbow (excitation mechanism) on the other hand, interplay of the different components of the installation in fluctuating régimes, effects of non-linearities.

It is also worth mentioning that, in alternative to the search for an "excitation mechanism", it might be more correct to reformulate the problem in terms of *"instability"* of the non-linear system, e.g. through the presence in the representative equations of terms having the nature of *"negative stiffness"* or of *"negative damping"*.

The final aim would be, needless to say, in line with the twofold goals of every theoretical model:

- on one hand, *a posteriori* interpretation of observations and correct guidelines for *identification* of non-measurable parameters. An obvious example is the identification of the excitation parameters keeping into account the effect of mutual interactions of the system components; indeed, what is measured is always a *local response* and, in particular, what is measured on a reduced scale model cannot be transposed directly to reality because the mutual interactions will in general produce different amplifications in the model and in the prototype;

- on the other hand, *a priori* forecasting of pulsating phenomena: correct transposition from the scale model to the prototype, i.e. forecasting (after the above said identification of excitation) of the amplitude of local responses in the real system; the provision of reliable guidelines for setting up experiments and for locating measuring instruments in the "right" positions; the forecasting of the effects of mitigating measures (such as e.g. air introduction or *"active control"* devices); etc.

Important progress is under way, of course, and more is to be expected, in the perfection of *experimental techniques*. Velocity fields can now be investigated in detail

2. Background

An excellent and thorough review of fluid transients in liquid-filled piping systems has been presented by Tijsseling (1996). His paper relates the historical contributions that lead to the present physical and mathematical understanding of unsteady liquid-pipe interaction. In addition, the paper describes various interactive fluid-piping numerical algorithms that provide predictions in the time domain and discusses their validation by physical experiments. Tijsseling's paper is recommended reading for those who desire an historical perspective on fluid-structure interaction in liquid filled piping, as well as an accounting of the significant contributions made by many investigators. The brief historical review that follows is excerpted from Tijsseling (1996).

Some type of pipe motion has always been considered when formulating the water-hammer problem; for example, Korteweg in 1878 accounted for the elasticity of pipe walls in the acoustic wave speed in a pipe. Lamb, in his treatment of acoustic velocities in tubes in 1898, recognized that an acoustic wave propagating in the liquid is modified by the tube wall motion, and that both axial and radial vibrations occur in the tube. Several researchers extended Lamb's work to include bending and rotation of the pipe element. Frequency-domain analyses that incorporated dynamic coupling between contained liquid and junction-like structural components were developed for simple systems such as a short single-elbow pipe.

With the advent of high-speed computing in the 1960's, water hammer analysis followed a traditional approach, progressing with the basic assumption that the piping and support structures are rigid and do not affect the predicted wave forms. The one exception was the adjustment of the acoustic wave speed based on static-elastic interaction between the liquid pressure and the pipe circumferential and axial strains. The possible vibration of a pipe component due to a water hammer pressure loading has been recognized in textbooks, and early attempts at coupled time-domain analysis incorporated, for example, a lumped mass-spring to approximate a portion of the piping system.

The nuclear power industry in particular has been required to analyze large-scale piping systems to maintain their integrity while being subjected to thermal stresses and seismic loads. Typically, standard structural finite element and modal analysis procedures have been used to predict piping vibrations without regard to acoustic waves propagating in the contained liquid. Water hammer loads had been considered secondary and often were not the major focus of a dynamic design analysis of the piping. From 1970 to 1980 in both Europe and North America, water hammer incidents in nuclear power plants kindled an interest in water hammer induced piping vibration; as a result, fluid transient and structural dynamics disciplines began to merge. Work was carried out simultaneously by a number of institutions and agencies, resulting in several computational methodologies, and more significantly, an improved understanding of the intricate and subtle interaction between fluid transient and pipe motion. The characteristics of systems studied and analytical methodologies employed to date are summarized in Table 1.

TABLE 1. Various approaches to fluid-structure interaction in piping systems.

System component characteristics:
    Piping: elastic or inelastic
    Fluid: continuous or discontinuous (liquid slug or column separation)
Analytical approach:
    Coupled versus uncoupled
    Time domain versus frequency domain

Uncoupled analysis consists of first obtaining the dynamic pressures from a water hammer calculation without regard to any piping motion, and then using them as loads to predict the piping response in a separate structural dynamics analysis. By contrast, coupled analysis requires that the water hammer and structural motion be solved simultaneously. In the sections that follow, we will focus on elastic piping, and on both continuous and discontinuous fluid components. Time-domain solutions will be emphasized, and the validity of coupled versus uncoupled analysis will be evaluated.

3. Essentials of Liquid-Piping Interaction

Three coupling mechanisms can be identified in liquid-piping interaction. *Poisson coupling* is associated with the radial stress perturbations produced by liquid pressure transients that translate to axial stress perturbations by virtue of the Poisson ratio coefficient. The axial stress and accompanying axial strain perturbations travel as waves in the pipe wall at approximately the speed of sound in the pipe. Typically its magnitude is three to five times greater than the acoustic velocity in the contained liquid in the pipe. *Friction coupling* is created by the transient liquid shear stresses acting on the pipe wall; usually it is insignificant when compared to the other coupling mechanisms. Both Poisson and friction coupling are distributed along the axis of a pipe element.

The third, and often the most significant, coupling mechanism is *junction coupling*, which results from the reactions set up by changes in liquid momentum at discrete locations in the piping such as bends, tees, valves, and orifices. In flexible piping, water hammer waves impacting at junctions may set up a vibration which in turn may translate to a variety of structural responses (bending, rotation, shear, axial stresses) at locations away from the junction. In addition, the vibrating junction will induce fluid transients in the contained liquid column, with acoustic waves traveling upstream and downstream from the junction. Clearly, the net effect is a complex motion in both the piping and the contained liquid, with resulting waveforms being highly dependent upon the geometry of the pipe system. Sources of excitation include not only those associated with liquid motion, but may also come from the structural side. Figure 1 relates some of the various excitation mechanisms.

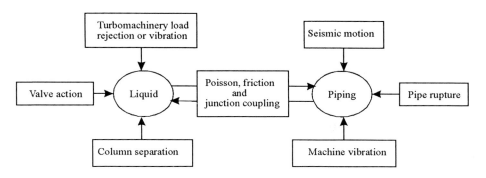

FIGURE 1. Sources of excitation in fluid-piping interaction.

Most large-scale industrial piping systems are comprised of liquid-filled, thin-walled piping elements, so that the transient behavior of the liquid can be described as one-dimensional wave phenomena. Consider a fundamental liquid-filled prismatic pipe element, with the z-coordinate positioned along the pipe axis, and the x- and y-coordinates normal to the z-axis. The pipe can transmit torsional motion θ_z in the axial direction, lateral motion u_x and u_y normal to the axis, and longitudinal motion u_z and stress σ_z along the axis. In addition, the contained liquid can exhibit transient velocity V and pressure p. Neglecting radial inertia of the pipe wall, a set of six partial differential equations can be derived that relate the liquid pressure and velocity, the axial displacement and stress of the pipe, the lateral pipe displacement, and the tortional rotation of the pipe (Heinsbroek and Kruisbrink, 1993):

$$\frac{\partial V}{\partial t} + \frac{1}{\rho_f}\frac{\partial p}{\partial z} + \frac{fV_r|V_r|}{2D} = 0 \tag{1}$$

$$\frac{1}{\rho_f c_f^2}\frac{\partial p}{\partial t} + \frac{\partial V}{\partial z} - \frac{2v}{E}\frac{\partial \sigma_z}{\partial t} = 0 \tag{2}$$

$$\rho_t A_t \frac{\partial^2 u_z}{\partial t^2} - EA_t \frac{\partial^2 u_z}{\partial z^2} = \frac{vDA_t}{2e}\frac{\partial p}{\partial z} + \frac{\rho_f A_f f V_r|V_r|}{2D} + \rho_t A_t g\sin\gamma \tag{3}$$

$$\left(\rho_t A_t + \rho_f A_f\right)\frac{\partial^2 u_x}{\partial t^2} + EI\frac{\partial^4 u_x}{\partial z^4} = -\left(\rho_t A_t + \rho_f A_f\right)g\cos\gamma \tag{4}$$

$$\left(\rho_t A_t + \rho_f A_f\right)\frac{\partial^2 u_y}{\partial t^2} + EI\frac{\partial^4 u_y}{\partial z^4} = 0 \tag{5}$$

$$\rho_t J\frac{\partial^2 \theta_z}{\partial t^2} - GJ\frac{\partial^2 \theta_z}{\partial z^2} = 0 \tag{6}$$

Additional parameters in the equations are: $c_f = 1 / \sqrt{\rho_f (1 / K + D / Ee)}$, the liquid wave speed; f = Darcy-Weisbach friction factor; V_r = liquid velocity relative to the axial pipe motion; D = pipe diameter; E = Young's modulus of the pipe material; v = Poisson's ratio; K = bulk modulus of liquid; ρ_f = fluid density; A_f = cross-sectional flow area; ρ_t = density of pipe material; A_t = cross-sectional pipe area; γ = pipe angle; I = moment of inertia of pipe; G = pipe material shear modulus, and J = polar moment of inertia of pipe. The extended liquid and axial displacement equations are coupled by friction and Poisson coupling along the axis of the pipe element. At pipe junctions, flow, liquid pressure forces, moments, displacements and torsion are translated from one pipe to another. Standard boundary conditions for fluid transients are employed; these include valve motion, turbomachinery transients such as load rejection or pump trip, and column separation. An example of a structural boundary condition is seismic excitation (Hatfield and Wiggert, 1990).

4. Methods of Solution

Solutions can be formulated in either the time domain or in the frequency domain. In the time domain, the most successful technique is to solve the water hammer equations by the method of characteristics and the structural equations are either by finite element representation (Lavooij and Tijsseling, 1991) or by modal component synthesis (Hatfield and Wiggert, 1991). Under special circumstances the method of characteristics has been used to solve the Poisson-coupled axial pipe motion (Budny, *et al.*, 1991, Heinsbroek, *et al.*, 1991) as well as lateral pipe motion (Wiggert, *et al.*, 1985a; Tijsseling, Vardy, and Fan, 1996). Frequently it is advantageous to simplify the structural motion to account for only those modes that interact significantly with the liquid transient. Under such circumstances, for small degree-of-freedom structural motion the method of lumped parameters (Wiggert, *et al.*, 1985b) can be employed, and for larger degrees of freedom, modal component synthesis is useful.

5. Water Hammer Interaction with Piping Structure

A simple example of how a water hammer load interacts with a vibrating pipe (Budny, 1988) is demonstrated in Fig. 2. The pipe is horizontally suspended so that it can move unrestrained in the axial direction. At the downstream end a valve was rapidly closed, and the resulting pressure waveform immediately upstream of the valve is shown. For the numerical analysis, the method of characteristics was utilized for both the liquid column and the axial mode of piping vibration. Two fundamental frequencies are superposed in the wave form: 1) water hammer $c_f / 4L \cong 6.6$ Hz, and pipe wall $c_t / 4L \cong$ 19 Hz, where $c_t = \sqrt{E / \rho_t}$, the acoustic speed in the pipe wall. Even though both Poisson coupling and axial pipe vibration are taking place, the axial vibration

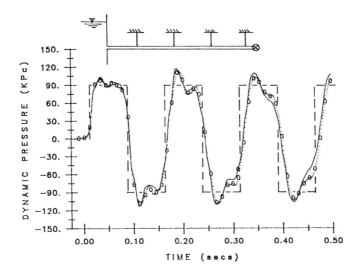

FIGURE 2. Axially-coupled motion in the liquid and structural modes (Budny, 1988).
Legend: o·····o experimental; ——— numerical; – – – Joukowsky

dominates the feedback into the pressure wave form. One noteworthy observation is that the Joukowsky pressure rise of 90 kPa is exceeded because of the dynamic interaction.

A more complicated interaction takes place when the piping system has an intricate geometry (Kruisbrink and Heinsbroek, 1992; Heinsbroek and Kruisbrink, 1993). For example, consider the system shown in Fig. 3. From a water hammer viewpoint, the system is identical to that shown in Fig. 2: a single liquid column with a reservoir at the upstream end and a rapid valve closure excitation at the downstream end, resulting in one primary mode of fluid motion. However, because of the complexity of the piping layout, there may be three or more significant modes of vibration in the piping excited by the valve closure. The pressure waveform at the valve, Fig. 4, reflects the feedback due to bending and translation of the piping. The pressure rise predicted by the Joukowsky approximation is approximately 3.8 bar, whereas the recorded peak dynamic pressure is about 5 bar. In the computation, the authors used a combined method of characteristics/finite element solution.

Both Figs. 2 and 4 demonstrate that in many similar industrial systems, when the piping is relatively compliant and not rigidly supported, the water hammer waveforms will deviate from the ideal Joukowsky predictions. Quite often, with higher frequency components the peak pressures in the waveforms will exceed the Joukowsky prediction, but normally not by a significant amount. What may be more important from an integrity viewpoint are the dynamic pipe strains and stresses resulting from the water hammer activity. It has been found that measured pipe displacements and strains are accurately predicted by coupled analysis, but are not accurately predicted using uncoupled analysis (Heinsbroek, 1993; Heinsbroek and Tijsseling, 1994).

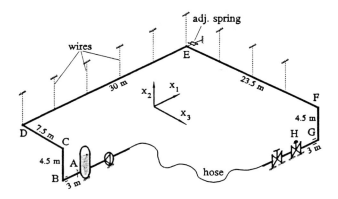

FIGURE 3. Experimental compliant pipeline system (Kruisbrink and Heinsbroek, 1992).

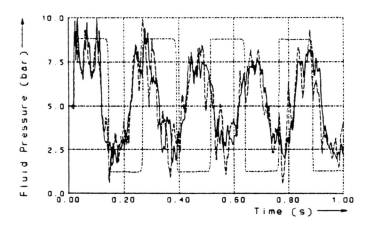

FIGURE 4. Dynamic pressure at the shutoff valve (Kruisbrink and Heinsbroek, 1992).
Legend: ——— measured; – – – computed; ·········Joukowsky

The significance of piping restraints has received little attention. The interaction of a pipe rack with the axial translation of a pipe was recently investigated by Tijsseling and Vardy (1996). They linked the interaction between the moving pipe and the rack by means of dry Coulomb friction, and successfully predicted the pipe axial velocity and the water hammer in the contained liquid.

6. Two-Component Flows

A series of studies related to fluid structure interaction and cavitation have been conducted in a novel manner by impacting a solid metal rod at the end of a freely suspended liquid-filled horizontal pipe (Tijsseling and Fan, 1991a,b; Fan and Tijsseling, 1992). In addition, Tijsseling, Vardy and Fan (1996) applied the same experimental procedure to a single-elbow pipe arrangement. The authors obtained experimental verification of the interaction between vapor cavities, the liquid, and the axial pipe wall motion. They utilized a discrete cavity model combined with fluid-structure interaction models to obtain predictions that compared well with experiments. These investigations show that some of the differences reported in earlier water hammer literature between measured and predicted cavitation can be attributed to fluid structure interaction.

A significant cavitation phenomenon that can lead to pronounced fluid-structure interaction is condensation-induced water hammer (Chou and Griffith, 1989). It is known to occur in piping when subcooled water comes in to contact with steam, and it has also been identified in refrigeration piping systems. Because of the stochastic nature of the void collapse, it is difficult to isolate an individual event and accurately estimate the loads that occur on the piping structure.

Liquid slug motion in voided lines leading to pipe support failure has been documented to occur in power plant piping. A slug, driven by a pressure gradient, can accelerate to high velocities and transfer significant loads when it impacts at a pipe junction, Fig. 5. There have been very few studies conducted that focus on this phenomenon (Fenton and Griffith, 1990; Bozkus and Wiggert, 1992). Experiments have shown that the impact waveforms and resulting peak pressures are highly stochastic, thereby increasing the uncertainty of a numerical prediction. Because of the

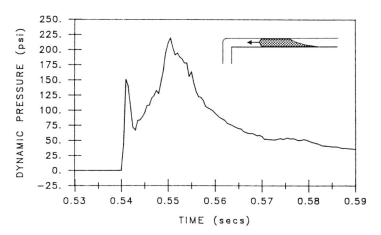

FIGURE 5. Pressure waveform at a pipe elbow due to liquid slug impact (Bozkus and Wiggert, 1992). Slug velocity = 17.3 m/s, slug length = 2.1 m, pipe diameter = 50 mm.

short duration of the impact pulse, when estimating piping stresses it is reasonable to uncouple the fluid transient from the structural motion. Additional experimental and analytical research is recommended, both for condensation-induced water hammer and liquid slug impact in full scale piping systems.

7. Discussion

Frequency domain studies have not been highlighted herein. Component synthesis techniques have been successfully applied to oscillatory flows that induce vibrations in small slender piping systems (Lesmez, Wiggert, and Hatfield, 1990). Recently, there has been interest in adapting these to large piping systems (Charley and Caignaert, 1993; Svingen, 1994, 1995; Gajic, et al., 1995). In hydropower and piping systems, vibrations in the flow may be induced by turbomachinery, leading to possible resonance and significant fluid-structure interaction. Additional research and development in this area is likely to occur.

A fundamental question often raised is: Can uncoupled analysis be used to determinine whether a piping system can withstand a water hammer load without substantial vibration or possible support failure? One must deal with an amount of uncertainty when estimating water hammer loads on piping. Dynamic pressures larger than those predicted by the Joukowsky approximation are known to exist when the piping is compliant, and coupled analysis has been shown to accurately predict such situations. For most commercial piping situations, the deviations in dynamic pressure are not significant, so that uncoupled analysis may be adequate, especially when safety factors are incorporated. However, with the present understanding, coupled analysis is possible without undue additional effort. Component synthesis enables one to combine conventional structural dynamic analysis techniques with extended water hammer algorithms, producing solutions that accurately predict the true dynamic interaction.

Acknowledgment

Appreciation is extended to Arris Tijsseling and Bjørnar Svingen for their assistance during preparation of the manuscript.

References

Bozkus, Z., and Wiggert, D.C. (1992) Hydromechanics of slug motion in a voided line, in R. Bettess and J. Watts (eds), *Unsteady Flow and Fluid Transients*, A.A. Balkema, Rotterdam, pp. 77-86.

Budny, D.D. (1988) The influence of structural damping on the internal fluid pressure during a fluid transient pipe flow, Ph.D. dissertation, Michigan State University, Department of Civil and Environmental Engineering, E. Lansing.

Budny, D., Wiggert, D.C., and Hatfield, F.J. (1991) The influence of structural damping on internal pressure during a transient flow, *ASME Journal of Fluids Engineering*, **113**, 424-429.

Charley, J., and Caignaert, G. (1993) Vibroacoustical analysis of flow in pipes by transfer matrix with fluid-structure interaction, *Proc. Work Group on The Behavior of Hydraulic Machinery Under Steady Oscillatory Conditions, 6th International Meeting*, IAHR, Lausanne.

Chou, Y., and Griffith, P. (1989) Avoiding steam-bubble collapse-induced water hammers in piping systems, EPRI Research Project NP-6647.

Fan, D., and Tijsseling, A. (1992) Fluid-structure interaction with cavitation in pipe flows, *ASME Journal of Fluids Engineering*, **114**, 268-274.

Fenton, R.M., and Griffith, P. (1990) The forces at a pipe bend due to the clearing of water trapped upstream, *Transient Thermal Hydraulics and Resulting Loads on Vessel and Piping Systems-1990*, American Society of Mechanical Engineers PVP Vol. 190, New York, pp. 59-67.

Gajic, A., Pejovic, S., Stojanovic, Z., and Josef, K. (1995), *Proc. Work Group on The Behavior of Hydraulic Machinery Under Steady Oscillatory Conditions, 7th International Meeting*, IAHR, Ljubljana.

Hatfield, F.J., and Wiggert, D.C. (1990) Seismic pressure surges in liquid-filled pipelines, *ASME Journal of Pressure Vessel Technology*, **112**, 279-283.

Hatfield, F.J., and Wiggert, D.C. (1991) Water hammer response of flexible piping by component synthesis, *ASME Journal of Pressure Vessel Technology*, **113**, 115-119.

Heinsbroek, A.G.T.J., Lavooij, C.S.W., and Tijsseling, A.S. (1991) Fluid-structure interaction in non-rigid piping: a numerical investigation, *SMiRT 11 Transactions*, B12/1, Tokyo, pp. 309-314.

Heinsbroek, A.G.T.J. (1993) Fluid-structure interaction in non-rigid pipeline systems: comparative analysis, *Proc. ASME/TWI 12th International Conference on Offshore Mechanics and Arctic Engineering*, Glasgow, pp. 405-410.

Heinsbroek, A.G.T.J., and Kruisbrink, A.C.H. (1993) Fluid-structure interaction in non-rigid pipeline systems: large scale validation experiments, *SMiRT 12 Transactions*, J08/1, Stuttgart, pp. 205-210.

Heinsbroek, A.G.T.J., and Tijsseling, A.S. (1994) The influence of support rigidity on waterhammer pressures and pipe stresses, *Proc. Second International Conference on Water Pipeline Systems*, BHR Group, Edinburgh, pp. 17-29.

Kruisbrink, A.C.H., and Heinsbroek, A.G.T.J. (1992) Fluid-structure interaction in non-rigid pipeline systems: large scale validation tests, *Pipeline Systems*, Kluwer Academic Publishers, Dordrecht, pp. 151-164.

Lavooij, C.S.W., and Tijsseling, A.S. (1991) Fluid-structure interaction in liquid-filled piping systems, *Journal of Fluids and Structures*, **5**, 573-595.

Lesmez, M.W., Wiggert, D.C., and Hatfield, F.J. (1990) Modal analysis of vibrations in liquid-filled pipe systems, *ASME Journal of Fluids Engineering*, **112**, 311-318.

Svingen, B. (1994) A frequency domain solution of the coupled hydromechanical vibrations in piping systems by the finite element method, *Proc. IAHR International Symposium on Hydraulic Machinery and Cavitation*, Beijing.

Svingen, B. (1996) Fluid structure interaction in slender pipes, *Proc. Pressure Surges and Fluid Transients in Pipelines and Open Channels*, Mechanical Engineering Publications Ltd., Suffolk.

Tijsseling, A.S., and Fan, D. (1991a) The response of liquid-filled pipes to vapour cavity collapse, *SMiRT 11 Transactions*, J10/2, Tokyo, pp. 183-188.

Tijsseling, A.S., and Fan, D. (1991b) The concentrated cavity model validated by experiments in a closed tube, *Proc. International Meeting on Hydraulic Transients with Water Column Separation: 9th Round Table of the IAHR Group*, Session A-3, Paper 2, Valencia.

Tijsseling, A.S. (1996) Fluid-structure interaction in liquid-filled pipe systems: a review, *Journal of Fluids and Structures*, **10**, 109-146.

Tijsseling, A.S., and Vardy, A.E. (1996) Axial modeling and testing of a pipe rack, *Proc. Pressure Surges and Fluid Transients in Pipelines and Open Channels*, Mechanical Engineering Publications Ltd., Suffolk, UK.

Tijsseling, A.S., Vardy, A.E., and Fan, D. (1996) Fluid-structure interaction and cavitation in a single-elbow pipe system, *Journal of Fluids and Structures*, to be published.

Wiggert, D.C., Hatfield, F.J., and Stuckenbruck, S. (1985a) Analysis of liquid and structural transients in piping by the method of characteristics, *Fluid transients in fluid-structure interaction–1985*, American Society of Mechanical Engineers, New York, pp. 97-102.

Wiggert, D.C., Otwell, R.S., and Hatfield, F.J. (1985b) The effect of elbow restraint on pressure transients, *ASME Journal of Fluids Engineering*, **107**, 402-406.

SYMPOSIUM CONTRIBUTORS

A DECISION AID SYSTEM FOR HYDRAULIC POWER STATION REFURBISHMENT PROCEDURE

Francis and Kaplan Turbine

BELLET L., PARKINSON E.* AND AVELLAN F.
IMHEF/LMH/EPFL - Av. de Cour,33
1007 LAUSANNE - Switzerland

AND

COUSOT T. AND LAPERROUSAZ E.
EDF-CNEH - Savoie Technolac
73373 LE BOURGET DU LAC - France

1. Project

Maintenance policies for hydrogenerating equipment vary considerably from one operator to another, from a least cost strategy to the greatest care and tightest scheduling. However, whatever policy is adopted, sooner or later maintenance is no longer enough to uphold acceptable levels of performance or safety. When that happens, renovation or refurbishment takes over.

In general, the decision to refurbish a scheme is inspired by a combination of causes, which can be grouped in three types: age, downgraded performance, and unsuitable or costly operating techniques.

It is fairly difficult to evaluate the effect of age, although it is well known that, as time passes, the risk of a serious accident increases. Obviously, refurbishment consecutive to an accident cannot be scheduled and can therefore not take place at the same cost nor with the same results as a scheduled refurbishment that is correctly integrated into the generating program.

Even the most painstaking maintenance works are not enough to preserve initial performance levels. Energy losses increase inexorably everywhere in the system: hydraulic, mechanic and electric losses.

Finally, operation becomes increasingly difficult. The original design requirements, such as operating limits imposed by cavitation or instability, are compounded by constraints due to wear and permanent deformations (watertightness of guide vanes or gates and valves) and the safety of the plant in general. Operating losses due to the frequency of maintenance must also be considered, and even, within a cascade of plants, the unsuitability of one scheme versus those upstream and downstream from it.

E. Cabrera et al. (eds.), Hydraulic Machinery and Cavitation, 71–80.
© *1996 Kluwer Academic Publishers. Printed in the Netherlands.*

Furthermore, although the *raison d'être* of power plants remains the generation of energy at least cost, other needs or requirements have grown up over the years such as:

— the safety of facilities;
— the match between performance levels and network requirements;
— compliance with new legislation;
— new operating requirements (irrigation, low stage support, recreational uses).

A well-managed refurbishment must take all these new data into account, as well as technological and scientific advances. Depending on those new requirements, a refurbishment operation can take many forms, from replacement of elements with identical equipment to the construction of a whole new plant. Compromises also exist, such as replacing a runner with one of modern design. In general the objective is to minimise the cost of the operation or guarantee that it is cost-effective because of improved generating performances.

Improving performance by replacing the runner alone, and when necessary modifying the generator, means extra cost and risk in comparison to replacement with identical equipment, extra cost because more study is required (and even model tests), and risk because it can never be absolutely guaranteed that performance will actually improve.

The IMHEF and EDF have been working together for several years on various subjects relating to hydraulic machinery. Since 1995, the two have joined forces to develop a procedure for the evaluation of older turbines. This joint work was inspired by the realisation of two facts:

1- EDF is confronted with an ageing stock of hydro facilities and the renewal of many of its concession agreements. EDF's engineers working in hydro generation are reflecting on how to enhance the value of French facilities, as part of a project called "Hydraulique Demain" ("The Hydraulics of Tomorrow"). This project is aimed at implementing the technico-economic tools and methods required to rationalise expenditures and to analyse the potential of hydro facilities.

2- The IMHEF and EDF have long worked for industrial groups, operators or water resources managers as consulting engineers. With the resulting awareness of those entities' needs in refurbishment projects, they have decided to adapt modern techniques to that field.

Our work is part of the global process required to prepare for a refurbishment, which among other things includes technical, operating and economic studies. Our objective is to ascertain how suitable a machine is for its site and to evaluate the energy potential of the stationary elements of an existing turbine. We therefore first focus on those stationary parts (spiral casing, distributor and where necessary headrace pipes) and then on the runner if necessary.

Such a study can be scheduled in various stages of a project, and can have various goals. When it is scheduled before design, it is a decision-making tool that helps in establishing the technical and economic balance sheet on a facility and in orienting the choice of what type of refurbishment to plan. If the decision is made to change the runner, it will serve in setting up the technical specifications for the new runner on solid bases and when necessary in predicting whether changes will be needed to stationary parts.

2. Refurbishment procedure

There are a multitude of components in a hydroelectric scheme, and therefore the engineer in charge of study of a refurbishment is faced with a wide and complex range of options. To guide his choices, he must take into account both technical and economic aspects, without neglecting legal aspects (the concession agreement) and environmental aspects, which may also play a decisive role. The procedure described herein concerns only technical aspects and is intended to help in establishing a rapid and accurate balance sheet on the site under study in order to be able to conduct an economic study. This procedure is divided into three stages, progressing from a global approach to a specific approach.

2.1. PHASE 1:RENOVA

The first stage in a refurbishment study consists in gathering all the available information on the project at hand concerning the hydraulic elements in the scheme, with the purpose of drawing up a balance sheet on the following main elements: water intakes, penstocks, gates and valves, turbine, headrace and tailrace channels.The information collected is used to identify critical zones in the scheme that the study will focus on. A complete and systematic summary can be transcribed in computer files of characteristics to create a data base, called *Renova*. This data base was created as part of the present project in order to:

- inventory existing machines described in the literature, whether refurbished or not;
- run a statistical study based on those machines;
- establish a data base for refurbishment projects.

In the statistical study, the machine at hand is compared to other machines, whether modern or older, and where appropriate a new type of turbine can be defined that is better suited to the site's present characteristics. This data base is for the moment essentially concentrated on the turbine, but can easily be extended to the other elements in a scheme.

2.2. PHASE 2:RENOVATURB

The second phase is more specific, as it concerns the high pressure parts of the turbine, i.e., the spiral casing and distributor. Its objective is to precisely determine conditions upstream from the runner. The *Renovaturb* computer program, which was partially developed during this project, allows geometrical analysis of those components:

- verification and computation of geometrical quantities (sections laws, skeletons lines, distributor opening law, opening, etc.);
- prediction of flow angles at the spiral casing outlet;
- compatibility between the estimated angles and actual angles of stay vanes;
- compatibility of geometrical angles between stay vanes and guide vanes;
- automatic generation of a structured mesh of the distributor and semi-automatic of the spiral casing in preparation for phase 3.

2.3. PHASE 3:COMPUTER MODELLING

Knowledge of hydraulic conditions in the high pressure parts of the turbine (spiral casing and distributor) is very important, and a computer model allows a more in-depth analysis of each component. Phase 2 of this procedure gives an estimation of the average flow angle, defined as $\alpha = atan[\frac{C_m}{C_u}]$ where C_m and C_u are respectively the meridional and tangential velocity, at the spiral casing outlet. If a more precise idea of the phenomena in the spiral casing is desired, a numerical calculation can be done, although it must be borne in mind that considerable time is needed for calculation and generation of the mesh (as the tong is specific to each spiral casing, generation of it must be modified in the *Renovaturb* program).

Calculation of the distributor gives the changes in flow angle as well as those in kinetic moment, so that the maximum energy available at the runner inlet and the flow angle are known. The runner computation will indicate whether it is capable of transforming this available energy. The geometry of the blades and the meridian channel are defined on the basis of drawings or from site measurements (articulated arm or theodolite) if no drawings are available or if the runner's geometry has been changed.

A study of the draft tube has not been envisaged for the moment, although it is a major element in the turbine, because of the complexity of the unsteady phenomena that take place in it. Calculation of this element with a Navier-Stokes program has still not been validated experimentally, and so cannot be systematically employed. Two types of computation have been envisaged for this procedure:

- Calculation using a Euler type code: the EULER-IMHEF [5]. The CALECHE © code (Metraflu), a finite element code based on a structured mesh, will be used for future calculation of the runner;
- Calculation with a Navier-Stokes type code: the N3S © [1] code (EDF-Simulog) in its turbomachine version.

3. Application to a real-life case

The Pinet hydropower plant on the Tarn river in France was chosen to validate this procedure. It was commissioned in 1929, with five vertical axis Francis turbines, each outputting maximum mechanical power of 8 MW at $250 \ rpm^{-1}$ for a net head of 32 m and a discharge of $Q = 30.4 \ m^3/s$. The Thoma number σ, or cavitation factor, defined as being the ratio between net positive suction specific energy and specific energy is close to 0.3.

3.1. PHASE 1

Data gathered on the site, including some of the original drawings, were used in this study. A statistical comparison of the existing turbine with the Francis turbines inventoried in the data base is presented in Figures 1 and 2. The dimensionless discharge and energy coefficients are defined as $\varphi = \frac{Q}{\pi.\omega.R^3}$ and $\psi = \frac{2.E}{\omega^2.R^2}$, where ω is the rotating speed and R the raduis. The following observations can be made:

- Specific speed, defined as $\nu = \frac{\omega.\sqrt{\frac{Q}{\pi}}}{(2.E)^{3/4}}$, which is representative of the type of turbine used, is well suited to the available head.
- The reference runner diameter, representative of the turbine's characteristic dimensions, is well suited to the discharge through the turbine.
- The reference diameter, on the other hand, is not suited to the mechanical power output by the runner. It is therefore possible, with a better runner, to supply more mechanical power to the shaft.
- The coefficient φ^2/ψ, that is representative of the specific kinetic energy at the runner outlet versus head, is within the statistical average. This coefficient is also proportional to energy losses in the draft tube, and therefore would seem to be suited to the runner being used.
- The Thoma number σ, or cavitation factor, is not isolated from the other values for Francis turbines (the number of values is reduced, since σ is not routinely given in the literature). However, this factor alone cannot give information on turbine elevation setting, which is why the coefficient κ, defined as $\kappa = \frac{\psi_{\overline{1}}}{tg\beta_{\overline{1}}}.\sigma_{min}$ where $\beta_{\overline{1}}$ is the relative flow angle at the runner trailing edge, has been calculated.
- The lower turbine setting acceptable for this machine requires that coefficient $\kappa \simeq 1$. A considerable margin of safety was taken by using a value of 3.5.

On the basis of these observations, it was proposed that a Francis runner with a smaller diameter would be better suited to the present operating conditions. A statistical comparison with the Kaplan turbines in the inventory (figure 1) shows that the use of a Kaplan (or Impeller) turbine with this reference diameter could also be envisaged.

It should be noted that almost none of the original equipment has been replaced (generators, transformers, runner, etc.), and, as there is no reg-

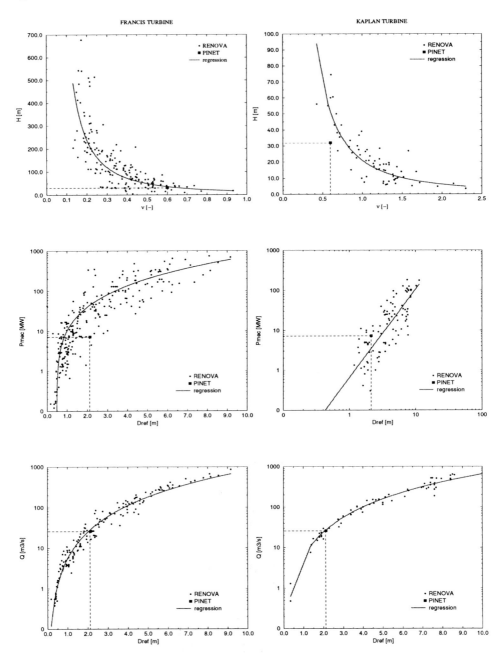

Figure 1.   Head/Specific speed - Mechanical power/Diameter - Discharge/Diameter

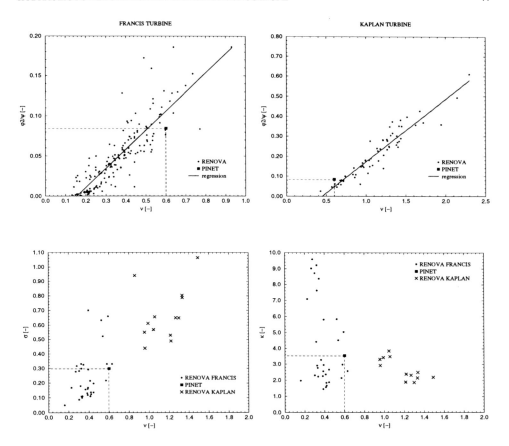

Figure 2. Coefficient $\frac{\varphi^2}{\psi}$/Specific speed - Thoma number/Specific speed - Coefficient κ/Specific speed

ulating device, the turbines always run at the same operating point. The essential information gathered on the turbine was the following:

— Major losses were measured in the inlet valves, as well as in the five-branch manifold upstream from those valves.
— The double curve spiral casing has two by-passes that divert flow directly downstream from the turbine in the event of any problem, and leakage has been observed at the outlet from those elements.
— The runner is suffering from relatively major erosion due to cavitation at the pressure-side inlet of the blades near the runner band.

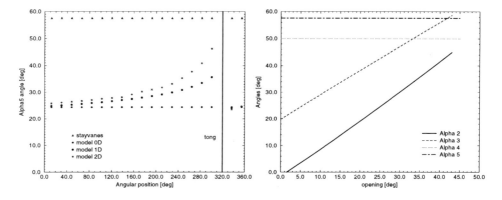

Figure 3. Estimated angles at the spiral casing outlet

Figure 4. Geometrical angles in the distributor

3.2. PHASE 2

Index 5 & 4 and 3 & 2 are corresponding respectively to the stay vane leading & trailing edge and guide vane leading & trailing edge. Three models for evaluation of average hydraulic flow angle at the spiral casing outlet have been programmed into *Renovaturb* and are shown in figure 3. The assumptions applied are the following:

- model 0D: $C_u = cste$ at the inlet and $C_r = cste$ at the outlet.
- model 1D: discharge is absorbed evenly around the machine.
- model 2D: the spiral casing is considered as a coiled pipe where headloss is assumed constant per unit of length.

It can be noted that the skeleton angle at the leading edge of the stay vanes is 57.6 [deg] while the models used give an average angle close to 28 [deg]. We may therefore expect major incidence losses at this location. On Figure 4, we note an intersection of the curves representing angles α_3 and α_4 for an opening of 33.18 [deg], where losses would be at a minimum in the distributor. As the opening used was 33.5 [deg], the setting of the stay vane is well suited to the setting of the guide vane.

3.3. PHASE 3

The distributor is of the radial type with adjustable guide vanes. It is composed of symmetrical stay vanes and guide vanes. For the Euler calculation, a structured mesh of a channel between blades, shown in Figure 5, was created with 29,600 nodes: $I_{max} = 74$ $J_{max} = 20$ $K_{max} = 20$, where direction i corresponds to the main direction of flow, and k to the height of the distributor. 10,000 iterations, i.e. one hour of computation time on an IBM Risc 6000-3AT work station (196 Mb) were required. The mesh for the

N3S computation was obtained from a structured hexahedral mesh measuring 70x16x15 divided into tetrahedrons to give an unstructured mesh of 125,941 nodes. The computation was done on EDF's Research and Development Department's Cray X-MP, with 1000 iterations and 3.5 hours of computation time.

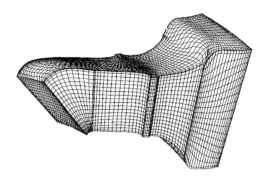

Figure 5. Structured mesh of the computational domain

3.3.1. Results of the Euler computation

For each side of the mesh i the average value of kinetic moment and hydraulic flow angle α was calculated (Figure 6), using the following formula:

$$\overline{r.c_u} = \frac{\int_{A_i} r c_u \vec{c}.\vec{n} dA}{\int_{A_i} \vec{c}.\vec{n} dA} \qquad \overline{tg\alpha_i} = \frac{\overline{r.c_m}}{\overline{r.c_u}} = \frac{\int_{A_i} r c_m \vec{c}.\vec{n} dA}{\int_{A_i} r c_u \vec{c}.\vec{n} dA} \qquad (1)$$

Two calculations, one with an initial angle of 28 [deg] and the other with 24.37 [deg], were run. Despite the difference in the flow angle at the distributor inlet, the average kinetic moment and the average flow angle at the outlet remained unchanged. It was observed that the average flow angle at the distributor outlet α_2 is about 37.25 [deg], while the geometric angle of the camber line α_{2geom} is 34 [deg] (see Figure 4). Maximum dimensionless energy available at the runner inlet, assuming that we are close to the top point and that $U.C_u \simeq 0$, is $\psi_{disp} = 0.628$ $(2.r.c_u = 0.7)$.

3.3.2. Results of the Navier-Stokes computation

For the same flow inlet conditions, the average flow angle at the distributor outlet is about 33.0 [deg], so 4 [deg] lesser than the Euler computation. The viscosity effect on the outlet flow angle is significant. Maximum dimensionless energy available at the runner inlet is $\psi_{disp} = 0.73$. Since the net dimensionless energy available for the turbine is $\psi_{\overline{1e}} = 0.806$ and the efficiency (measured) is close to 87%, the dimensionless transformed energy is about $\psi_{mec} = 0.701$, which is in accordance with the computational ψ_{disp} value.

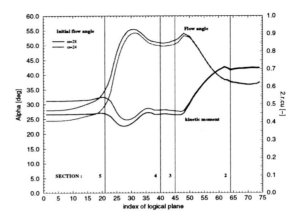

Figure 6.   Flow angle and kinetic moment

4. conclusion

This first study was the opportunity to set up an overall balance sheet on the hydraulic elements of the scheme and to emphasis sensitive points. The next stages in the project will be to measure runner geometry on site, in order to analyse it and to have all the elements needed to define, from a technical standpoint, the modifications required.

5. Acknowledgments

The authors wish to thank J-F Combes and E. Dueymes of EDF-DER for their help in achieving the N3S computations. They wish, also, thank Hydro-Vevey S.A and Sulzer Hydro for their contribution of the development of *Renovaturb*.

References

1.   J.-P. Chabard, B. Metivet, G. Pot, and B. Thomas. An efficient finite element method for the computation of 3d turbulent incompressible flows. *Finite Elements in Fluids*, vol. 8, 1992.
2.   EPRI. *Hydropower Plant Modernization Guide-Hydroplant modernization*, volume 1. Electric Power Research Institute, 1989.
3.   EPRI. *Hydropower Plant Modernization Guide-Turbine Runner Upgrading*, volume 2. Electric Power Research Institute, 1989.
4.   P. Henry. *Turbomachines hydrauliques*. Presses polytechniques et universitaires romandes, 1992.
5.   E. Parkinson, P. Dupont, R. Hirschi, J. Huang, and F. Avellan. Comparison of flow computation results with experimental flow surveys in a Francis turbine. In *XVII IAHR International Symposium*, Beijing, China, September 1994.

(*)Now at Hydro-Vevey

A THREE DIMENSIONAL SPIRAL CASING NAVIER-STOKES FLOW SIMULATION

A comparative study

Ch. Bruttin
J.-L. Kueny

IMHEF-LMH / EPFL
Av. de Cour, 33
1007 Lausanne
Switzerland
christophe.bruttin@imhef.dgm.
epfl.ch

B. Boyer
K. Héon
T. C. Vu
General Electric Hydro
795 First Avenue
Lachine, Québec H88 258
Canada
boyerb@hydro.ge.com

E. Parkinson

Hydro-Vevey S.A
Av. des deux gares, 6
1800 Vevey
Switzerland
hydro@hydro.ch

Abstract

Navier-Stokes flow simulations are now commonly used in the process of hydraulic turbomachinery design. Comparison with experiments and careful understanding of the behavior of such tools is however very important to the correct interpretation of the results. Following this trend, Navier-Stokes flow simulations, using three solvers, applied to a pump turbine spiral casing in turbine mode, are presented and discussed in this paper. The physics of the flow and guidelines for the interpretation of the results are emphasized

Figure 1. Pump-turbine model on test-rig (Hydro-Vevey SA)

1. Introduction

The understanding and the prediction of the flow behavior in a turbine spiral casing can be of crucial importance in a refurbishment project where the slightest efficiency

E. Cabrera et al. (eds.), Hydraulic Machinery and Cavitation, 81–90.
© 1996 *Kluwer Academic Publishers. Printed in the Netherlands.*

increase in any component of the turbine can make the difference. Even more, non classical flow patterns can occur when dealing with old designs. Therefore, sole local analysis provide accurate information to design and/or optimize hydraulic components [1, 2, 3, 4]. Three dimensional flow simulations appear then as the optimum solution, not talking yet of experimental analysis. Navier-Stokes simulations, when interested in energetic loss predictions and local vane optimization, short cutting all mesh generation problems, suffer from the amount of required mesh nodes and boundary definitions in such components. Indeed, the spiral casing outlet flow is strongly coupled with the stay vanes' hydraulic definition. Therefore, either the numerical domain of the simulation includes the stay vanes, implying large mesh sizes and mesh generation complexity [5], either the outlet boundary condition simulates the stay vane ring influence. This solution requires reasonable mesh sizes and is therefore affordable on standard workstations. Such an option requires extensive validations.

This paper examines a pump turbine spiral casing in turbine mode and various codes are applied. Following a brief description of the geometry, different outlet boundary conditions are detailed and tested with three codes. The influence of the turbulence modeling, either mixing length or k-epsilon types, is also analyzed. Both global and local analysis outline the possibilities and the limits of such flow simulations in spiral casings, with regards to reasonable CPU time and memory sizes.

2. Spiral casing description

2.1 GEOMETRY

The geometry used for this study a pump turbine spiral casing, illustrated when operated on the test rig at Figure 1. The geometry of the spiral casing, see Figure 2, is defined with 21 sections along the azimuth direction.

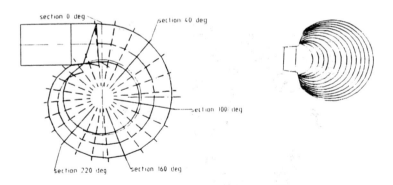

Figure 2. Geometry of the spiral casing

2.2 MESH GENERATION

Two different meshes are generated. In the first case, a block structured grid was created with FIMESH© of FDI. The blocks are structured with a butterfly description of the computing domain, see Figure 3, for the inlet and spiral pipes. A single block is added for the inlet section of the runner. This method ensures a correct grid distribution near the wall. The main hypothesis made for this grid generation is that the stay vane connected to the spiral tongue is not taken into account in the mesh generation, i.e. no solid walls boundaries exist in this part. Consequently, discharge can cross over from the end of spiral casing pipe to the inlet sections. Mesh dimensions vary, according to the considered solvers, from 70'000 to 90'000 nodes.

As a second case, a 3-D structured single block body fitted grid system of the casing is defined, see Figure 3, using GE Hydro's CFD-based computer aided engineering system. On the opposite to the previous mesh definition, no fluid connectivity exists in the tongue domain. Therefore, no discharge exists in the tongue domain between the inlet pipe and the spiral end. The structured mesh dimensions are approximately of 60'000 nodes.

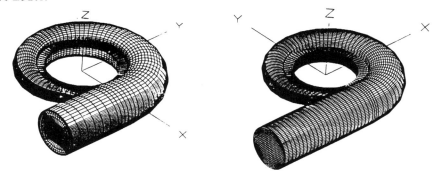

Figure 3. Butterfly and single bloc spiral casing meshes

3. Flow simulation strategy

3.1 TWO DIFFERENT APPROACHES

Three turbulent Navier and Stokes packages were used to perform the simulations: FIDAP© of FDI, TASCflow© of ASC, both commercial packages and GE Hydro's proprietary software. A short description of their main features is provided below. According to these different features, two different approaches are applied to calculate the spiral casing's flow. The spiral casing flow is calculated as an isolated turbine component with FIDAP and TASCflow, whereas with GE Hydro's solver, applied as a two stage approach for the casing / distributor assembly.

3.2 ISOLATED CALCULATIONS WITH FIDAP AND TASCFLOW

FIDAP is a finite element Navier-Stokes software, with k-ε and mixing length turbulence modeling capabilities. TASCflow is a multiblock finite volume code using a standard k-ε turbulence modeling. In all cases treated with FIDAP and TASCflow, the boundary condition at the inlet is given as an uniform normal velocity profile at the inlet section. At the outlet, different options are used to simulate the stay vanes. The first option is to impose a constant pressure at the outlet of the domain. These cases are referenced as F1, F3 and F4 in table 1, where all tested configurations are listed. « F » stands for FIDAP. A second option is to impose an open boundary condition, cases T1 and T2, where « T » stands for TASCflow, and last an open boundary with a drag coefficient (1), case F2.

$$f_n = 0{,}5\rho C_l u_n^2 \qquad (1)$$

As the simulated flow is turbulent, a zero velocity including logarithmic laws is imposed for all walls. A constant intensity of 5% and an eddy length of one tenth of the inlet pipe diameter are imposed at the inlet section for cases F3, F4, T1 and T2, for which k-ε turbulence modeling is applied.

3.3 CASING / DISTRIBUTOR CALCULATIONS WITH GE HYDRO'S SOLVER

As stated previously, a casing/distributor assembly is analyzed in a two-level flow analysis approach with GE Hydro's solver. A global flow analysis is performed for the combined geometry where the distributor region is treated as a porous medium in order to simulate the flow resistance or head loss produced by the distributor. On the other hand, to provide input to the porous medium treatment, a series of 2D viscous flow analysis is conducted to determine the individual distributor head loss where the incoming attack flow angle is obtained by the global flow analysis. Therefore, near the outlet portion of the casing flow domain, where the presence of the distributor is simulated by a porous medium, Darcy's coefficient has to be specified. For this specific application, the Reynolds-averaged Navier-Stokes equations with an extra inclusion of the porous medium treatment based on the Darcy's law are adopted. Validation of the computational model against experimental data and its application as a hydraulic design tool are explained in [6, 7, 8, 9]. The casing / distributor flow analysis is performed with the GE Hydro CFD-based computer aided engineering system developed specifically for hydraulic turbine application [10, 11]. The solution, referenced as G1, where « G » stands for GE Hydro. This information is obtained with a series of 2D flow analysis for the distributor at 20° of wicket gate opening.

TABLE 1. Characteristics of considered flow simulations cases

Case	Solver	Nodes	Turbulence	Outlet Condition	CPU time
F1	Fidap	70'000	mixing-length	Pressure	8 hours
F2	Fidap	70'000	mixing-length	Drag coefficient	8 hours
F3	Fidap	70'000	k - ε	Pressure	20 hours
F4	Fidap	90'000	k - ε	Pressure	35 hours
T1	TASCflow	70'000	k - ε	Pressure	6 hours
T2	TASCflow	90'000	k - ε	Pressure	8 hours
G1	GE Solver	60'000	k - ε	Free Outlet Porous media	5 hours

4. Global flow analysis

4.1 DISCHARGE DISTRIBUTION

The first analysis to be performed when considering the amount of data generated by the flow simulations is the discharge distribution along the main flow direction. It is indeed a key factor for the runner inflow conditions, in turbine mode. It is illustrated at

Figure 4, considering a single case per solver, where the discharge in radial cross sections of the spiral is presented versus the azimuth direction. As observed, the discharge remains constant in the inlet pipe. The chart shows the influence of the assumption made for the spiral tongue in the FIDAP and TASCflow computations. In these two cases, the discharge decreases faster than in GE's computations due to the lack of the stay vane. The opposite appears at the casing end.

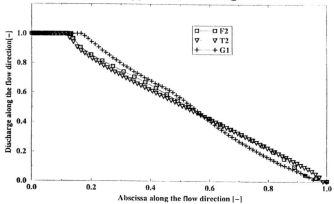

Figure 4. Discharge distribution along the main flow direction

4.2 AVERAGED FLOW ANGLES

The knowledge of flow angles in the spiral casing is of a great importance for stay vane optimization. Two flow angles can be extracted from the flow computations: a mean flow angle based upon the velocity integration along the stay vane height and an energetic flow angle. This mean flow angle is defined as

$$\overline{\alpha} = \frac{\int \alpha C_r dz}{\int C_r dz} \qquad (2)$$

Its distribution along the azimuth abscissa for all calculated cases is illustrated at Figure 5. This chart is directly related with Figure 4 and calls for similar comments. Indeed, F2 and T2 profiles cross over G1's profile at the same position, i.e. roughly 140°. The comparison between F and T computed angles shows two different patterns:
• TASCflow and FIDAP with mixing length turbulence modeling where flow angles remain within a margin of 4°.
• FIDAP computations with k-ε.
The energetic transfer between C_r and C_u components is analyzed using an energetic flow angle (3).

$$\overline{tg\alpha}_e = \frac{\overline{C_r U}}{\overline{U C_u}} \qquad (3)$$

It is also used to defined an optimum stay vane angular setting, with regards to incidence losses. In such a case, the boundary layer domain should be eliminated from the integration domain, as shown later, within the local flow analysis. Its azimuthal distribution is presented at Figure 6.

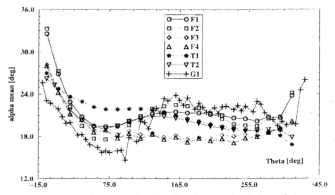

Figure 5. Mean flow angle distribution at the inlet of the stay vanes

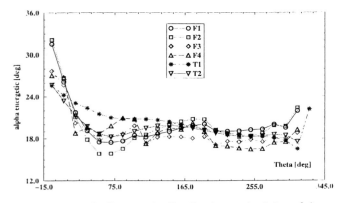

Figure 6. Energetic flow angle distribution at the inlet of the stay vanes

5. Local flow analysis

To analyze the different results given by the three solvers, 4 vertical survey sections are chosen at the stay vanes' inlet section. Their respective azimuth positions are 40, 100, 160 and 220 degrees, see Figure 2.

5.1 TANGENTIAL AND RADIAL VELOCITIES

Both tangential C_u and radial C_r velocity components for all four survey sections are plotted versus the vertical z axis in Figures 7 and 8 respectively. All solutions present similar quantitative values, forgetting the boundary layers. Sole G1's results in sections 160 and 220 degrees show a clear difference. It is related to the discharge distribution, itself function of the tongue treatment, considered as a solid wall for G1. The thinnest boundary layers are calculated by both finite volume solves (T1, T2, G1), independently of FIDAP's turbulence model. It will thus directly influence the predicted energetic losses.

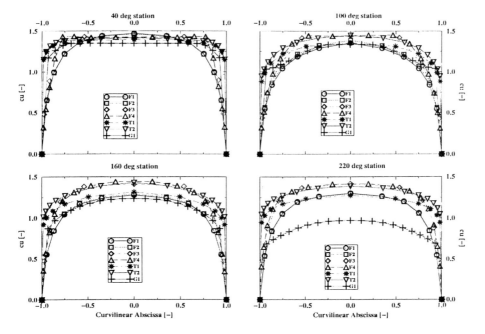

Figure 7. Cu distribution for 4 reference stations

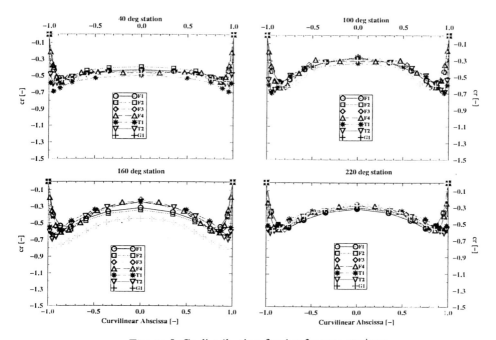

Figure 8. Cr distribution for 4 reference stations

5.2 LOCAL PRESSURE TRANSFER

To illustrate the pressure to kinetic energy transfer, considering for example case T2, the momentum equation in the radial direction is considered in (4).

$$c_z \frac{\partial c_u}{\partial z} + c_r \frac{\partial c_r}{\partial r} + \frac{c_u}{r} \frac{\partial c_r}{\partial \theta} + \frac{c_u^2}{r} = -\frac{1}{\rho} \frac{\partial p}{\partial r} + \upsilon (\nabla^2 c_r - \frac{c_r}{r^2} - \frac{2}{r^2} \frac{\partial c_u}{\partial \theta}) \qquad (4)$$

In a first approximation, it can be simplified as (5) :

$$c_r \frac{\partial c_r}{\partial r} + \frac{c_u^2}{r} \approx -\frac{1}{\rho} \frac{\partial p}{\partial r} \qquad (5)$$

Considering in the following the azimuth position $\theta= 270°$, vertical iso-pressure values are observed in the meridian section, see Figure 9.

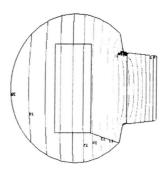

Figure 9. Iso-pressure in meridian section $\theta= 270°$

To analyze this pressure gradient source terms, the momentum equation terms are estimated along a line in the middle plane of the spiral casing. Referring to Figure 10,

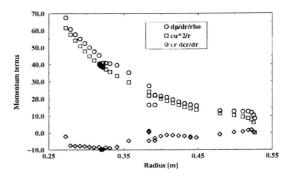

Figure 10. Momentum equation terms along a mid-line in the meridian plane

the pressure transfer is completely due to the centrifugal term along this line, as pressure and C_u terms profiles are equivalent. On the opposite, the bottom near wall pressure gradient is mainly due to the radial component of the velocity. It induces, as seen in Figure 8, higher values of C_r near the wall than at the middle of the section

6. Conclusion

Following the analysis of all cases from the various codes in use for this study, conclusions arise on the geometry handling, turbulence modelling and two stage calculations. Correct modelling of the tongue geometry is a key factor for such flow analysis. Indeed, it will directly influence the discharge distribution and therefore all main flow parameters. The blockage effect of the distributor, which creates a non uniform azimuth pressure distribution, as a second order influence, is also unavoidable and justifies all calibrated porous medium type analysis. Indeed, when comparing these simplified approaches to a complete geometric modeling of casing / stay vanes / wicket gate, limited CPU time and memory sizes remain very important factors if such analysis is to be used in the design process. As observed with the butterfly type meshes, used for FIDAP and TASCflow, a high sensitivity of the flow angles to nodes distribution is illustrated. Care is thus needed when using such meshes. Moving to turbulence modeling, different behaviors are observed between k-ε and mixing length models with FIDAP. Further study is needed to complete this analysis.

These first numerical steps of an on-going development work on high pressure turbine components, as discussed in the paper, raise many points. Future work should include comparisons with a complete geometric model (fully modelling casing, stay vanes and wicket gate) and of course, experimental results.

7. Acknowledgments

The authors wish to thank Hydro-Vevey SA for allowing the publication of this study and GE Hydro for its contribution to this work.

8. List of symbols

Symbol	Name	Unit
x, y, z	Cartesian coordinates	m
r, θ, z	Cylindrical coordinates	m, °, m
A_{ref}	Inlet reference area	m^2
Q_{ref}	Inlet reference discharge	$m^3.s^{-1}$
U_x, U_y, U_z	Cartesian velocity components	$m.s^{-1}$
C_r, C_u, C_z	Cylindrical velocity components	$m.s^{-1}$
C_{ref}	Reference velocity (Q_{ref}/A_{ref})	$m.s^{-1}$
p	Pressure	Pa
k	Turbulent kinetic energy	$m^2.s^{-2}$
ε	Turbulent dissipation rate	$m^2.s^{-3}$
α	Flow angle	°
$α_e$	Energetic flow angle	°

9. Bibliography

1. Vu, T.C., Héon, K., Coulson, S., Neury, C., Winkler, S. - *A comparative study of computational methods for a high specific speed Francis runner flow analysis* - XVII IAHR Symposium, Beijing, China, 1994.
2. Parkinson, E., Neury, C., Vuilloud, G., Walther, W. - *An optimum combination of numerical and experimental tools for pump-turbine developments* - Modelling, testing & Monitoring for Hydro-Powerplants - II, Lausanne, Switzerland, 1996.
3. Avellan, F., Dupont, Ph., Fahrat, M., Gindroz, B., Henry, P., Parkinson, E., Santal, O. - *Numerical and experimental analysis of the flow in a Francis turbine* - XV IAHR Symposium, Belgrade, 1990.
4. Parkinson, E., Dupont, Ph., Hirschi, R., Huang, J., Avellan, F. - *Comparison of flow computational results with experimental flow surveys in a Francis turbine model* - XVII IAHR Symposium, Beijing, China, 1994.
5. Soares Gomez, F., Mesquita, A., Ciocan, G., Kuény, J.L. - *Numerical and experimental analysis of the flow in a pump-turbine spiral casing in pump operation* - XVII IAHR Symposium, Beijing, China, 1994.
6. Vu, T.C., Héon, K., Shyy, W. - *A comparative study of three dimensional viscous flows in semi and full spiral casings* - XV IAHR Symposium, Belgrade, 1990.
7. Vu, T.C., Héon, K., Desbiens, E. - *Development of a new hydraulic design tool for spiral casing / distributor geometry* - Water Power Conference, Denver, Colorado, USA, 1991.
8. Shyy, W., Vu, T.C. - *Modeling and computation of flow in a passage with 360° turning and multiple airfoils* - Journal of Fluid Engineering, Vol. 115, pp. 103-108, 1993.
9. Vu T.C., Shyy, W. - *Viscous Flow Analysis as a Design Tool for Hydraulic Turbine Components* - Journal of Fluid Engineering, Vol.112, March 1990.
10. Vu, T.C., Héon, K., Shyy, W. - *An integrated CFD tool for hydraulic turbine efficiency prediction* - 5th International Symposium on Refined Flow Modeling and Turbulence Measurements, Paris, France, 1993.
11. Vu, T.C., Héon, K., Shyy, W. - *A CFD based computer aided engineering system for hydraulic turbines* - XVII IAHR Symposium, Beijing, China, 1994.

TIP CLEARANCE FLOW IN TURBOMACHINES - EXPERIMENTAL FLOW ANALYSES

Gabriel Dan CIOCAN
LEGI - ENSHMG - INPG
Laboratoire des Ecoulements Géophysique et Industriels
B.P. 53 - 38041 Grenoble Cèdex 9, FRANCE

Jean Louis KUENY
LEGI - ENSHMG - INPG
Laboratoire des Ecoulements Géophysique et Industriels
B.P. 53 - 38041 Grenoble Cèdex 9, FRANCE

1. Abstract

To analyse the tip leakage flow, it is considered as necessary the set up of an experimental database. For that, a test's water tunnel, without rotation, that represents the blade-to-blade canal of a rocket inducer, has been constructed and qualified at Centre de Recherches et d'Essais de Machines Hydrauliques de Grenoble (CREMHyG - France). For these configurations the flow evolution in the water tunnel has been analysed, and specially the vortex of the leading edge by the 3D velocity measurements, performed by LDV. The tip clearance flow analysis has been carried out by velocity measurements: to the inlet and the exit of the tip clearance by LDV and into the tip clearance by PIV. The different levels of the turbulence kinetic energy has been measured at the different positions. The analysis of the phenomenon has been thus achieved.

2. Introduction

Among the rotating machines, a large number operates with an open runner and a more or less important tip clearance, between the extremity of blades and the annulus wall. Many pumps, inducers, Kaplan turbines, bulb turbines and compressors operate on this principle, similarly as the propeller's boats. At the level of this tip clearance a leakage flow develops from the pressure side to the suction side and this flow generates efficiency losses that can be very important, up to 15-20 % of total power. These phenomena of the tip clearance flow are badly known and modelled, and, if it want to advance from a practical viewpoint this machines type, it is important to minutely analyse the mechanisms and to research influence parameters, to find construction solutions which also avoid to the maximum their pernicious effects.

E. Cabrera et al. (eds.), Hydraulic Machinery and Cavitation, 91–100.
© *1996 Kluwer Academic Publishers. Printed in the Netherlands.*

Owing to the development of the experimental investigation means and the numerical modelling, the analysis and the understanding of the tip clearance flows have largely progressed.

From the phenomenology viewpoint, the pressure gradient from the pressure surface to suction surface generates the transverse flow that crosses the endwall. The geometry and dimensions of the tip clearance can determine a complex tip clearance flow. Thus one can find three mechanisms: a separation, a reattachment and a zone of the flow mixing due to a strong leakage vortex.

The separation is due to the flow acceleration in the pressure side with linestreams that are convergent to the endwall, and therefore linked to the form the blade extremity. This behaviour induces a locally increased velocity by passage section decrease (Venturi effect).

According to Bindon [2], for the forms well-adapted forms of the blade extremity profile this phenomenon can be avoided. An other solution consists in the utilisation of the blades supply tablets coming to prolong the pressure surface (Morphis and al. [13]), but this system puts problems of the mechanical resistance viewpoint. This mechanism is important because it is a furnace of cavitation for liquids and it is the place of the acoustic emission and the erosion of the blade extremity.

The reattachment in the endwall is an other very sensitive phenomenon of the geometry. In this case, besides the form of the blade extremity, an important parameter is the proportion between the thickness of the endwall τ and its length e. Thus according to Wadia and al. [15] for a proportion $\tau / e < 0.25$ the reattachment takes place on the blade. This behaviour type is often the case of turbines and more rarely of pumps and compressors. Moore and al. [12] has measured the pressure distribution on the blade extremity of a subsonic turbine. One notices a rapid decrease of the pressure on the pressure surface, with a minimum from the edge. In the separation zone the pressure is nearly constant and increases abruptly in the reattachment zone, while being constant for the rest of the endwall. This separation zone favours thermal changes which can damage the blade. It can be eliminated by the adjustment of the profile form.

A flow mixture due to a strong leakage vortex at the exit of the endwall is generated in the interaction zone between the tip clearance flow and the main flow. According to several authors the leakage vortex is separate from the blade in a well precise position. This position and the trajectory of this vortex, the centre of a strong turbulent dissipation, and the separation point are a function of the geometry and of the blade loading, as well as the pressure gradient sense, direct (turbine) or reverse (pump) - see Bindon [2], Dishart [6].

At the tip clearance exit, the tip clearance flow is in interaction with the main flow and with the other secondary flows: the annulus wall boundary layers and the blade boundary layers. The annulus wall boundary layers represent the 3D boundary layers which develop on the annulus and that induces a supplementary viscous perturbation according to the machine type. In the turbine one can even suppress the tip clearance flow for weak value of the tip clearance. In the pump or in the compressor the effect of this phenomenon comes to increase the leakage flow.

Lakshminarayana and al. [9] has shown both experimentally and numerically the existence of the blades boundary layers deviation from the hub to the annulus. Thus, into an inducer with a wedging angle of 80°, radial velocities in the boundary layers reach 20% the upstream main velocity, they cannot therefore be neglected for a correct representation of the phenomenon in interaction. In the pressure surface, this blade boundary layers feeds the tip clearance flow and in the suction surface, this centrifugal boundary layer will roll with the leakage flow in the leakage vortex.

All these mechanisms cause important losses, and for this reason they have become the object of the numerous studies. The most recent ones Dishart and al. [6], Yaras and al. [18], Storer and al. [14] the different mechanisms of the losses generation for compressors have been identified and separated: the reattachment viscous effects in the tip clearance, the losses by mixture in the tip clearance, the viscous effects due to the interaction of the tip clearance flow with blade boundary layers, the viscous effects due to the interaction of the tip clearance flow with annulus wall boundary layers and the losses by mixture of the tip clearance vortex with the main flow. The authors are generally agreed from a qualitative point of view. There remain however important differences to the level of the qualitative estimation of these phenomena. According to Dishart and al. [6], for a tip clearance value of 2.5% of the chord, 40% of losses are generated in the tip clearance and mainly in the separation bulb. Yaras and al. [18], finds 17% of losses to the exit of the tip clearance for a turbine and with a tip clearance of 2.1% of the chord. Others authors estimate to 10-15%.

These phenomena have been studied both experimentally that numerically. The complete calculation of the flow in a turbomachine, taking into account the tridimensional effect and interaction runner-stator is not again possible with the modern computers. Knowing the importance of the tip clearance flow for a correct phenomena's representation on the whole calculation, at this phase, the introduction of the simplified patterns of the tip clearance appears as a necessity. To this end an experimental database has been constituted. This database will serve to validate tip clearance flow calculation. From the numerical and experimental results, a physical pattern of the tip clearance flow will be introduced in the complete calculations of the turbomachines.

Further on we present the constitution of the database and the first results obtained.

3. Experimental facility and methods

A test tunnel simulating, without rotation, the blade-to-blade channel of a rocket inducer has been conceived to CREMHyG laboratory. This tunnel has the runner blades reduced to two fixed profiles allowing to analyse the tip clearance effects. The final form of the tunnel is presented in Le Devehat [5] - see the figure 1.

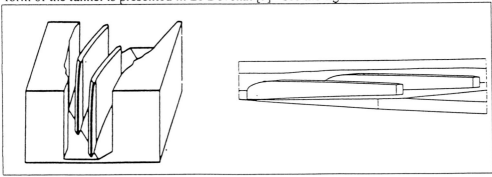

Figure 1. Experimental facility

The definition of the section and profiles has been obtained by a meridian calculation in accordance with Katsanis and al [7]. This has been performed for the nominal flow rate of the inducer. To obtain a similitude of pressure coefficients on the blade and the profiles, a blade-to-blade calculation, quasi 3D, S1-S2 type, in finished element, has been performed -see Kueny et al. [8].

The Reynolds number based on the chord is 4×10^6 and the tip clearance can vary from 1 to 5 mm, for a constant thickness of the profile of 10 mm.

For the tunnel validation and to determine the pressure evolution on the tip of the profiles and on the profiles the followings have been provided: 60 pressure tapes on the suction surface, 48 on the pressure surface and 48 in the tip clearance and on the wall. A water gate cupboard with electrically driven gates was used for the connection between the static pressure taps and two Desgranges and Huot balances. The electro-gates' cupboard was driven by a micro-computer. The precision of the pressure measurements is of +/- 1mb.

The test tunnel has been performed in plexiglas to allow the carrying out of the velocity measurements by LDV and PIV. The quality performance of walls has allowed the carrying out of the measurements.

The LDV system used is a DANTEC system whit two components, in the back scatted light and transmission by optical fibre, with a laser of 5W argon - ion source. Optical characteristics of the system are found in the table 1.

Laser wave lengths	488 / 514.5 nm
Probe diameter	60 mm
Beam expander	80 mm
Beam spacing probe	38 mm
expander	70.5 mm
Focal length	310 mm
Fringe spacing	2.16 /2.277 nm
Volume of measurements $\sigma_x = \sigma_v \sim$	0.1 mm
$\sigma_z \sim$	1 mm

Table 1 Characteristics of the LDA system

The signal processing for the Doppler frequency was accomplished by a spectrum analysis. The measurements precision has been estimated following Mofat and al. [11] of 4%. The displacement of the laser probe has been made by a carriage. The precision of displacement is 0.1 mm. The reference position for the decreasing of the measurement volume has been obtained by the crossing of beams on the internal part of the wall.

For the PIV part a cooperation with DANTEC company has been performed. The stroboscopic lightsheet (thickness of ~1 mm) has been performed with a double-cavity Nd:YAG laser that performed 200 mJ during 9 ns. An optical high power light guide has allowed the performance of measurements on all the length of the chord. The images have been obtained with a CCD camera developed by DANTEC, in a succession of a pair of images at 100 μs, that constitutes a measure. The interval between two measures has been 100 ms. The camera resolution has been 768x480 pixels for a spacial resolution of 30x40 mm. The synchronisation of the shot segment with luminous flashes, the acquisition and the image processing have been performed with the

The two LASER measurements systems have been performed with the spherical particles, in silver covered glass, empty inside to have the water density. The size of these particles is 20 μm.

4. Results of measurements

After the qualification of the test tunnel from the pressure similitude viewpoint, a control of the inlet section has been performed. Thus the secondary velocity in this section does not exceed 3% of the main velocity.

4.1. PRESSURE MEASUREMENTS

The static pressure measurements in the profiles have allowed to validate the tunnel in similitude viewpoint - see Le Devehat et al.[4]. The pressure distribution on the profile are shown in figure 2.

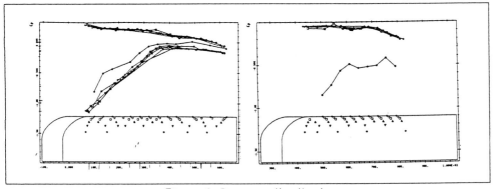

Figure 2. Pressure distribution

4.2. VELOCITY MEASUREMENTS BY LDV

To perform the 3D measurements, one carries out two orthogonal measurements. The calculation of the optical arrangements is presented in Mesquita and al [1]. For the flow description 6 sections considered representative have been chosen - see figure 3 and 4.

The measurements have been carried out for three different tip clearance values: 1 mm; 3 mm and 5 mm. In figures 3, 4, 5 and 6 have been represented the comparisons between the tip clearance flow of 1 mm and 5 mm.

The velocities have a-dimensioning by the average velocity in the inlet section and the turbulence rate by the square of this velocity. The length has been divided whit the characteristics dimensions of the inlet section: height - ho, width - lo and the length of the chord - lch.

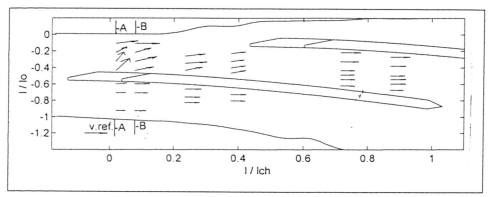

Figure 3. Longitudina! section to 3 mm of walls to a gap clearance of 1 mm

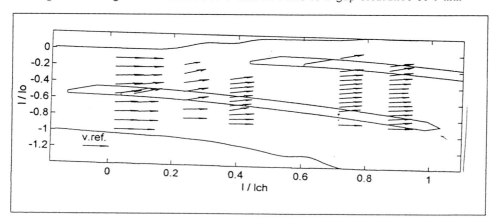

Figure 4. Longitudinal section to 3 mm of walls to a gap clearance of 5 mm

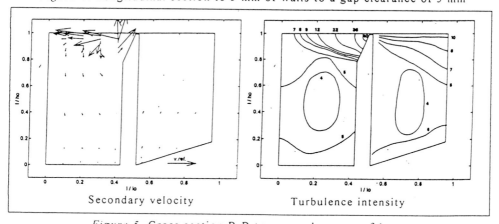

Figure 5. Cross section B-B to a gap clearance of 1 mm

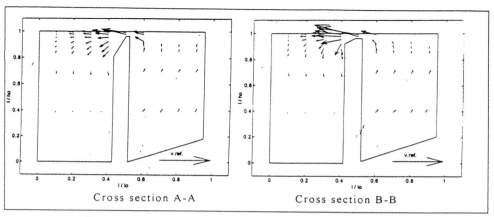

Cross section A-A   Cross section B-B

Figure 6. Secondary velocity to a gap clearance of 5 mm

4.3. VELOCITY MEASUREMENTS BY PIV

The measurements presented in the figure 7 and 8 are accomplished at half of the tip clearance of 5 mm. The figure 7 corresponds of the leading edge and 8 to the half of the chord. The lack of the data in the figure 8 is due to the reflections produced by taps of pressure.

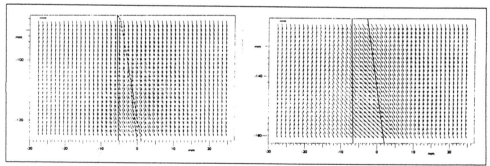

Figure 7. Velocity toleadingedge to 3 mm of walls to a gap clearance of 5 mm

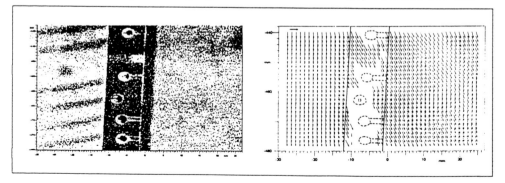

Figure 7. Velocity to mi-chord to 3 mm of walls to a gap clearance of 5 mm

In the figure 9 one presents a comparison between a measurements' profile performed by LDV and by PIV. A direct quantitative comparison of the different LDA and PIV measurements gives excellent agreement - less 4% .

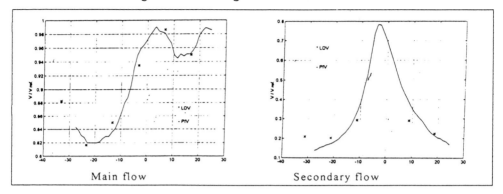

Main flow   Secondary flow

Figure 9.LDV - PIV comparison

5. Discussion

For a tip clearance of 1 mm a flow mixing due to a strong leakage vortex develops to the superior limit of the leading edge. The reasons for the formation of this vortex are both the effect of theleadingedge and the leakage flow generated by the strong pressure gradient in this area, as well as the interaction of this flow with main flow. Vortex develops on 25% of the section (in depth) and is deviated to the second profile, without penetrating between profiles. For the rest of the cord, the tip clearance flow is less important, even if it is always present. The lacks of the kinetic energy prevents the development of a new vortex flow to the exit of the tip clearance and it is convected by the main flow. This is linked also to the decrease of the pressure gradient along the chord.

For the tip clearance of 5 mm one does not observe the same behaviour. The vortex disappears and it is replaced by a tip clearance flow that develops in the blade-to-blade channel but that no longer rolls in the vortex. The increase of tip clearance (5% of the height of the profile) has reduced the velocity values and the height of the tip clearance has allowed the development of the flow which crosses the tip clearance (the same gradient of pressure by for the tip of 1 mm). The tip clearance flow intensity decreases along the cord in the direction of the decrease of the blade loading.

This behaviour - disappearing of the leading edge vortex for a increasing in the gap, was found out also in the experiments performed on a prototype inducer at the KSB-GUINARD Company. This is favorable for the off-design flow rate, with a view to decreasing the NPSH - see Ciocan and al. [3].

The turbulence intensity, calculated in the isotope hypothesis, has increased values upon coming out of the tip clearance and in the vortex action area (up to 40%) for lowers values than 5% in the rest of the section. This distribution is not very different for several cross sections and tip clearance.

6. Conclusion and Perspectives

The study performed on the tunnel permitted to obtain the evolution of the leakage phenomenon and the estimation of the leakage flow rate for several values of the tip clearance. A database has been carried out on this pattern. The behaviour observed has been confirmed by the tests performed with inducer prototype.

To be able to understand the phenomena interaction, a more complete running of the possibilities offered by this pattern is therefore necessary: the study of the the the interaction of the boundary layers limit with the tip clearance flow, the influence and the appearance of the cavitation. The instrumentation of the profile with 8-10 unsteady miniature pressure sensors, placed on the profile, will allow to quantify the level of depression on the suction side corresponding to the starting point of the leakage vortex. The utilisation of an unsteady pressure probe for the turbulent losses determination is envisaged.

Using these experimental data as boundary conditions and the validation data bank, a numerical simulation, using a uses 3D Navier-Stokes numerical code with a k-ε model of the turbulence, will be performed. A simplified geometry allowing to preserve the pressure gradients from the pressure to the suction surface will be used to represent the studied phenomena.

The experimental and numerical approaches will allow to describe the tip clearance flow and the mixing area formed by this flow and the main flow. Finally we shall try to specify the influence of each parameter in the losses' generation process.

7. Acknowledgements

The authors take this opportunity to thank both DANTEC Company for their participation to PIV measurements and the Région Rhônes-Alpes, KSB-GUINARD Company and SEIM-MGI COUTIER for their financial support and participation in this project.

8. References

1. Amarante Mesquita, A. L., (1993) Experimental Techniques for the Flow Characterisation in Hydraulic Turbomachines Models, *Proceedings of the 2nd Meeting of the Latin-American Division of the International Association of the Hydraulic Research* - IAHR, Vol. 1, pp. 193-209, Ilha, Solteira Brazil
2. Bindon, (juin, 1988) Mesurement and formation of tip clearance loss, *Gas turbine and Aeroengine Congress*, Amsterdam,
3. Ciocan, GD, Kueny, JL, Analyse experimentale de l'ecoulement de jeu dans les inducteurs, *PEH 96 Conference*, Bordeaux, France, 1996.
4. Le Devehat, E., Kueny, J.L., Dellby, J., (decembre, 1992) Tip leakage flow in blade tip clearance of a 3D test-facility, *Seminar on 3D Turbomachinery Flow Prediction*, Val d'Isere, France,
5. Le Devehat, (1993) Ecoulement de fuite dans l'entrefer d'une turbomachine. Analyse expérimentale et numérique - *Thèse de doctorat à l'INPG*
6. Dishart, Moore, (1989) Tip leakage losses in a linear turbine cascade - *ASME paper*, Brussels, Belgium,
7. Katsanis, « Katsanis, revised FORTRAN program for calculating velocities and streamlines on the hub-shroud midchannel stream surface of axial, radial or mixed flow turbomachines of annular duct »,

8. Kueny, J.L., Le Devehat, E., (january, 1994) Tip clearance flow calculation - 3D channel with tip clearance flow, *Seminar and workshop on 3D turbomachinery flow prediction II*, Vol. III, pp. 200-223, Val d'Isere, France,

9. Lakshminarayana, Sitaram, Zhang, (janvier, 1986) End-wall and profil losses in a low-speed axial flow compressor rotor - *Journal of Engineering for Gas turbines and Power*, vol. 108,

10. McCluskey, D.R., (july 1995) Obtaining high-resolution PIV vector maps in real time - *International Workshop on PIV*, Fukui, Japan,

11. Mofat, R. J., (1985) Using Uncertainty Analysis in Planning of an Experiment, *Journal of Fluid Engineering*, Vol. 107, pp. 173-178,

12. Moore, Le Fur, (juillet, 1990) Computational study of 3D turbulent air-flow in a helical rocket pomp-inducer - 26th *Propulsion Conference*, Orlando,

13. Morphis, Bindon, (juin, 1988) The effects of relative motion, blade edge radius and gap size on the blade tip pressure distribution in annular turbine cascade with clearance - *Gas turbine and Aeroengine Congress* - Amsterdam,

14. Storer, Cumpsty, (1991) Tip Leakage flow in axial compressors - *Journal of Turbomachinery*, vol. 113

15. Wadia, Booth, (janvier, 1982) Rotor-typ leakage. Part II: Design optimization through viscous analysis and experiment - *Journal of Engineering for Power* - vol.104,

16. Yaras, Sjolander, (1989) Losses in the tip leakage flow of a planar cascade of turbine blades - Secondary Flows in Turbomachines, Luxembourg

NUMERICAL PREDICTION OF HYDRAULIC LOSSES IN THE SPIRAL CASING OF A FRANCIS TURBINE

PETER DRTINA
Sulzer Innotec AG
CH-8401 Winterthur, Switzerland

AND

ANDREAS SEBESTYEN
Sulzer Hydro AG
CH-8023 Zürich, Switzerland

1. Introduction

Power plants that have been in operation for several decades show a considerable potential for improvement in hydraulic performance by bringing the design up to modern standards. This situation has led to numerous rehabilitation projects for existing hydraulic power plants.

Up to now most of the modernisation of Francis turbines has been aimed at replacing the runner while retaining all of the existing stationary parts of the machines, e.g. spiral casing, stay vanes, guide vanes and draft tube. In order to improve the hydraulic performance of a Francis turbine as a whole, attention should also be paid to the non-rotating elements of the machine.

For a long time the design of stay vane profiles has been influenced mainly by the considerations of safety, stress analysis and manufacturing technology. Indeed, the high pressure part of the Francis-machine is one of the most critical elements from the point of view of material loading. The blade profiles of stay vane rings of many existing machines mainly reflect the result of mechanical and not hydraulic optimization: they are robust and of very simple geometrical shape. Modern safety engineering routinely uses three-dimensional finite element stress calculus for this part of the machine. This allows arbitrary shaped stay vane profiles to be applied with sufficient safety margin. Hence it is now possible to design stay vane profiles of high performance which optimize both hydraulic and stress considerations.

E. Cabrera et al. (eds.), Hydraulic Machinery and Cavitation, 101–110.
© *1996 Kluwer Academic Publishers. Printed in the Netherlands.*

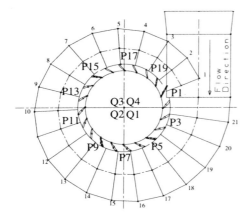 

Figure 1.   Draft of spiral casing in the   *Figure 2.*   Entire computational grid for
symmetry plane.   the spiral casing.

The spiral casing transforms the rectilinear penstock-flow into a uniformly distributed swirling inflow over the circumferential inlet surface of
the stay vane ring. The stay vane passages turn the flow to the guide vanes
under a given flow angle. The requirement from the hydraulic point of view
is that the flow turning in the annulus cascade of the stay vanes produces
a very low level of hydraulic loss. The use of an appropriate 3D viscous
calculation procedure enables the hydraulic loss to be minimized by the
choice of the best possible hydraulic profiles for the stay vane ring.

The present paper shows that the application of a sophisticated calculation method is capable of evaluating the difference in hydraulic loss
between two differently shaped stay vane rings. Consequently, it is possible
to estimate theoretically the gain in energy terms which can be reached by
the modification and improvement of an existing stay vane ring.

Shyy and Vu (1993) presented a simplified two-level approach to simulate the non-rotating components upstream of a hydraulic turbine runner.
In a first step a flow calculation for the spiral casing has been performed
with the details of the distributor smeared out by applying a porous medium
treatment instead of modelling each stay vane. In a second step an individual cascade is investigated with the inlet conditions extracted from the
global spiral casing analysis. Lipej et al. (1994) describe numerical results
for all components of a Francis turbine obtained by applying FIDAP and
TASCflow. For the spiral casing less than 7000 nodes were used and neither
the stay vane ring nor the guide vanes were taken into account, which cer-

tainly leads to a weak prediction of the flow. Parkinson et al. (1994) point out that the circumferential discharge distribution obtained from their spiral casing computation does not correspond correctly to the experimental data due to the absence of the stay vanes in their calculation. Instead of including them into the computational domain, which they claim to be very time consuming and node increasing, they tend to prefer the application of a loss model to include the effects of the distributor. Ruprecht et al. (1994) also performed separate computations for the spiral casing and the the distributor. The inlet condition for their single passage computations are extracted from the spiral casing results. They propose an iterative procedure to account for the coupling between spiral casing and distributor. DeHenau and Markovich (1995) and DeHenau (1995) use a combination of coarse grid flow simulations for the complete spiral casing including the distributor and a refined calculation on a selected stay vane/wicket gate passage with inlet boundary conditions taken from the coarse grid simulation. This approach yields a reliable loss prediction enabling the authors to compare the original stay vane design with an optimised design.

In the present case a flow analysis for a spiral casing including 20 stay vanes is performed by solving the incompressible Reynolds-averaged Navier-Stokes equations closed with a standard k-ε turbulence model and logarithmic wall functions. The results obtained for an original stay vane design are used to optimise a new design with respect to losses. The second key issue of the present investigation is to prepare the inlet velocity profile for the downstream components (wicket gate, runner, draft tube) of the Francis turbine. Stage calculations carried out for all components from the stay vane inlet to the draft tube outlet leading to a numerically predicted hill chart are described in detail by Keck et al. (1996). A description of the basic numerical procedure as well as an extensive comparison between stage calculation results and experimental data is published by Sick et al. (1996).

2. Computational method, grid generation and boundary conditions

Early potential flow models (Athanassiadis, 1984; Kubota et al., 1984) and 3D finite element models (Ruprecht et al., 1994; Soares-Gomes et al., 1994) have been developed for studying the flow in spiral casing assuming the flow to be inviscid. An accurate study requires today the solution of the laws of conservation for viscous flow by using the finite element method or the finite volume method in the discretization scheme.

The flow in the spiral casing has been considered as incompressible, steady and viscous, leading to the need of applying a Navier-Stokes solver. The commercially available CFD code TASCflow (ASC Canada) which solves the conservation equation for mass and the Reynolds-averaged Navier-Stokes equations has been used. This code is nowadays very often used in the industrial environment (Casey *et al.*, 1995). Validation of the code (Sick *et al.*, 1996) promises high reliability and therefore its use deserves to win confidence as a design tool.

For the generation of the computational grid the ICEM-CFD mesh generator P^3 (PCUBE) has been applied. In combination with its advanced CAD capabilities the model setup time could be reduced remarkably compared to the previously applied grid generation technique. The translator interface of the mesh generator allows the transformation of the constructed mesh to most of the well-known commercial CFD solvers. For the case under investigation a block-structured grid has been generated which consists of 190 blocks. In order to avoid highly skewed grid cells the main body of the mesh which fills the casing region has been constructed using a butterfly arrangement of subblocks. In order to capture all significant flow phenomena the flow field resolution had to be sufficiently high. On the other hand the available computer resources restrict the mesh size. In the past spiral casing grids have been set up with different refinement levels ranging from 7000 nodes (Lipej *et al.*, 1994), 31185 nodes (Shyy and Vu, 1993) up to 240000 nodes (DeHenau and Markovich, 1995). Even unstructured grids have already been applied (Soares-Gomes *et al.*, 1994). With the grid shown in figure 2 the above mentioned constraints are sufficiently met. It consists of about 280,000 nodes for one half of the spiral casing and all stay vane passages.

At the spiral casing inlet the mass flow and appropriate turbulence conditions for the k,ϵ-equations are defined, while at the outflow boundary only the average pressure is set. This allows the pressure distribution to be a result of the computations even in the outlet area. Due to geometrical symmetry only half of the spiral casing and the stay vane ring had to be modelled. At all solid surfaces the logarithmic law of the wall was applied.

3. Simulations

Numerical simulations have been performed in a three-step design procedure. First, flow calculations for a spiral casing including a set of original stay vanes have been carried out. Five different operating points were considered ranging from $0.75\dot{Q}_1$ to $1.25\dot{Q}_1$ with $\dot{Q}_1$ being the design volume flow rate. Overall and local flow phenomena as well as loss sources have been identified and studied. In the second step an improved set of stay vanes has been developed based on the computational results and on the engineering knowledge available. In the third step, the new design was transfered to the grid generator via the IGES data interface. After performing moderate

Figure 3.   Distribution of hydraulic losses for the original design.   *Figure 4.*   Distribution of hydraulic losses for the improved design.

modifications, the new grid has been generated and the calculations were repeated for three different operating points. To achieve a satisfactorily converged solution (maximum residuals $\leq 10^{-4}$) the CPU time required for a single run is in the range of 20h - 40h on a SGI Power Challenge. A restart run from a previous solution may converge within 5h - 10h.

4. Computational results and analysis

For an industrial engineer involved with the design and optimisation of hydraulic turbine components the most important criteria to judge a design is the overall component efficiency. In order to see the effect of modifying the stay vane profiles, the overall component loss defined by

$$L_{global} = \frac{\int_{A_{inlet}} \dot{m} p_{tot} dA - \int_{A_{interface}} \dot{m} p_{tot} dA}{\int_{A_{inlet}} \dot{m} p_{tot} dA} \cdot 100 \qquad (1)$$

has been plotted in figure 5 as a function of the volume flow rate. A parabola which fits to the original stay vane data has been added. The numerical results show the expected behaviour. With increasing flow rate (increasing Reynolds number) the loss increases quadratically. Comparing the original and the optimised design shows clearly the improvement which has been achieved by the geometrical modification of the vanes. In the design point the loss is reduced from 3.72% to 3.20% that is a reduction of 14%. Thus the global goal of the optimisation procedure has been reached. The next question which arises automatically is: How is the loss distributed in the spiral casing and the stay vane ring and what are the reasons for the improvement?

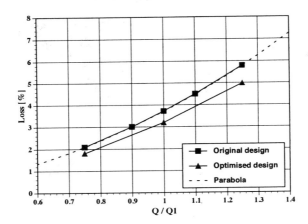

Figure 5.   Global loss of spiral casing and stay vane ring as a function of volume flow rate.

Figure 3 (original design) and figure 4 (optimised design) indicate the regions of high losses which are located, as expected, on the suction sides of the stay vanes. The grey scale patterns range from 0% loss (light grey) to 10% loss (black) based on:

$$L_{local} = \frac{p_{tot,local} - \overline{p}_{tot,inlet}}{\rho g H} \cdot 100 \qquad (2)$$

Comparing both loss distributions reveals some interesting aspects.

- At each of the 20 stay vanes the losses have been reduced.
- For both designs losses are high within the first and the second quarter (Q1 and Q2, see figure 1). A loss minimum occurs in the third quarter (passages P15-P17).
- The location of the major loss changes from the suction to the pressure side for the last 3 vanes.

Figure 6 shows quantitatively the circumferential variation of the normalised total pressure $p_{tot}(\Theta)/\overline{p}_{tot,inlet}$ (averaged over the span) up- and downstream of the stay vanes. $\Theta = 0^{\circ}$ is located at the 'leading edge' of the tongue and increases in main flow direction. Note: While the R1-lines start with the first passage (P1), the R2-lines (downstream of the stay vanes) start within passage P16 (same circumferential position Θ!). The peaks indicate the circumferential position of the stay vanes wakes. For the optimised design - marked with symbols - the size of the wakes is clearly reduced. Even upstream of the stay vanes the losses decrease slightly. Some

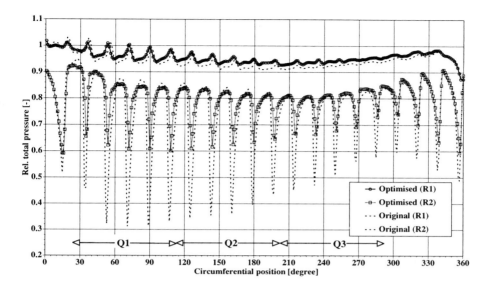

Figure 6. Circumferential variation of normalised total pressure.

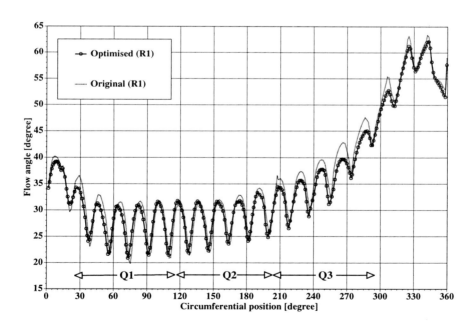

Figure 7. Circumferential variation of flow angle.

more insight can be gained by looking onto the vector plots in figures 8 to 11.

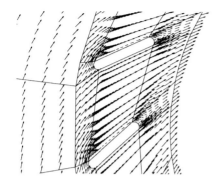 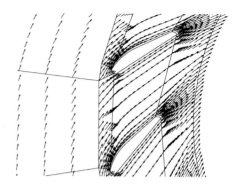

Figure 8.   Velocity vectors in passage P7 at $z = 0$ for the original design.
        Figure 10.   Velocity vectors in passage P7 at $z = 0$ for the improved design.

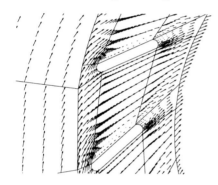 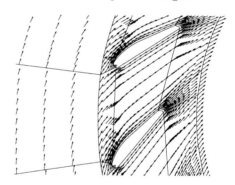

Figure 9.   Velocity vectors in passage P7 at $z = 0.8 * H$ for the original design.
        Figure 11.   Velocity vectors in passage P7 at $z = 0.8 * H$ for the improved design.

The original design shows a separation zone close to the head cover, which extends far downstream. For the optimised design this separation zone is removed by the improved alignment of flow and stay vanes. From figure 12 the same conclusion can be drawn. While the velocity and flow angle profiles upstream of the stay vanes only change very slightly from the original to the optimised design, significant differences occur in the downstream profiles. The scatter (difference of minimum and maximum values) is reduced remarkably with the mean profiles being only marginally affected. Velocities and flow angles are non-dimensionalised with the average values at the corresponding radius. Figure 7 shows that in the circumferential sectors Q3/Q4 the mean flow angle is in the range of about $40° - 50°$ which corresponds very well to the inclination of the stay vanes which is $\approx 53°$

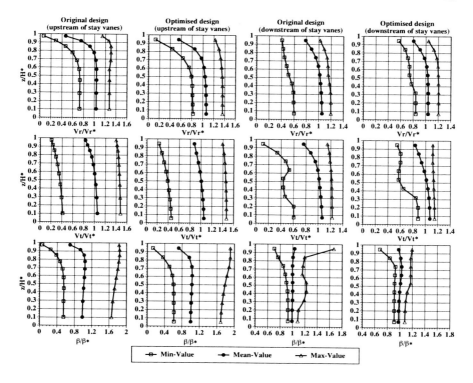

Figure 12.   Velocity and flow angle profiles. First and second columns show velocity and flow angle profiles upstream of the stay vanes for the original and optimised design, respectively. Third and fouth columns show downstream conditions for the original and optimised design, respectively. Rows: Non-dimensional radial velocity component (top), non-dimensional tangential velocity component (middle) and non-dimensional flow angle (bottom).

(original) and $\approx 48°$ (optimised) respectively, and accounts for the low loss in passages P15 to P17. Note the weak design of this particular spiral casing which does not lead to constant flow angle around the circumference.

5. Conclusion

The results presented in this paper show that numerical flow simulations can be integrated successfully into a design procedure for hydraulic turbine components which can dramatically shorten the time needed for a new development. The visualisation of the flow (pressure, velocity vectors, flow angles) and derived quantities (losses) helps the designer to analyse the flow and to decide which modifications may be most appropriate for improvements to important local flow phenomena. In the present case the position and strength of the separation zones, and the detailed loss distribution, made it possible to adjust the stay vane profiles very quickly leading to the demonstrated improvements.

The described flow problem could only be solved with the help of a sophisticated CAD/grid generation software and an efficient robust flow solver. Both are available as commercial software packages offering a wide range of capabilities to tackle complex industrial problems.

In particular, the spiral casing calculations presented here show the important effect of flow angle distribution into the stay vanes and the potential for improvements with stay vane redesign.

References

Athanassiadis, N. (1960) Potential Flow through Spiral Casings, Diss. ETH Zürich, Mitteilungen aus dem Institut für Aerodynamik Nr.30

Casey, M.V., Borth, J., Drtina, P., Hirt, F., Lang, E., Metzen, G., Wiss, D. (1995) The Application of Computational Modeling to the Simulation of Environmental, Medical and Engineering Flows, *SPEEDUP Journal*

DeHenau, V. Markovich, M.S. (1995) Optimization of the Sir Adam Beck II Turbine Distributor using Computational Fluid Dynamics, *Updating and Refurbishing Hydro Power Plants*, International Water Power and Dam Construction and National Hydropower Association, Nice

DeHenau, V. (1995) Turbine Rehabilitation: CFD Analysis of Distributors, *Waterpower'95*, Proc. Intl. Conf. Hydropower

Sick, M., Casey, M.V., Galpin, P. (1996) Validation of a Stage Calculation in a Francis Turbine, *Proc. XVIII IAHR Symposium*

Keck, H., Drtina, P., Sick, M. (1996) Numerical Hill Chart Prediction by means of CFD Stage Simulation for a Complete Francis Turbine, *Proc. XVIII IAHR Symposium*

Kubota, T., Takimoto, S., Kawashima, M. (1984) Effect of Stay Vane Shape on the Performance of Hydraulic Turbine, *Proc. 12th IAHR Symposium*

Lipej, A., Jost, D., Oberdank, K., Velensek, B., Jamnik, M. (1994) Numerical and Experimantal Flow Analysis in a Francis Turbine, *Proc. XVII IAHR Symposium*

Parkinson, E., Dupont, P.,Hirschi, R.,Huang, J., Avellan, F. (1994) Comparison of Flow Computation Results with Experimental Flow Surveys in a Francis Turbine, *Proc. XVII IAHR Symposium*

Ruprecht, A., Bauer, C., Riedelbauch, S. (1994) Numerical Analysis of Three-Dimensional Flow Through Turbine Spiral Casing, Stay Vanes and Wicket Gate, *Proc. XVII IAHR Symposium*

Shyy, W., Vu, T.C. (1993) Modeling and Computation of Flow in a Passage with 360-Degree Turning and Multiple Airfoils, *J.Fluids Engineering*, **115**

Soares-Gomes, F., Mesquita, A.A., Ciocan, G., Kueny, J.L. (1994) Numerical and Experimental Analysis of the flow in a Pump-Turbine Spiral Casing in Pump Operation, *Proc. XVII IAHR Symposium*

IMPROVEMENTS OF A GRAPHICAL METHOD FOR CALCULATION OF FLOW ON A PELTON BUCKET

MORTEN HANA
Ph.D. student
Norwegian University of Sience and Technology
Trondheim, Norway

Abstract

A graphical method for calculation of particle paths over Pelton buckets is presented. This method involves a laborious graphical derivation. This paper presents a new method for deriving particle velocities and accelerations. This method is based on cubic splines, and is calculated on a computer. A comparison with a graphical method is performed.

Resume

Nous présentons une méthode graphique de calcul de la trajectorie des particules liquides dans une aube de turbine Pelton qui implique une dérivation graphique trés laborieuse. Une nouvelle méthode de calcul par ordinateur d'obtention des vitesses et des accélérations est présentée, utilisant des fonctions splines cubiques. Nous comparons ses résultats avec ceux de la méthode graphique.

E. Cabrera et al. (eds.), Hydraulic Machinery and Cavitation, 111–119.

1. Introduction

At the Hydropower laboratories at the Norwegian University of Science and Technology, there has been a tradition for graphical calculations of flow on a Pelton bucket. This tradition dates back to 1918, when Henrik Christie wrote the first diploma thesis at the Laboratory. In this thesis he introduced the method, which is based on the simple formulation that a particle on the surface always has a acceleration that is perpendicular to the surface.

Later Hermod Brekke continued this tradition at Kværner, where Christie was his superior. Brekke modernized the method by introducing reduced quantities, which made the method more universal. In 1984 he presented his work in a paper called "A general study on the design of vertical Pelton turbines" [1].

As part of a Ph.D. thesis the graphical method has been investigated. Improvements done to the method will be presented in this paper. The scope of this work has been to make part of the graphical analysis into a numerical analysis.

2. Theoretical Background

The flow over Pelton buckets is non-stationary. Thus making it very difficult to establish a thorough theoretical analysis of the flow over Pelton buckets.

By using a graphical method that traces particles, a analysis of the flow is possible. This graphical method is based on the following facts [1]:

1. The acceleration on a particle on the jet surface is zero before entering the buckets. And in the same way is the acceleration zero on the particle after the outlet of the buckets.
2. The acceleration on a particle must always be normal to the water surface.
3. The water volume on a bucket included the outlet water behind the buckets is equal to the water volume cut out of the jet.

These three simple facts makes a graphical analysis possible.

2.1. REDUCED VALUES

All the values used for velocity, acceleration and time in this paper is reduced values. The use of reduced quantities in the graphical method was introduced by Brekke [1].

Reduced values are calculated the following way:

$$\underline{V} = \frac{V}{\sqrt{2gH}}$$

Where $\underline{V}$ is the reduced dimensionless velocity, V is the velocity, g gravity and H is the net head.

The reduced angular velocity is defined as:

$$\underline{\omega} = \frac{\omega}{\sqrt{2gH}}$$

Which gives the redusec angular velocity the dimension $[m^{-1}]$.

The reduced acceleration is found by examining equation (1), which is presented later, which gives that the acceleration should have the same dimension as $2\omega V$ which is $[m^{-1}]$:

$$\underline{a} = \frac{a}{2gH}$$

If we consider that $a = \frac{dV}{dt}$, the reduced time become:

$$\underline{t} = t \cdot \sqrt{2gH}$$

In the remaining of this paper all quantities will be reduced as described above

3. Procedure for the graphical analysis

The first step in analyzing flow over a Pelton bucket is to establish the geometrically description of the water surface as it moves over the bucket. One way to do this is by obtaining stroboscopic photos of the surface.

Modern visualization tools like CCD-cameras and multiflash stroboscopic light together with computer tools for photographic analysis will in the future make this part of the process faster and more precise.

It is assumed that a geometrical description of the water surface is established for a Pelton bucket for the following calculations.

The next step is to trace particle paths and derive both velocities and acceleration for the particles. The particle path is at first guessed, but experience soon makes it good guesses. Acceleration is calculated by these basic equations:

$$
\begin{aligned}
a_x &= \frac{d^2x}{dt^2} - \omega^2 R \cos\varphi - 2\omega V_y \\
a_y &= \frac{d^2y}{dt^2} + \omega^2 R \sin\varphi + 2\omega V_x \qquad (1) \\
a_z &= \frac{d^2z}{dt^2}
\end{aligned}
$$

These equations are for a relative coordinate axis on the Pelton wheel, as shown in figure 1. They demonstrates that when moving to a reference system fixed on the Pelton wheel one has to take into account the centrifugal and Coriolis accelerations.

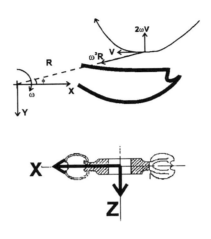

Figure 1.   Coordinate axis on a Pelton bucket

The different signs on the Coriolis and centrifugal forces is deducted when considering the coordinate axis, velocities and spin axis, as shown in figure 1. The different terms in the equations are defined in the nomenclature list at the end of this paper.

When the acceleration has been calculated it is checked according to the second statement that said that it must always be perpendicular to the surface. If this is not fulfilled the particle path is adjusted and acceleration calculated again. This iteration process is repeated until the acceleration is perpendicular to the surface.

In figure 2 two shapes of water is showed along with some of the particle paths. The particle paths is named a, b, c and so on. Figure 2 also shows the graphical calculations and derivations of first and second derivatives. Figure 2 are from a calculation done by Hermod Brekke[1].

4. Cubic Splines

When deriving the velocities and accelerations for the particle moving along a path, the laborious and time consuming method of graphical derivation has been used in the past, as shown in figure 2.

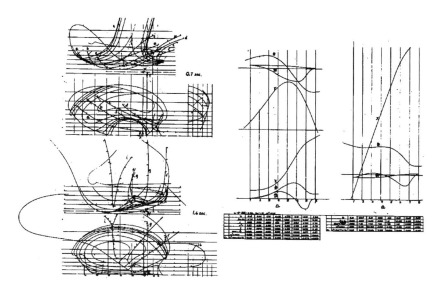

Figure 2. Example of water surfaces and graphical derivation

By using cubic splines to approximate the path, one gets the velocities and acceleration directly from the splines calculations. The cubic spline method is described in numerous numerical books, for instance the book by Cheney and Kinciad [2], on which this method is based.

Cubic Spline fitting is based upon Taylor series expansion. Taylor series expansion with regard to the time t may be written [2]:

$$S_i\left(t_i\right) = A_i + B_i t_i + C_i t_i^2 + D_i t_i^3 \qquad (2)$$

Where the index i is a counting variable that indicates the internal splines between the particle points. Thus: $i = 1, 2, \ldots, (n-1)$. Where $n =$ number of particle points. This gives $(n-1)$ equations with $4(n-1)$ unknown constants. These unknowns can be found by using boundary conditions. These boundary conditions are for x-direction:

$$S_i\left(t_i\right) = x_i \qquad (3)$$

$$\lim_{t \to t_i^-} S^{(k)}\left(t_i\right) = \lim_{t \to t_i^+} S^{(k)}\left(t_i\right), \text{ where } k = 0, 1, 2 \qquad (4)$$

$$S''\left(t_1\right) = S''\left(t_n\right) = 0 \qquad (5)$$

The last condition (5) makes this a Natural Cubic Spline (NCS) approach. The NCS condition is reasonable if we consider that the second derivative of S, in either x, y or z direction, is the acceleration in that direction. According to the first statement made earlier the acceleration is zero on a particle on the jet surface before entering and after leaving the buckets, which makes (5) a reasonable boundary condition.

The two first conditions (3) and (4) are internal boundary condition ensuring continuity in the spline fitting. The boundary conditions gives:

Boundary condition	No. of conitions
1	$2(n-1)$
2	$2(n-1)$
3	2

Totally this gives $4(n-1)$ boundary conditions, and makes the system of equations solvable. This leads to a simple and fast method that substitute the laborious method of manual derivatives.

5. Computer Implementation

When the cubic splines system of equations is established, there are several ways to implement the solution procedure. Solving can simply be done by solving each spline separately, matrix solving or commercial available numerical mathematics programs. Since the cubic spline computer algorithms as described in text books like Cheney and Kinciad [2] is quite simple, the solution algorithm was implemented by programming a C/C++ code.

By programing the algorithm it was possible to have control of the adding of Coriolis and centrifugal acceleration according to equation (1).

The code has a simple input/output format. The input reads the x, y and z-coordinates, the code then calculates the respective first and second derivatives, and then writes the result to a file. The result file can be plotted.

The input file can contain one, several or all particle paths that is to be calculated. It also reads start time, time increment and how many time steps it shall calculate.

6. Test case: Skjåk Pelton turbine

In 1963 Hermod Brekke graphically calculated particle paths on the Skjåk Pelton turbine. These calculations has been used for comparing the cubic spline method with the graphical method.

The calculation began by reading the different paths, velocities and accelerations, from drawings as showed in figure 2. The coordinates was send as input to the program, and the result was compared to the read velocities and accelerations.

Figure 3 shows the results from the calculation of particle path B. This is a particle path which enters the bucket early and hence leaves at the back of the bucket. Other more complicated paths, which enters later, was also calculated.

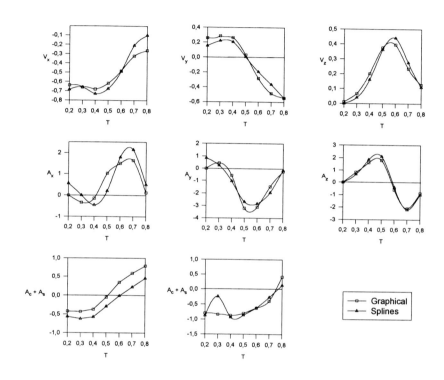

Figure 3.   Example of results from Spline calculations

The results shows very good agrement between the graphical and spline calculations. The spline calculations show a tendency of oscillation around the graphical calculations.

By examining the spline approach one can see that a small change in the coordinate gives a relative big change in acceleration. When the acceleration was used as a guideline to make small changes in the coordinate, the particle path, velocity and acceleration became smoother. The changes in the particle path was made within the reading error.

7. Further improvements

As mentioned adjustment of the particle path were done manually, but it would be possible to implement a smoothing function on the acceleration path and integrate numerically the spline function to obtain a new particle path. And by allowing a magnitude of displacement of the particle path, smoother results will be obtained. A smoothing scheme that could be used could be it Gauss-Seidel iteration. Which is similar to relaxation schemes known from multigrid methods.

For future calculations of new paths, velocities and accelerations it is necessary to check that the sum of the acceleration always is perpendicular to the surface. If the surface can be described by surface vectors, a numerical comparison can be obtained. By developing certain rules for changing the particle paths depending the difference in acceleration and surface vector a almost complete numerical calculation can be achieved.

The surface description can, as mentioned earlier, be obtained by a CCD-camera. Numerical manipulation of these results can give a files for surface vectors.

8. Concluding remarks

The Natural Cubic Spline algorithm has been successfully implemented and tested. The splines show a oscillating nature, which could be smoothed manually. Future development will include numerical smoothing schemes.

The development of modern visualization tools will make a closer link between experiments and numerical analysis of flow over Pelton buckets.

Nomenclature

a_x, a_y, a_z	$[m^{-1}]$	Reduced acceleration in x, y and z directions, respectively
$V_x = \frac{dx}{dt}$		Reduced velocity in x direction, respectively
$V_y = \frac{dy}{dt}$		Reduced velocity in y direction, respectively
$R = \sqrt{x^2 + y^2}$	$[m]$	Radius
ω		Reduced angular velocity
$\varphi = \tan^{-1}\left(\frac{x}{y}\right)$		Angle
n		Number of particles in one path
g	$\left[\frac{m}{s^2}\right]$	Gravity $= 9.81 m/s^2$
H	$[m]$	Net head

References

1. Brekke, Hermod: "A general study on the design of vertical Pelton turbines", 1984, presented at 25th anniversary Turbuinstitut, Lubljana.
2. Kinciad, Ward and Cheney, David: "Numerical mathematics and computing", 2nd edition 1985, Brooks/Cole Publishing Company.

FLOW BEHAVIOR AND PERFORMANCE OF DRAFT TUBES
FOR BULB TURBINES

T. KANEMOTO
M. UNO
Kyushu Institute of technology
Sensui 1-1, Tobata, Kitakyushu 804, Japan
M. NEMOTO
Kanagawa Institute of Technology
Shimo-ogino 1030, Atsugi 243-02, Japan
T. KASHIWABARA
Koochi Technical College
Shinkai-otsu 200-1, Monobe, Nankoku 783, Japan

1. Introduction

To cope with the warming global environment, the development of the ultra-low head energy is active by welcoming as the clean and cool energy resources. Bulb turbines, which are suitable to the ultra-low head, have the very high velocity head at the runner outlet, relative to the net head. Since the role of the draft tube on the hydraulic performance is extremely high, the bulb turbine normally uses the straight draft tube which is composed of the inlet annular diffuser with the hub followed by the diffuser having the transitional cross section from conical to rectangular shape. The performance of such a draft tube can not be predicted successfully by usual data for two-dimensional, conical and/or annular diffusers. Thus, the pressure recovery of the straight draft tube for the bulb turbine and the internal flow in the.S-shaped draft tube for the tubular turbine have been investigated [1]-[3].

The internal flow and performance of the straight draft tube equipped with the conically shaped hub with the cusp end were also investigated by us, under the free vortex type swirling flow at the draft tube inlet [4]. The flow condition at the draft tube inlet is determined from the turbining operation point, and the above free vortex type swirling flow corresponds to the flow condition discharged from the runner at the partial load operation point. As making the load increase, the swirling direction on the tip side of the runner blade becomes to differ from that on the hub side (for instance, clockwise on the tip side and counter-clockwise on the hub side). Besides, the profile of the prototype hub is not the conical shape with the cusp end but the trapezoid shape, in general.

Accordingly, this paper discusses the effects of the inlet swirling flow type and the hub shape on the internal flow structure and the performance of the straight draft tube for the bulb turbine, through the experimental results.

E. Cabrera et al. (eds.), Hydraulic Machinery and Cavitation, 120–129.

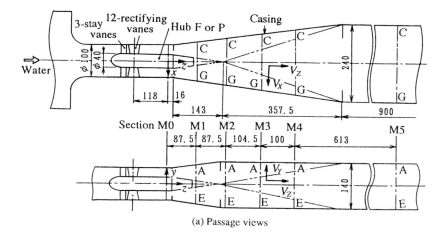

(a) Passage views

2. Model Draft Tube and Inlet Flow Condition

2.1. MODEL DRAFT TUBE

Figure 1(a) shows the tested model passage of the draft tube with the inlet approach and the downstream diffuser. Figure 1(b) illustrates the cross sections on which the traverse lines for the flow measurements are displayed with dot and dash lines. The twelve adjustable rectifying vanes can generate the necessary swirling flow in place of the rotational runner, and Section M0 which corresponds to the runner outlet was considered as the inlet datum section for the model.

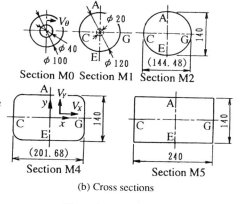

(b) Cross sections

Figure 1. Model draft tube

Figure 2 is the tail shape of the hub. The tail of Hub F is the conical shape with the cusp end (a), and the tail of Hub P is the trapezoid shape with the flat end (b). The distance from the tail end of Hub P to Section M1 is 31.5mm.

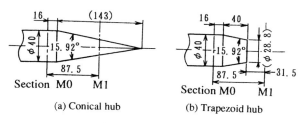

(a) Conical hub   (b) Trapezoid hub

Figure 2. Model hubs

2.2. INLET FLOW CONDITION

Figure 3 shows the flow conditions at Section M0, under which the effects of the inlet swirling flow type on the internal flow and the performance were investigated with Hub

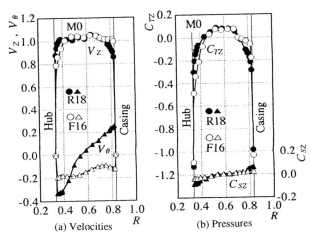

TABLE 1. Swirling intensity at Section M0 (R18, F16)

Flow type	R18	F16
$\lvert m \rvert$	0.183	0.159
m	0.0952	-0.159

(a) Velocities   (b) Pressures

Figure 3. Inlet Flow condition at Section M0 (R18, F16)

F. The abbreviations V_z and V_θ give the dimensionless velocities in the z and θ directions (Fig.1), divided by the sectional avaraged velocity v_{zo} at Section M0. The abbreviations C_{TZ} and C_{SZ} are the total and static pressure coefficients estimated with the difference from the mass flow avaraged pressures and v_z at Section M0, and R is the dimensionless radius divided by double the inlet width B_0. These flow conditions are axissymmetrical and the Reynolds numbers are 5.6×10^5 estimated with the mean abso- lute velocity and the equivalent pipe diameter at Section M0.

The inlet swirling intensities $m = \Sigma r V_\theta V_z \, dA \ / \ B_0 \Sigma V_z^2 \, dA$ and $\lvert m \rvert = \Sigma r \ \lvert V_\theta \rvert V_z \, dA \ / \ B_0 \Sigma V_z^2 \, dA$ are presented in Table 1, where r is the radius and dA is the segmental area. Above inlet flow conditions are hereafter described as R18 and F16. The inlet swirling flow type is distinguished with alphabet R or F and the numerical value alludes to the inlet swirling intensity. The swirling intensity of R18 on the casing side is slightly larger than that on the hub side, as recognized from value of m in Table 1. The intensities $\lvert m \rvert$ estimated from the absolute swirling velocity component $\lvert V_\theta \rvert$, however, are

nearly the same irrespective of the inlet swirling flow type.

Figures 4, 5 and Table 2 show the flow conditions at Section M0,under which the effects of the hub shape are investigated. The flow type with alphabet P means the case equipped with Hub P, the flow type with alphabet F means the case equipped with Hub F, and numerical value alludes to the inlet swirling intensity. These flows are also axissymmet-

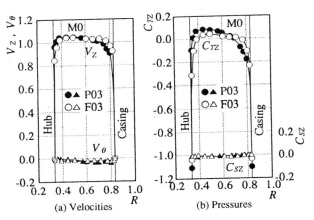

(a) Velocities   (b) Pressures

Figure 4. Inlet flow condition at Section M0 (P03, F03)

rical and the Reynolds numbers are the same as those of R18 and F16 mentioned above.

3. Effect of Inlet Swirling Flow Type on Flow Behavior

In this chapter, the effects of the inlet swirling flow type on the flow behavior in the draft tube equipped with Hub F are discussed, under the inlet swirling flow types of R18 and F16 given in Fig.3 and Table 1.

3.1. FLOW IN ANNULAR DIFFUSER

The velocity distributions at Section M2, just behind the cusp end of Hub F, are shown in Fig.6. The abbreviations V_X, V_Z are the dimensionless velocities divided by the sectional avaraged inlet velocity v_{ZO}, and Y is the dimensionless vertical distance divided by $2B_0$ (see Fig.1). The axissymmetrical flow condition of R18 at Section M0 already collapses in the annular diffuser composed of the axissymmetrical cross section, because the flow condition near the boundary where the swirling direction changes from clockwise to counter-clockwise (near $Y= -0.6$ and $Y=0.7$) is very unstable. On the contrary, the axissymmetrical flow condi-

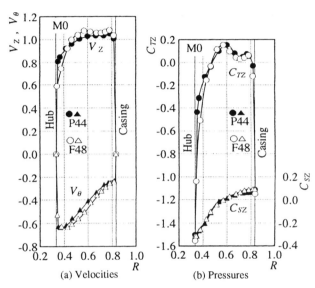

(a) Velocities   (b) Pressures

Figure 5. Inlet flow condition at Section M0 (P44, F48)

TABLE 2. Swirling intensity at Section M0 (P03, F03, P44, F48)

Hub	P		F	
Flow type	P03	P44	F03	F48
| m |	0.0301	0.442	0.0284	0.478

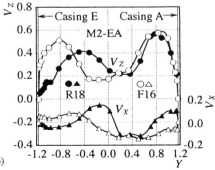

Figure 6. Velocity distributions on E-A at Section M2 (R18,F16)

tion of F16 at Section M0 is maintained moderately, but the velocity defect due to the hub is large as compared with one of R 18. Such a velocity defect is directly induced from the boundary layer along the hub wall [4]. By means of the boundary layer analysis [5] in the main flow predicted with the steramline curvature method [6], it was clear that the boundary layer flow of R18 does not separates but the flow of F16 separates at the first half of the hub wall.

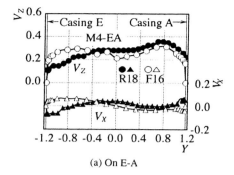

(a) On E-A

3.2. FLOW IN DOWN-STREAM DIFFUSER

The flows at Section M4 are shown in Fig. 7, where V_Y and X are the dimensionless velocity and horizontal distance (see Fig.1). The velocity deffects due to the hub come to disappesr considerably, and the flow of F16 tends to incline toward the one of the casing wall. Besides, it is noticeable that the swirling component V_X or V_Y, of R18 decreases conspicuously from Section M2 to M4 owing to the shearing stress. Such a flow mixing induced from the clockwise and counter-clockwise swirling components may make the hydraulic loss increase and the total pressure drop from Sections M2 to M4 of R18 is also larger than that of F16, as recognized in Fig.8.

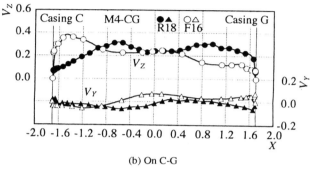

(b) On C-G

Figure 7. Velocity distributions at Section M4 (R18, F16)

At Section M5 in the straight passage composed of the rectangular cross section connected to the model draft tube outlet, the velocity V_Z distributions become uniform except for the thin boundary layer region on the casing wall,

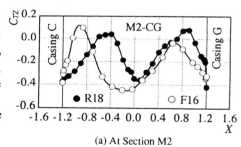

(a) At Section M2

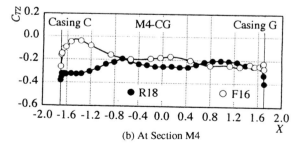

(b) At Section M4

Figure 8. Total pressure distributions on C-G (R18, F16)

and the swirling velocity components V_Y are nearly the same value, irrespective of the inlet swirling flow type (Fig.9).

4. Effect of Hub Shape on Flow Behavior

In this chapter, the effects of the hub shape on the flow behavior in the draft tube is discussed, under the inlet flow conditions given in Figs.4, 5 and Table 2.

4.1. SWIRLLESS FLOW

The inlet flow conditions of P03 and F03 given in Fig.4 were here regarded as the swirless flow, because of tiny swirling velocity components. The velocity distributions at Section M1, just behind the tail end of Hub P with the trapezoid shape, are shown in Fig.10, under the inlet flow conditions given in Fig.4. When Hub P is installed, the dead water region whose wide is nearly equal to the tail end diameter is observed obviously as marked with P03, and the axissymmetrical flow condition at the inlet already collapses in spite of the axissymmetrical passage. The flow of F03, however, is axis-symmetrica and stable because Hub F makes the annular and narrow cross section.

The dead water region due to Hub P disappears at Section M2, as confirmed in Fig.11 (P03). Such a flow mixing makes the jet flow along the casing, surrounding the dead water region, diffuse suddenly into the whole passage and causes the unexpected pressure rise in the meridional direction.

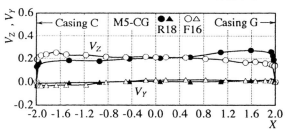

Figure 9. Velocity distributions on C-G at Section M5 (R18,F16)

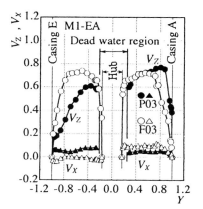

Figure 10. Velocity distributions on E-A at Section M1 (P03, F03)

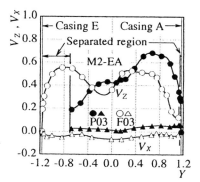

Figure 11. Velocity distributions on E-A at Section M2 (P03, F03)

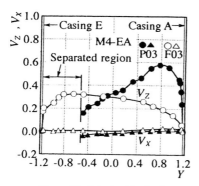

Figure 12. Velocity distributions
on E-A at Section M4 (P03, F03)

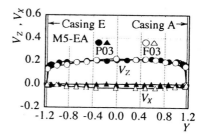

Figure 13. Velocity distributions
on E-A at Section M5 (P03, F03)

This pressure rise promotes remarkably the undesirable boundary layer separation on the casing wall. The separation is naturally observed in the flow of F03, but its region is very narrow and the separated flow reattaches soon. On the contrary, the separation of P03 never reattaches and the dead water region spreads undoubtedly at downstream Section 4 close to the draft tube outlet, as shown in Fig.12. Such a separation affects markedly the pressure recovery of the draft tube. Accordingly, this result should be taken into account, in the design of the prototype draft tube whose outlet is directly opened in the dischrging water way. As our model draft tube outlet is connected to the straight passage composed of the rectangular cross section, however, the separation reattaches finally and velocity distributions become uniform irrespective of the hub shape (Fig. 13).

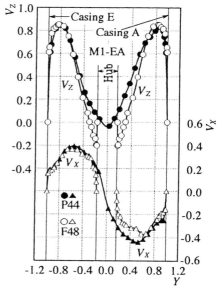

Figure 14. Velocity distributions
on E-A at Section M1 (P44, F48)

4.2. SWIRLING FLOW

The flow conditions at Section M1, under the inlet flow with strong swirling velocity compnent, are shown in Fig.14. Contrary to the flow behavior without the swirling velocity component (P03), the flow of P44 never stagnates and runs somewhat backward accompanying with the forced vortex type swirling velocity component, in the center of this section just behind Hub P. This reason may be that the flow mixing is promoted well by the strong pressure gradient in the radius direction due to the swirling velocity component. Moreover, the flow condition of P44 is axissymmetrical and its axissym-

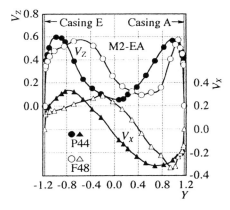
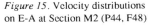

Figure 15. Velocity distributions
on E-A at Section M2 (P44, F48)

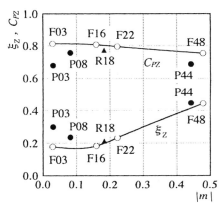

Figure 16. Model draft tube performances

metry is maintained as far as Section M2 (Fig. 15). On the other hand, the axissymmetrical flow condition of F48 at Section M1 becomes to collapse at Section M2 owing to the comparatively large reverse flow occured on the hub wall (see Fig. 14).

Except for the difference of the axissymmetrical flow conditon, the velocity profiles of P44 are almost the same as those of F48 at Section M2. Thus, the more downstream velocity profiles are also scarcely affected by the hub shape.

5. Hydraulic Loss and Pressure Recovery

Figure 16 shows the hydraulic loss coefficient ξ_z and the pressure recovery C_{PZ} for the inlet swirling intensity $|m|$, whose values are estimated with the difference of the mass flow avaraged pressures between Sections M0 and M5 and the dynamic pressure with the sectional avaraged velocity v_{zo} at Section M0. The abbreviation F with the numerical value on the full line is the data for Hub F, under the free vortex type swirling flow at Section M0 [4]. As compared with these values, the performances of R18, P03, P08, and P44 deteriorate at the same inlet swirling intensity $|m|$. These interesting results are discussed below.

5.1. EFFECT OF INLET SWIRLING FLOW TYPE

For R18 and F16 with Hub F, the hydraulic loss coefficient ξ_z and the pressure recovery C_{PZ} in the segmental passage are presented in Table 3. In the annular passage from Sections M0 to M2, ξ_z of F16 is larger than one of R18, owing to the difference of the boundary layer condition on the hub wall as discussed in the heading 3.1. On the contrary, ξ_z of R18 becomes conspicuous in the downstream passage from Sections M2 to M4, as compared with one of F16. That is, the flow mixing accompanying with the decline of the swirling velocity component with the differnt direction may contribute

TABLE 3. Effect of swirling type

(a) Loss coefficient ξ_Z

Flow type	R18	F16
M0~M2	0.080	0.115
M2~M4	0.110	0.058
M4~M5	0.023	0.010

(b) Pressure recovery C_{PZ}

Flow type	R18	F16
M0~M2	0.665	0.646
M2~M4	0.053	0.110
M4~M5	0.065	0.051

TABLE 4. Effect of hub shape

(a) Loss coefficient ξ_Z

Flow type	P03	F03	P44	F48
M0~M1	0.090	0.012	0.271	0.100
M1~M2	0.134	0.072	0.074	0.256
M2~M5	0.076	0.094	0.102	0.087

(b) Pressure recovery C_{PZ}

Flow type	P03	F03	P44	F48
M0~M1	0.466	0.553	0.315	0.348
M1~M2	0.041	0.114	0.173	0.155
M2~M5	0.173	0.145	0.201	0.256

considerably to the increase of the hydraulic loss.

The static pressure recoveres mostly in the annular passage from Sections M0 to M2 and the quantity of the pressure recovery C_{PZ} in each segmental passage corresponds conversely to the hydraulic loss, irrespective of the inlet swirling flow type.

The effect of the inlet swirling flow type on the draft tube performance is small in comparison with the same |m|, but the performance of R18 cannot be helped estimating low in comparison with another inlet swirling intensity m defined in the heading 2.2, as confirmed in Fig.16.

5.2. EFFECT OF HUB SHAPE

Table 4 shows the effect of the hub shape on the hydraulic loss coefficient ξ_Z and the pressure recovery C_{PZ} in each segmantal passage. Under the swirlless flow condition, ξ_Z of P03 is larger than one of F03, in the segmental passage from Sections M0 to M2. It may be the reason why the mixing loss hehind Hub P with the trapezoid end is larger than the friction loss on Hub F with the cusp end. Such a flow mixing undoubtedly contributes to the increase of the hydraulic loss in the segmental passage from Sections M0 to M1, too.

Under the swirling flow condition, the hydraulic loss coefficient ξ_Z of P44 is large in the segmental passage from Sections M0 to M1 owing to the mixing loss just behind Hub P, while ξ_Z of F48 is large in the segmental passage from Sections M1 to M2 owing to the friction loss on the hub wall and the mixing loss induced from the reverse flow discussed in Fig.14. The increase of the hydraulic loss of P44, however, is almost the same as that of F48. This result means that the effect of the hub shape on the hydraulic loss in the segemntal passage from Sections M0 to M2 becomes comparatively small as increase of the inlet swirling intensity.

The pressure recovery C_{PZ} is affected by not only the hydraulic loss but also the displacement effect of the velocity defect in the wake flow and the boundary layer separation. That is, the pressure recovery C_{PZ} of P03 can not be expected from Sections M1 to M2 but expected from Sections M2 to M5 where the boundary layer separation reattaches.

6. Conclusion Remark

The effects of the swirling flow type at the inlet section and the hub shape on the flow behavior and performance of the model draft tube are summarized as follows.

(1) The swirling flow with the clockwise and counter-clockwise velocity components make the internal flow unstable, and the axissymmetrical inlet flow condition collapses at the first half of the annular duffuser.

(2) The clockwise and counter-clockwise swirling velocity components decrease obviously behind the annular passage owing to the shearing stress, and that makes the hydraulic loss increase.

(3) The trapezoid shaped hub accompanies with the dead water region behind the tail end, under the swirlless inlet flow condition. The diffusion of the jet flow surrounding this dead water region makes not only the mixing loss increase but also the flow along the casing separate. On the contrary, the water behind the trapezoid shaped hub hardly stagnates, under the strong swirling inlet flow condition.

Aknowledgement

The authors would like to express them thanks to Professor T. Kubota of Kanagawa University, Professors M. Nishi and H. Tsukamoto of Kyushu Institute of Technology for kindly giving valuable guidance and advice, and thanks to Messrs. H. Akamatsu, Y. Nagai, S. Kurahashi, Y. Yamane, A. Fujishima and K. Horikawa for devotedly helping these experiments

References

1. Kikuyama,K. et al., The Swirling flow effect on low head water turbine draft tube, *Proc.JSME Symp.*(1994),411=413(in Japanese).

2. Shimizu,Y. et al., Study on S-shaped draft tube for Tubular turbines, *Trans.JSME*,**52**-474,B(1986), 585-592, and **53**-492,B(1987),2500-2506(in Japanese).

3. Loeffler,A. Jr., A navier-Stokes code for S-shaped diffusers-review, *Int. J. for Numerical Methods in Fluids*, **8**(1988),463-474.

4. Kanemoto,T. Kubota,T. and Ishibashi,H, Swirling flow in straight draft tube with annuala and rectangular cross section, *11th Australasian Fluid Mechanics Conf.* **2**(1992),845-848.

5. Schlichting,H.,*Boundary Layer Theory*, 4th Ed.(1960),574.

6. Kanemot,T. and Toyokura,T., Flow in annular diffusers(2nd Report*), Bull. JSME*,**26**-218(1983), 1323-1329.

PERFORMANCE ANALYSIS OF DRAFT TUBE FOR GAMM FRANCIS TURBINE

T. KUBOTA
Kanagawa University
Yokohama, Japan

F. HAN
Huazhong University
Wuhan, China

F. AVELLAN
IMHEF, EPFL
Lausanne, Switzerland

1. Introduction

To precisely analyze the hydraulic losses in the various components of a Francis turbine, it is necessary to know the internal flow velocity distributions in the respective components. Since, however, the hydraulic energy loss $\Delta \psi$ and the performance characteristics such as energy coefficient ψ and discharge coefficient ϕ etc. are all the one-dimensional information for a whole turbine, it is sufficient to know the sectional mean velocities along a single representative mid-streamline through the turbine. Thus, the one-dimensional flow theory is decisively useful for the loss analysis corresponding to the performance diagrams. Performance diagrams acquired by the precise model test are so reliable that utilizing the diagrams to the loss analysis is highly recommendable. Two of the authors have presented, applying the one-dimensional flow theory, a new algorithm of extracting the various component losses in bulb turbines from the performance diagrams measured with the model tests [1].

Hydraulic loss in a draft tube is strongly affected by the intensity of swirl flow at the runner outlet. Since the swirl intensity of runner outflow varies with the operating conditions of ψ and ϕ, the draft tube loss also depends on ψ and ϕ. So far, the effect of swirl intensity on the flow behavior in the draft tube is well-investigated relating to the pressure pulsations [2]. Little is reported, however, concerning its effect on the hydraulic energy loss in the draft tube. To analyze the draft tube losses versus ψ and ϕ is essential to improve the hydraulic performance of Francis turbines, and to convert the model performance to its prototype considering the reliable scale effect.

GAMM Francis model turbine was designed and manufactured by IMHEF in EPFL to provide the experimental data on the hydraulic performances and flow distributions in the turbine for the 1989 GAMM Workshop on 3D-computation of incompressible internal flows [3]. The model test results revealed that the efficiency hill-diagram has the unusual two optimum peaks with the unsatisfactory efficiencies of 0.905, respectively. To investigate the unusual performance of the turbine, a *special* test was added by relocating

130

E. Cabrera et al. (eds.), Hydraulic Machinery and Cavitation, 130–139.
© 1996 *Kluwer Academic Publishers. Printed in the Netherlands.*

the low pressure measuring section for the determination of the specific hydraulic energy of turbine, SHET from the draft tube outlet to its *inlet*. The hill-diagram of the special test showed a single optimum peak with the improved efficiency of 0.920 [4].

The aim of this investigation is to extract the hydraulic energy loss in the bend draft tube for the GAMM Francis turbine from the above two performance diagrams, the one acquired by the *normal* model test with the draft tube *outlet* section, and the other by the *special* test with the draft tube *inlet* section.

2. Specific Hydraulic Energy of Turbine SHET

The specific hydraulic energy of turbine SHET shall, by nature, be defined as the difference of the true specific hydraulic energies between the inlet and outlet of the turbine. The true flow energy in a cross section can be determined by integrating the distributions of mass-averaged pressures and kinetic energies including the swirl velocity under the assumption of the same potential energies between inlet and outlet. Practically, however, the IEC code stipulates to calculate SHET E_{nor} with the following equation (cf. Fig. 1);

$$E_{nor} = E_{Ci} - E_{Do} = \left(\frac{P_{Ci}}{\rho} + \frac{V_{Ci}^2}{2} \right) - \left(\frac{P_{Do}}{\rho} + \frac{V_{Do}^2}{2} \right) \tag{1}$$

where P_{Ci} and V_{Ci} is the *wall* pressure and the *sectional mean* velocity at the spiral case inlet, and P_{Do} and V_{Do} at the draft tube outlet, respectively. Equation (1) would be approximately true only when the flows at both the inlet and outlet sections are uniform without swirl. In general, when the flow has a swirl component, it is necessary to correct the swirl energy to the normal SHET E_{nor} of Eq. (1).

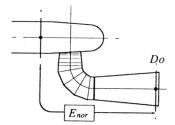

Fig. 1 *Enor* at normal test

Generally, a model test stand supplies rather uniform flow to the model turbine from the upper tank with rectifier via short straight inlet pipe. The flow in the high pressure measuring section is, therefore, nearly uniform without swirl irrespective of the operating conditions. The SHE (specific hydraulic energy) E_{Ci} at turbine inlet calculated with the *wall* pressure and the *sectional mean* velocity is approximately close to the true energy. On the other hand, the flow at runner outlet greatly varies with the operating conditions, and the bend draft tube distorts the flow toward the outlet. The flow at draft tube outlet is not uniform with strong distortion, swirl and/or reverse flows. As a result, the SHE E_{Do} at the draft tube *outlet* calculated with the wall pressure and the sectional mean velocity differs from the true energy.

Under the operating condition of swirl-free flow at the runner outlet, the SHE E_{Di} at draft tube *inlet* obtained from the wall pressure and the sectional mean velocity is rather close to the true energy. If we measure the *special* SHET E_{spc} by relocating the low pressure measuring section to the draft tube *inlet* as shown in Fig. 2 that differs from the IEC code specified, then, the SHET is close to the true energy at the swirl-free condition.

$$E_{spc} = E_{Ci} - E_{Di} = \left(\frac{P_{Ci}}{\rho} + \frac{V_{Ci}^2}{2} \right) - \left(\frac{P_{Di}}{\rho} + \frac{V_{Di}^2}{2} \right) \tag{2}$$

This E_{spc} reaches maximum at an operating point where the E_{Di} becomes minimum under the given E_{Ci}, and where the Euler energy of runner becomes maximum. Since E_{Di}

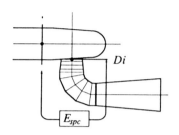

reaches minimum at the point where the runner outflow has no swirl, the energy efficiency obtained by E_{spc} reaches maximum at this point. The special SHET E_{spc} does not include the hydraulic loss in the draft tube, so the efficiency based on E_{spc} must be higher than the one based on E_{nor}. At the point where the runner outflow has swirl velocity, the swirl flow energy shall be corrected to both E_{nor} and E_{spc} for getting the true energy.

Fig. 2 *Espc* at special test

3. Model Test Results

3.1. MODEL TURBINE AND TEST PROCEDURE

The tested model Francis turbine has the discharge specific speed n_{sQ} of 76, the number of runner vanes Z_R 14, the runner outlet radius R_{ref} 0.200 m and the runner vane outlet radius R_2 on the representative mid-streamline of 0.1394 m. The adopted draft tube has an inlet conical diffuser with half cone angle θ_{Di} of 6.6 deg., a 90 deg.-bend of circular cross section with the constant area, an outlet conical diffuser with half cone angle θ_{Do} of 5.0 deg and the area ratio of outlet to inlet AR of 3.23.

The *normal* model test was executed by the experts of IMHEF by selecting the draft tube *outlet* to the low pressure measuring section according to the IEC code for the ten guide vane opening angles α_G from 15 deg. to 35 deg. The tested energy E_{nor} was set to be a constant value of 98 J/kg according to Eq. (1).

After that, the *special* test followed by relocating the low pressure measuring section to the draft tube *inlet* with the same range of guide vane angles. The tested energy E_{spc} was also set to be a constant value of 98 J/kg according to Eq. (2).

3.2. DIMENSIONLESS REPRESENTATION OF PERFORMANCES

To find the corresponding operating points between the two tests, we need the dimensionless performances that are normalized to the identical reference conditions according to the similarity law. Since both the E_{nor} and E_{spc} are adjusted to be the same value of 98 J/kg during the tests, the actual energy for the special test is higher than the normal test, even under the same guide vane angle. Since the speed- and discharge-factor n_{DE} and Q_{DE} etc. based on the SHET E can not be used, therefore, discharge-, torque- and energy- coefficient ϕ, τ and ψ must be adopted based on the peripheral speed at runner outlet U_{ref} ($=\omega R_{ref}$) as follows;

$$\phi = \frac{Q}{\pi \omega R_{ref}^3} \tag{3}$$

$$\tau = \frac{2T}{\rho \pi \omega^2 R_{ref}^5} \tag{4}$$

$$\psi = \frac{2E}{\omega^2 R_{ref}^2} \tag{5}$$

The above ϕ and τ do not include E, so if the complete similarity to the internal flow through the model turbine is kept between the normal and the special test, the relation of τ versus ϕ shall coincide between the two tests irrespective of the tested energy E. As an example, Fig. 3 shows the correlation of τ and ψ versus ϕ under the guide vane angle of 30 deg. The blank marks illustrate the normal test results and the filled marks correspond to the special test results. The torque coefficients for the both tests well coincide against the discharge coefficients and demonstrate the dimensionless performances of Eqs. (3) to (5) are applicable.

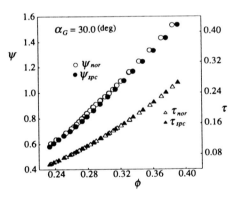

Fig. 3 Dimensionless Performances (α_G=30 deg.)

When the slight deviation is detected in the setting of α_G between the two tests that are executed separately, the value of ϕ is fine tuned to hold the consistency of ϕ versus α_G for both tests. The corresponding τ and ψ are corrected with the square of ϕ by confirming the coincidence of τ and ϕ. From the difference in ψ thus corrected, now we can get the specific hydraulic energy of draft tube SHED E_{spc}-E_{nor}.

3.3. NORMAL TEST RESULTS

The diagram of normal performances measured with E_{nor} are shown in Fig. 4. Strangely, the diagram has two efficiency peaks with the insufficient values of 0.905 under the lower discharge ($\phi_{opt1} = 0.26$, $\alpha_{Gopt1} = 22.5$ deg.) and the higher discharge ($\phi_{opt2} = 0.31$, $\alpha_{Gopt2} = 25$ deg.).

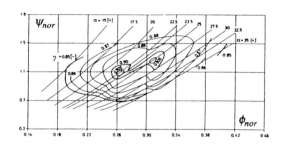

Fig. 4 Normal Performances based on E_{nor}

3.4. SPECIAL TEST RESULTS

The diagram of special performances measured with E_{spc} are shown in Fig. 5. The diagram has an improved single efficiency peak of 0.92 at the discharge coefficient of 0.29. As mentioned above, the efficiency based on E_{spc} reaches optimum at the operating condition of swirl-free runner outflow. We can designate the optimum discharge in the special performances as the swirl-free discharge ϕ_{cf} (=0.29).

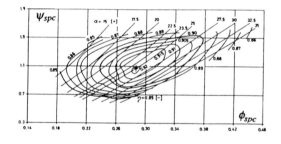

Fig. 5 Special Performances based on E_{spc}

4. Extraction of Draft Tube Energy

As confirmed with Fig. 3, the difference in ψ between the two performances is very small. To increase the sensitivity of detecting the small difference $\Delta\psi$ on the diagram, it is better to represent the diagram so as that the variation of ordinate is minimum against the variation of abscissa. For this reason, the energy ψ and the torque τ of the two performances are transformed by multiplying the square and the cube of the ratio of the swirl-free discharge ϕ_{cf} to the respective discharge ϕ, respectively. As the typical examples, Fig. 6 illustrates the magnified ordinates $\psi(\phi_{cf}/\phi)^2$ and $\tau(\phi_{cf}/\phi)^3$ versus ϕ for the guide vane angles of 25, 30 and 35 deg., respectively. Blank and filled mark corresponds to the normal and special performance, respectively. Both the τ_{nor} and τ_{spc} are coincide well within a wide range of ϕ irrespective of α_G. The magnified ψ_{nor} varies much against ϕ with increasing α_G, whereas the variation of ψ_{spc} is moderate.

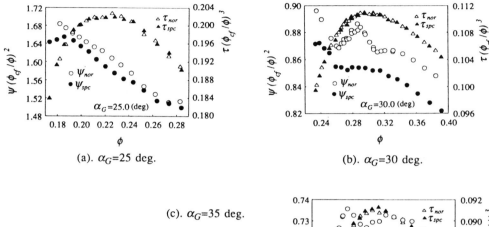

(a). α_G=25 deg.   (b). α_G=30 deg.

(c). α_G=35 deg.

Fig. 6 Magnified performances

The differential energy coefficient ψ_{nor}-ψ_{spc} under the identical discharge coefficient ϕ implies the specific hydraulic energy of draft tube SHED based on the IEC definition (wall pressures and sectional mean velocities) as follows;

$$\Delta\psi_{IEC} \equiv \psi_{nor} - \psi_{spc} = \frac{2(E_{Di} - E_{Do})}{\omega^2 R_{ref}^2} \qquad (6)$$

The SHED thus extracted from the both performance diagrams are illustrated in Fig. 7 with the entire guide vane angles. The curves of $\Delta\psi_{IEC}$ repeat up and down against ϕ irrespective of α_G except the minimum angle of 15 deg. The local maxima on the middle of curves appear at the swirl-free discharge ϕ_{cf} of 0.29 irrespective of α_G. According to the typical Fig. 6(b), this results from ψ_{nor} including the draft tube loss and not from ψ_{spc}. So, the local maxima is the evidence of sudden increase in draft tube energy necessary for the swirl-free flow at runner outlet.

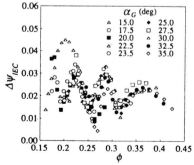

Fig. 7 Specific hydraulic energy of draft tube $\Delta\psi_{IEC}$

5. Correction of Swirl Flow at Draft Tube Inlet

As understood by Eq. (6), $\Delta\psi_{IEC}$ does not include the swirl energy at draft tube inlet. Actually, a Francis runner generates the swirl flow at runner outlet except the swirl-free operating conditions. The total kinetic energy of the swirl flow shall be the axial energy based on the sectional mean velocity plus the swirl energy. Also, the measured wall pressure includes the pressure rise due to the centrifugal force of swirl flow in addition to the static pressure. To obtain the actual energy at draft tube inlet, we have to correct the above effects of swirl flow as follows:

5.1. SWIRL VELOCITY AT RUNNER OUTLET

The velocity triangle at the intersection of a representative mid-streamline with the trailing edge line of runner vanes (subscript 2) forms the right triangle at the swirl-free discharge ϕ_{cf}. The relative angle β_{2cf} of runner outflow can directly be obtained from ϕ_{cf} as follows;

$$\tan\beta_{2cf} \equiv \frac{V_{m2cf}}{U_{2cf}} = \phi_{cf}\left(\frac{\pi R_{ref}^3}{A_2 R_2}\right) \tag{7}$$

where A_2 is the cross sectional area at runner vane trailing edge. The fact that the swirl-free discharge ϕ_{cf} does not depend on the guide vane angle as shown in Fig. 7, tells us that also the runner outflow angle β_{2cf} is not dependent on the guide vane angle. This is understandable because the solidity of Francis runner vanes is high and, in general, the runner outflow angle is approximately constant within the normal operating range.

Assuming the runner outflow angle β_2 is constant $(=\beta_{2cf})$ irrespective of ϕ, the swirl velocity V_{u2} of the runner outflow can be calculated for the arbitrary ϕ as follows;

$$V_{u2} = U_2 - \frac{V_{m2}}{\tan\beta_{2cf}} = U_2\left(1 - \frac{\phi}{\phi_{cf}}\right) \tag{8}$$

5.2. SWIRL ENERGY AT DRAFT TUBE INLET

The swirl energy coefficient $\Delta\psi_{vu}$ of the runner outflow can be obtained with the swirl velocity of Eq. (8) for the correction of the draft tube inlet energy as follows;

$$\Delta\psi_{vu} \equiv \frac{V_{u2}^2/2}{U_{ref}^2/2} = \left(1 - \frac{\phi}{\phi_{cf}}\right)^2\left(\frac{R_2}{R_{ref}}\right)^2 \tag{9}$$

5.3. WALL PRESSURE RISE DUT TO SWIRL FLOW

Assuming the flow field between the runner vane trailing edge (position 2) and the low pressure measuring section at draft tube inlet (for P_{Di}) is of free-vortex, the wall pressure rise coefficient $\Delta\psi_p$ due to the swirl velocity of Eq. (8) is calculated with the following equation for the low pressure measuring section;

$$\Delta\psi_p \equiv \frac{\Delta p/\rho}{U_{ref}^2/2} = \Delta\psi_{vu}\left(\frac{R_2}{R_{ref}}\right)^2 \tag{10}$$

6. Extraction of Draft Tube Loss

By correcting the swirl energy $\Delta\psi_{vu}$ of Eq. (9) and the wall pressure rise $\Delta\psi_p$ of Eq. (10) to the IEC draft tube energy $\Delta\psi_{IEC}$ of Eq. (6), the effect of swirl flow on the energy at draft tube inlet has been corrected. Theoretically, the similar correction would be necessary for the swirl flow at draft tube outlet. As mentioned in Chap. 2, however, the flow at draft tube outlet is distorted greatly with the complicated swirl and reverse flow. The maximum velocity in the outlet section with swirl component would be much higher than the sectional mean velocity. Nevertheless, the kinetic energy of the tube outflow is negligibly small than the runner peripheral speed energy, except the case of high specific speed machines. The wall pressure at tube outlet is close to the pressure in the tail water tank, and less affected by the pressure rise because the swirl flow is not dominant. According to the above consideration, the actual energy loss $\Delta\psi_D$ in the draft tube flow can be to calculated with the following formula;

$$\Delta\psi_D \equiv \frac{2\Delta E_D}{\omega^2 R_{ref}^2} = \Delta\psi_{IEC} + \Delta\psi_{vu} - \Delta\psi_p$$

$$= \Delta\psi_{IEC} + \left(1 - \frac{\phi}{\phi_{cf}}\right)^2\left(\frac{R_2}{R_{ref}}\right)^2\left(1 - \frac{R_2^2}{R_{ref}^2}\right) \tag{11}$$

Figure 8 shows the actual draft tube loss $\Delta\psi_D$ obtained from SHED $\Delta\psi_{IEC}$ in Fig. 7 by using Eq. (11). Roughly speaking, the extracted draft tube loss coefficients are aligned on a single line of W-shape except the minimum guide vane angle. The tube loss reaches maximum at the swirl-free discharge, and decreases once when the discharge leaves ϕ_{cf}. As understood by Eq. (8), when the discharge decreases from the swirl-free discharge, the positive swirl velocity increases, and vice versa. There are two optimum discharges ϕ_{opt} of 0.26 and 0.31

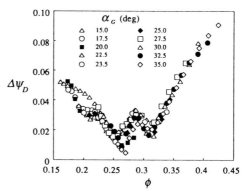

Fig. 8 Actual draft tube energy loss $\Delta\psi_D$

where the tube loss reaches minimum. The discharges can be designated as the optimum swirl discharge. The tube loss at the lower side of ϕ_{opt} is smaller than the loss at the higher side. At the lower side, the swirl energy at runner outflow is positively optimum, and vice versa. When the discharge decreases or increases apart from the optimum swirl discharges, the tube loss increases especially at high discharge range.

In general, if there is no-swirl flow at the inlet of a bend draft tube, the flow is used to separate from the inner bend wall, and the energy loss increases. The draft tube applied for this study has a circular bend of constant sectional area. The aspect ratio of the bend (the ratio of the radius of curvature at outer bend wall to that at inner bend wall) is high, and the secondary flow is apt to develop in the bend. This is the reason why the draft tube loss at the swirl-free discharge is large compared with the normal bend draft tube. The swirl energy at tube inlet increases when the discharge leaves the swirl-free discharge and suppresses the flow separation at the bend, resulting in the decrease of tube loss irrespective of the direction of swirl. Between the two minima of tube loss at the optimum swirl discharges, it is natural that the loss at the lower side of discharge is smaller. Since the tube loss at the swirl-free discharge is large, the contribution of swirl to the loss reduction is also large. Apart from the two optimum swirl discharges, with increasing positive or negative swirl energy, the tube loss becomes large.

Now we can understand the reason why the normal performance diagram has two efficiency peaks as follows: At the swirl-free discharge, the separation loss is large in the bend draft tube. At the two optimum swirl discharges smaller or larger than the swirl-free discharge, the separation loss is drastically suppressed by the optimum swirl. As a result, the two efficiency peaks appear at the two optimum swirl discharges. On the other hands, in the case of an ordinary Francis turbine with the well designed bend draft tube, since the separation loss in the bend is not so large, only a single efficiency peak will appear at the lower side of the two optimum swirl discharges.

The draft tube loss coefficient ζ_D based on the kinetic energy of the sectional mean velocity V_{mDi} at draft tube inlet can be deduced from $\Delta\psi_D$ in Fig. 8 as follows;

$$\zeta_D \equiv \frac{\Delta E_D}{V_{mDi}^{\ 2}/2} = \frac{\Delta\psi_D}{\left(V_{mDi}/U_{ref}\right)^2} \tag{12}$$

The obtained results are illustrated in Fig. 9. The extracted draft tube loss coefficients are also aligned on a W-shaped curve. The increase of coefficient is sharp at the lower discharge range, whereas the slope of curve is small at the higher discharge region with approaching to a constant value asymptotically.

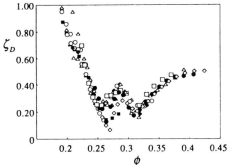

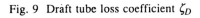

Fig. 9 Draft tube loss coefficient ζ_D

7. Conclusion

Applying the GAMM Francis model turbine, the normal model test based on the draft tube outlet for the low pressure measuring section and the special test based on the draft tube inlet were executed separately. The SHED (specific hydraulic energy of draft tube) $\Delta\psi_{IEC}$ is extracted from SHET (specific hydraulic energy of turbine) for both performances under the identical flow conditions of the model turbines. By correcting the swirl energy at draft tube inlet to $\Delta\psi_{IEC}$, tne draft tube loss coefficient ζ_D can be empirically extracted from the both performance diagrams

8. References

1. Han, F., Ida, T. and Kubota, T.: Analysis of Scalable Loss in Bulb Turbine in Wide Operating Range, 17tn IAHR Symposium - Beijing (1994), Vol.2, G5, 853 - 864.
2. Kubota, T. and Yamada, S.: Effect of Cone Angle at Draft Tube Inlet on Hydraulic Characteristics of Francis Turbine, 11th IAHR Symposium - Amsterdam (1982), 53, 1-14.
3. Sottas, G. and Ryhming, I.L.; "3D-Computation of Incompressible Internal Flows", Notes on Numerical Fluid Mechanics, Vol.39, (1993) Vieweg.
4. Avellan, F., Dupont, P., Farhat, M., Gindroz, B., Henry, P., Hussain, M., Parkison, E. and Santal, O.; "Flow Survey and Blade Pressure Measurements in A Francis Turbine Model", 15th IAHR Symposium - Belgrade (1990), Vol.2, I5.

MODELLING COMPLEX DRAFT-TUBE FLOWS USING NEAR-WALL TURBULENCE CLOSURES

Y. VENTIKOS
Postdoctoral Associate

F. SOTIROPOULOS
Assistant Professor

School of Civil and Environmental Engineering
Georgia Institute of Technology
Atlanta, Georgia 30332, U.S.A.

and

V.C. PATEL
Professor and Director

Iowa Institute of Hydraulic Research
The University of Iowa
Iowa City, Iowa 52242, U.S.A.

Abstract

This paper presents a finite-volume method for simulating flows through complex hydroturbine draft-tube configurations using near-wall turbulence closures. The method employs the artificial-compressibility pressure-velocity coupling approach in conjunction with multigrid acceleration for fast convergence on very fine grids. Calculations are carried out for a draft tube with two downstream piers on a computational mesh consisting of 1.2×10^6 nodes. Comparisons of the computed results with measurements demonstrate the ability of the method to capture most experimental trends with reasonable accuracy. Calculated three-dimensional particle traces reveal very complex flow features in the vicinity of the piers, including horse-shoe and longitudinal vortices and regions of flow reversal.

1. Introduction

Understanding hydroturbine draft-tube flows is a crucial prerequisite for addressing numerous operational and environmental challenges facing the hydropower industry today. From the operational standpoint, the draft-tube is of paramount importance for the overall efficiency and smooth operation of a hydraulic turbine, particularly at off-design conditions. Its importance is best demonstrated by the fact that the primary

140

E. Cabrera et al. (eds.), Hydraulic Machinery and Cavitation, 140–149.

consideration when designing turbine blades is to ensure that they deliver well-conditioned flow at the draft-tube entrance (Fisher, 1995). Poor inflow conditions are associated with several undesirable flow phenomena--flow reversal downstream of the runner, formation of rope vortices, and cavitation--which could induce large efficiency losses, devastating pressure pulsations in the entire system, and even failure. From the environmental standpoint, draft tube flows are important for understanding the causes of injury and/or mortality of passing fish as well as for developing effective strategies for improving the tailrace water quality. Regions of large flow gradients, intense streamwise vortices, cavitation, areas of flow recirculation, and formation of the so-called "back-roll" vortices at the draft-tube/tailrace interface may be responsible for injuring and/or disorienting passing fish. On the other hand, tailrace water quality, which is affected by the depletion of Dissolved Oxygen (DO) in the lower levels of the reservoir during warm months of the year (Bohac and Ruane, 1991), may be substantially improved using autoventing hydroturbines (AVT). AVT technology relies on turbulence mixing within the draft tube to ensure transfer of oxygen from the air-bubbles--injected into the water at strategic locations downstream of the runner--to the water at a rate sufficient to increase the tailwater DO concentration at environmentally acceptable levels (Carter, 1995).

A typical draft tube consists of a short conical diffuser followed by a strongly curved 90° elbow of varying cross-section and then a rectangular diffuser section. Its cross-sectional shape changes continuously from circular at the inlet, to elliptical within the elbow, and finally to rectangular at the exit. Additional geometrical complexities include the presence of one or more piers, downstream of the elbow, splitter blades, guide vanes, slots, etc. The flow that enters the draft tube--the wake of the turbine runner--is turbulent and three dimensional, with high swirl levels. This already complex inlet flow undergoes additional straining as it passes through the elbow, induced by the rapid area changes, the very strong longitudinal curvature, and the presence of various obstacles. The resulting flow is extremely complicated with regions of strong induced pressure gradients, intense longitudinal and horse-shoe vortices, regions of flow reversal, etc. These complexities make the numerical simulation of draft tube flows particularly challenging for even the most advanced numerical methods available today. Yet modern computational fluid dynamics (CFD) methods offer the most promising alternative for elucidating the physics of draft tube flows at a level of detail necessary for addressing the operational and environmental issues noted above.

Numerical simulations of draft tube flows have been reported, among others, by Vu and Shyy (1988), Agouzoul et al. (1990), Sotiropoulos and Patel (1993), and Reidelbauch et al. (1995). With the exception of Sotiropoulos and Patel (1993)--who employed a two-layer near-wall k-ε model and a moderately fine computational mesh (approximately 200,000 nodes)--all these studies adopted the standard, high Reynolds number, k-ε model with wall functions, and reported results on rather coarse meshes (40,000 to 100,000 nodes). Despite reproducing general physical trends, regarding the effect of inflow swirl on the flow development, none of these studies demonstrated their ability to quantitatively predict the flow details.

The objective of this work is to develop the computational framework that would enable accurate quantitative predictions of turbulent flows through complex draft-tube geometries over a range of operating conditions. An efficient, finite-volume numerical method is presented for solving the three-dimensional Reynolds-averaged

Navier-Stokes equations in conjunction with near-wall turbulence closures on very fine, highly stretched and skewed computational meshes. Numerical solutions are obtained for one of the Norris Power Plant draft tubes (Tennessee Valley Authority) at model-scale Reynolds numbers. The two-layer k-ε model of Chen and Patel (1988) is employed for turbulence closure. The computed solutions are compared with available mean velocity measurements at several locations downstream of the elbow (Hopping, 1992) and analyzed in terms of three-dimensional particle traces.

2. The numerical method

The numerical method of Sotiropoulos and Lin (1996) is modified and used in the present study. This method solves the three-dimensional Reynolds-averaged Navier-Stokes (RANS) equations, in conjunction with two-equation, near-wall, turbulence closures, formulated in generalized curvilinear coordinates in strong conservation form. Pressure-velocity coupling is achieved using the artificial compressibility approach. The governing equations are discretized on a non-staggered computational mesh using finite-volume discretization schemes. Three-point central differencing is employed for the viscous fluxes and source terms in the turbulence closure equations. The method features a number of options for approximating the spatial derivatives of the convective flux-vectors. These include second-order, central--with scalar and matrix valued fourth-difference artificial dissipation terms--and flux-difference splitting upwind (ranging from first to fifth-order accuracy) differencing schemes. The spatial resolution of these schemes has been carefully evaluated in both laminar (Lin and Sotiropoulos, 1996a) and turbulent flow simulations (Sotiropoulos and Lin, 1996b).

The discrete mean flow and turbulence closure equations are integrated in time using a four-stage explicit Runge-Kutta algorithm (Jameson, 1983) enhanced with local time-stepping, implicit residual smoothing, and multigrid acceleration. A three-grid level V-cycle algorithm with semi-coarsening in the transverse plane (that is, coarse grids are constructed by doubling the grid spacing only in the transverse directions) is employed in the present study. One, two, and three iterations are performed on the first, second, and third grid level, respectively. The present multigrid method is capable of solving the turbulence closure equations in both loosely and strongly-coupled fashion. In the first approach, multigrid is applied only to the mean-flow equations while the turbulence closure equations are solved only on the finest mesh (the eddy-viscosity values are injected to the coarser meshes and held constant during the cycling process). In a strongly coupled strategy, on the other hand, multigrid is applied simultaneously to both the mean and turbulence closure equations and the eddy-viscosity values are updated at each grid level (see Sotiropoulos and Lin (1996) for a detailed discussion and comparison of the various methods). All subsequently presented calculations have been obtained using the loosely coupled algorithm with three iterations performed on the turbulence closure equations per multigrid cycle.

The present method features a number of isotropic and non-isotropic (non-linear) two-equation turbulence models (see Sotiropoulos and Ventikos (1996) for details). In the present study, however, only the isotropic two-layer k-ε model of Chen and Patel (1988) is employed. Work is currently underway to implement and validate the various non-linear models for draft-tube geometries.

3. Test case and computational details

The draft tube configuration, used for the present computations, is one of the TVA Norris Autoventing Power Plant (Norris, Tennessee) draft tubes designed to operate with 66,000 HP hydroturbines. The area expansion ratio for this draft tube (ratio of the exit to inlet cross-sectional area) is approximately 4.4:1 while the radius of curvature of the elbow is 1.34 diameters of the inlet circular cross-section. Two vertical piers, symmetrically placed about the centerline, support the downstream rectangular diffuser.

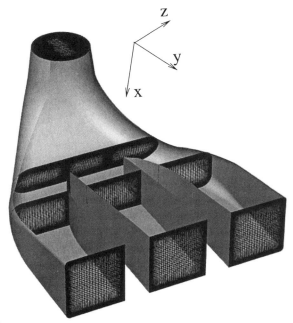

Figure 1. Cross-sectional views of the computational grid

The computational grid for every cross-section is generated using an efficient algebraic grid generation method which employs linear and third-order spline interpolation. The grid lines are concentrated near the walls using the hyperbolic tangent stretching function. The cross-sectional grids are then stacked along the centerline of the tube to complete the three-dimensional grid. To accurately resolve the flow in the vicinity of the piers, the streamwise planes are clustered around the pier leading edges also using hyperbolic tangent stretching. Typical cross-sectional views of the computational mesh and the relevant coordinates are shown in figure 1. All the subsequently reported calculations were carried out on a grid with 85 x 73 x 193 nodes (a total of approximately 1.2×10^6 nodes), in the streamwise, vertical, and horizontal (ξ, η, and ζ) directions, respectively, which is the finest mesh to be used so far for draft-tube calculations. The near-wall coordinate surfaces are located everywhere such that $1 < n^+ < 5$, where $n^+ = u_\tau n / \nu$ (u_τ is the shear velocity, n denotes the distance from the wall, and ν is the molecular kinematic viscosity).

        Inlet conditions are specified using the experimental measurements (Hopping, 1992). These include both the axial and transverse mean velocity components at a plane downstream of the runner (see Figure 2). The measurements were carried out using a laser velocimeter for a range of operating conditions with and without air injection. The present simulations correspond to experimental run No. 1 (see Hopping, 1992) which was performed with the air off, runner speed 898 rpm, net head 24.8m, and water discharge 0.44cm^3/sec. These conditions correspond to a Reynolds number Re=1.1x10^6, based on the diameter D, and bulk velocity U_b at the inlet of the draft tube. It should be noted that inlet measurements were obtained along two mutually perpendicular radii (see Figs. 2a, and c), which suggest that the flow is not circumferentially symmetric. Due to lack of more detailed data, however, the calculations were carried out by arbitrarily choosing one of the two profiles and assuming that the inlet flow is axisymmetric. To facilitate the application of outflow boundary conditions, an artificial straight extension (of total length 10D) was added downstream the end of the draft tube. The flow quantities at the downstream end of this extension were obtained by assuming zero streamwise diffusion. On the solid walls the velocity components and turbulence kinetic energy are set equal to zero.   The pressure at all boundaries is calculated by using linear extrapolation form the interior nodes.

        The computational domain is treated as a single block with the piers accounted for by using a blanking technique. This treatment necessitates the use of several two-dimensional arrays to store the Jacobian and metrics of the geometric transformation and the pressure field on each pier wall.   Converged solutions (four orders of magnitude reduction in residuals) are obtained after approximately 800 multigrid cycles.   The computational time per grid node per cycle is 2.2x10^{-4} secs on a single-processor Silicon Graphics, 90MHZ, R8000, Power Challenge workstation.

4. Results and Discussion

In this section we present comparisons of calculated mean streamwise velocity profiles with measurements at several locations within the three bays. The numerical solutions are also interrogated using particles traces to elucidate the structure of the three-dimensional flow separation and vortex formation phenomena within the elbow and the downstream diffuser.

        Figure 2 shows comparisons of measured (Hopping, 1992) and calculated streamwise mean velocity profiles at two streamwise locations, downstream the start of the piers, in all three bays. The velocity profiles, are plotted at two y = constant planes (see Figures 2a and 2c for axes definition) along the horizontal (Fig. 2a), and vertical (Figs. 2b, c, and d) centerlines of each cross-section. Figures 2a and 2c also include the measured streamwise and swirl velocity components at the inlet section, which, as discussed above, were used to provide inlet conditions for the calculations. All velocities in these figures have been scaled by the bulk velocity at the inlet of the draft tube.

        The measurements in Fig. 2 suggest that most of the flow passes through the left (with respect to an observer standing at the draft-tube inlet looking downstream) bay. This is evident by the overall larger velocities through that bay and is obviously associated with the clockwise direction of the inflow swirl. The calculations reproduce this flow feature and appear to capture reasonably well most experimental trends. Some

discrepancies are observed at the downstream location in the right bay (Fig. 2a), where
the calculated streamwise velocity profile indicates the presence of a small reversed flow

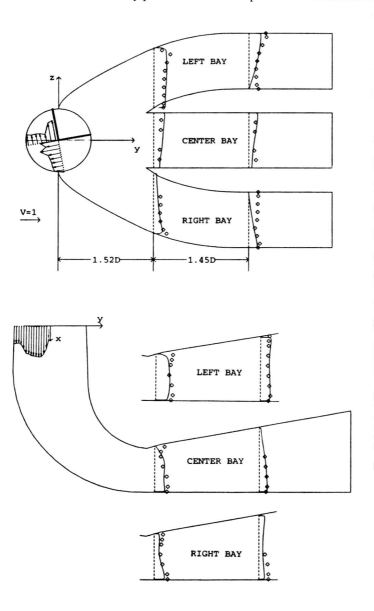

Figure 2. Comparisons of measured and computed streamwise mean velocity profiles

region near the inner wall.   Contour plots of the calculated streamwise velocity
component, not shown here due to space considerations, reveal a recirculating flow
region starting upstream of that section and ending immediately downstream.   The
measurements, on the other hand, suggest a fuller and almost uniform velocity profile
there which appears to have recovered very rapidly from its upstream distorted shape.
Similar discrepancies, albeit not as pronounced, are observed at the downstream section
in the left bay as well.   It should be noted, however, that the experimental measurements
are not detailed enough to allow a comprehensive assessment of the accuracy of the
numerical solutions. Given the continuous area expansion downstream of the elbow, it
is very likely that reversed flow does exist in the experiment, although may be not at the
same locations indicated by the calculations, but could not be resolved by the few
available velocity measurements. Yet another source of uncertainty is the lack of detailed
velocity measurements at the inlet. As discussed in the previous section, the inlet flow
was assumed axisymmetric, although the limited available measurements do not support
such an assumption (see inlet swirl profile in Fig. 2a). Given the complexity of the
draft-tube geometry, even small differences in inlet conditions could account for the
observed discrepancies.   Obviously, the present calculations can not offer positive
answers to all these questions.   They do, however, underscore the need for carefully
designed, very detailed laboratory experiments.

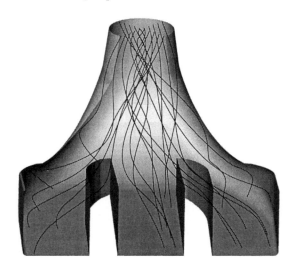

Figure 3. Three-dimensional particle traces: General view

Figures 3, 4 and 5 depict particle traces released at strategically selected
locations to clarify various three-dimensional flow features.   A global view of the
flowfield is given in Fig. 3, which shows the paths of particles originating along two
mutually perpendicular diameters at the inlet plane. It is seen that most of the flow
passes through the left bay and the left half of the center bay, which is consistent with
the trends exhibited by the velocity profiles discussed above. Particles released near the
center of the inlet section are seen to form a coherent, rope-like, vortical structure which
appears to pass through the left half of center bay. Significant secondary motion is also

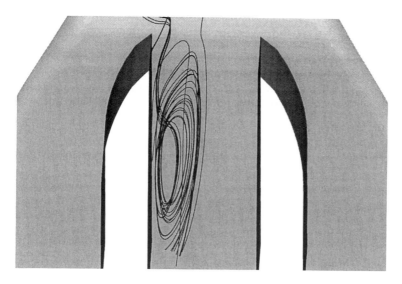

Figure 4. Three-dimensional particle traces: Reversed flow region

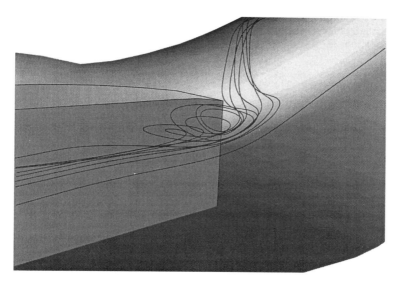

Figure 5. Three-dimensional particle traces: Horse-shoe vortex

present in the right bay as indicated by the twisting particle trajectories there. Figures 4 and 5 reveal some very complex three-dimensional flow patterns along the flat wall of the right pier. Figure 4 indicates the existence of a recirculation region which is located near the top (diverging) wall of the draft tube--although not shown herein due to space limitations, the particles that are trapped in this area originate from the near-wall region at the left side of the inlet section. Underneath this recirculating flow region there is a very intense longitudinal vortical structure which is shown in Figure 5. This structure is similar to horse-shoe like vortices known to form at wing-body junctions and is produced by lateral skewing of the vorticity vector. These flow patterns--which to the best of our knowledge have not been reproduced before numerically for draft-tube geometries--serve to demonstrate the enormous complexities of such flows, underscore the challenges for advanced CFD methods, and point, once again, to the need for very detailed laboratory experiments to provide data for numerical validation.

5. Summary and Conclusions

An efficient finite-volume method was presented for carrying out fine-grid calculations with near-wall turbulence models for complex draft-tube geometries. The computed solutions are compared with mean velocity measurements downstream of the elbow. The calculations reproduce most experimental trends with reasonable accuracy. Three-dimensional particle traces reveal, for the first time, the presence of very complex three-dimensional flow patterns around the piers. These include longitudinal and horse-shoe vortex formation, and regions of reversed flow. The present study underscores that detailed three-dimensional flow measurements are of crucial importance for further advancements in numerical modelling of real-life draft-tube geometries. Current work focuses on further improving the efficiency of the multigrid method, by implementing grid sequencing techniques, as well as implementation and testing of advanced turbulence models that account for turbulence anisotropy.

6. Acknowledgments

The initial phase of this work, while the second author was at the University of Iowa, was supported by a grant from the Tennessee Valley Authority. The continuation of this work at the Georgia Institute of Technology is funded by Voith Hydro, Inc. and the U.S. Department of Energy. The authors are most grateful to Richard K. Fisher, Jr., of Voith Hydro, Inc., and Patrick March and Paul Hopping, of TVA, for their advice and support.

7. References

Agouzoul, M., Reggio, M., and Camarero, R. (1990), "Calculation of Turbulent Flows in a Hydraulic Turbine Draft Tube," *ASME J. Fluids Eng.*, 112, pp. 257-263.
Bohac, C. E., and Ruane, RT. J. (1991) "Tailwater Concerns and the History of Turbine Aeration," ASCE National Hydraulic Engineering Conference.
Carter, J., Jr. (1995) "Recent Experience with Turbine Venting at TVA," *Water-Power 95*, ASCE Int. Conf. & Exposition on Hydropower, San Francisco, California, Vol. 2, pp. 1396-1405.
Chen, H. C. and Patel, V. C. (1988), "Near-Wall Turbulence Models for Complex Flows Including ," *AIAA J.*, Vol. 26, pp. 641-648.
Fisher, Jr., R. K. (1995), private communication.
Hopping, P. N. (1992), "Draft Tube Measurements of Water Velocity and Air Concentration in the 1:11.71 Scale Model of the Hydroturbines for Norris Dam," *Tennessee Valley Authority Engineering Laboratory*, Report No. WR28-2-2-116, Norris, Tennessee.

Jameson, A. (1983), "Solution of the Euler Equations by a Multigrid Method," *Applied Mathematics and Computation*, Vol. 13, pp. 327-356.

Lin, F., and Sotiropoulos, F. (1996), "Assessment of Artificial Dissipation Models for Three-Dimensional, Incompressible Flow Solutions," to appear in the *ASME J. of Fluids Engineering* (April 1995).

Riedelbauch, S., Fisher, Jr., R. K., Faigle, P., and Franke, G. (1995), "The Numerical Laboratory Gets Better," *Water-Power 95*, ASCE Int. Conf. & Exposition on Hydropower, San Francisco, California, pp. 1386-1395.

Sotiropoulos, F., Lin. F. (1996), "Strongly-Coupled Multigrid method for 3-D Incompressible Flows Using Near-Wall Turbulence Closures," to appear in the *ASME J. of Fluids Engineering*.

Sotiropoulos, F., and Patel, V. C. (1992), "Flow In Curved Ducts Of Varying Cross-Section," Iowa Institute of Hydraulics Research, IIHR report No. 358, The University of Iowa, Iowa City, Iowa.

Sotiropoulos, F., and Ventikos, Y. (1996), "Assessment of Some Non-Linear Two-Equation Turbulence Models in a Complex, 3-D, Shear Flow," to be presented at and appear in the proceedings of the *6th International Symposium on Flow Modelling and Turbulence Measurements*, September 8-10, Tallahassee, Florida.

Shyy, W. and Braaten, M. E., (1986), "Three-dimensional Analysis of the Flow in a Curved Hydraulic Turbine Draft Tube," *Int. J. Numerical Methods in Fluids*, Vol. 6, pp. 861-882.

Vu, T. C. and Shyy, W. (1990), "Viscous Flow Analysis as a Design Tool for Hydraulic Turbine Components," *ASME J. Fluids Engineering*, Vol. 112, pp. 5-11.

FLUID FLOW INTERACTIONS IN HYDRAULIC MACHINERY

T. ASCHENBRENNER, N. RIEDEL AND R. SCHILLING
Institute for Hydraulic Machinery and Plants
Technical University of Munich, Germany

Abstract.

The paper deals with the fluid flow interactions between rotor and stator as well as rotor and side spaces of hydraulic machinery. It is shown that the flow through the rotor strongly influences the leakage flow through the side space, the friction torque and the axial·forces acting on the rotor.

1. Introduction

In the past years the development of hydraulic machinery has made much progress due to very sophisticated numerical tools. After having defined the point of operation a recalculation of each component is started to find out the potential of improvement. Based on these flow analysis results each component may be optimized seperately so that they reach a high hydraulic quality. However, this strategy does not necessarily yield the optimum performance of the hydraulic machinery since the fluid flow interactions between different components is not yet considered.

Due to the increasing performance of modern workstations and parallel computation techniques the optimization of the whole hydraulic machinery becomes more realistic.

A potential to improve the efficiency of hydraulic machinery still exists by taking into account the fluid flow interactions between the components of the machinery in the design and optimization process. The present paper describes two main fluid flow interactions in hydraulic machinery.

2. Definition of problems

In order to take into account the fluid flow interactions in hydraulic machinery during the design process a setup of numerical tools has to be

150

E. Cabrera et al. (eds.), Hydraulic Machinery and Cavitation, 150–159.
© 1996 *Kluwer Academic Publishers. Printed in the Netherlands.*

developed. These tools have to be fast and accurate enough to give reliable results within some hours.

At the institute a 3D-Euler Code which is used for rotor-stator-interaction (RSI) -calculations has been developed, see /3/. Also a Navier-Stokes code for the recalculation of the flow in the side space between the rotor and the housing of hydraulic machinery has been developed, see /4/. Various turbulence models are beeing tested. Combining these codes both the interaction between rotor-stator (RSI) and rotor-side space (RSSI) may be considered as a coupled fluid flow problem. A research programme has been defined to investigate the effects in hydraulic machinery induced by these fluid flow interactions. The aim is to find out the best computational model for RSI and RSSI interaction, to calibrate the codes with experimental data and to study the interactions in different types of hydraulic machinery.

3. Rotor-Stator-Interaction (RSI)

3.1. THE 3D EULER CODE

Within the research programme a fast block structured 3D-Euler code has been developed. The governing equations for the incompressible fluid flow are modified using Chorins artificial compressibility concept yielding a hyperbolic set of equations. Since the rotor calculations are performed in the rotating frame of reference Coriolis and centrifugal forces are added to the right hand side. The governing equations are discretized by means of an implicit cell-centered finite volume scheme. The numerical fluxes in the cell faces are constructed using a second order upwind scheme similar to Roe's method, and an advanced limiter function is used. The system of algebraic equations is solved by means of a point-Gauss-Seidel scheme.

3.2. NUMERICAL MODELS

Four different approaches to perform a rotor-stator calculation were checked. First, an isolated runner calculation has been carried out taking the exit angle of the guide vane camberline as inflow boundary condition. This approach is used for a standard recalculation. Second, only the flow through the guide vanes has been analyzed without rotor using a computational domain from the guide vane inlet down to the draft tube. This calculation is performed using the massflux and the approximate outflow angle of the spiral casing as inflow boundary conditions. The stator solution then defines new boundary conditions for a rotor calculation. Using this approach the evaluated rotor inflow angle can be expected to be more accurate then in the first case. The third test case is defined to check whether the flow field downstream of the stator is influenced by the rotor. Using overlapping grids, the computational domain of the stator ends at the leading edge of the rotor and the rotor domain begins at the trailing edge of the stator. The basic idea is to start a stator calculation then to take this solution as rotor inflow boundary condition to recalculate the runner flow and to feed back the pressure field as boundary condition for the stator outflow. Thus, convergence may be ecpected after some iterations.

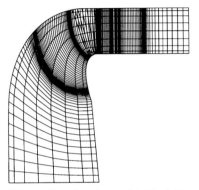
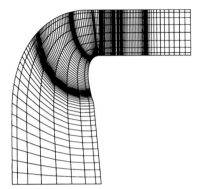

Fig. 1: Overlapping grid, Test Case 3   Fig. 2: Connected grid, Test Case 4

The last test case is a coupled rotor-stator analysis where rotor and stator are calculated in parallel on two workstations. The two computational domains are linked together at an intersection plane, exchanging the boundary conditions after each time step. Since the flow within each component does react directly on changes at the intersection plane a faster convergence may be optained.

3.3. BOUNDARY CONDITIONS

For all test cases periodic boundary conditions are prescribed in circumferential direction except at blade surfaces where the normal velocity is set to zero. At in-and outflow a special approach is used. Since the boundaries are very close to leading and trailing edges, a non-uniform flow field has to be assumed and only momentum averaged values can be specified. The distribution of a characteristic quantity f is momentum averaged in circumferential direction on a torus surface for line correction and over the whole plane for areal correction. The averaged value $\bar{f}$ is then compared with the boundary condition f_{bc} yielding a difference Δf.

$$\Delta f = f_{bc} - \bar{f}$$

This difference is then added to the discrete value of f in each boundary cell face so that the new momentum average exactly yields f_{bc} and the circumferential distribution of f remains unchanged. For stability reasons this correction can be reduced by an relaxation parameter α

$$f'_k = f_k - \alpha \Delta f, \qquad\qquad k = k_1 \cdots k_2$$

In cases where f is an integral value as the massflux, $\bar{f}$ has to be transformed to a velocity correction.

Boundary conditions for test case 1
Since the inflow cells are located in the radial part of the computational domain, the axial velocity is set to zero. Massflux and inflow angle β are specified by the velocities c_r and c_ϕ. The velocities are corrected using the integral method described above.

Boundary conditions for test case 2
Here the boundary conditions of the rotor inflow are evaluated from the stator calculation. The conditions specified are $\bar{c}_r, \bar{c}_\phi, \bar{c}_z$ and the corrections are added using the prescribed line method. For the rotor and stator calculation the radial equilibrium for the pressure is used at the outlet.

Boundary conditions for test case 3
In this test case a stator calculation is performed using an estimated pressure field at the outlet. Then a rotor calculation is started with the boundary conditions on the torus surface computed at the trailing edge of the present stator solution as prescribed for test case 2. After convergence the rotor solution yields a new pressure distibution for the next stator calculation. This iterative procedure is continued until the changes of the boundary values reach a convergence criterion.

Boundary conditions for test case 4
For this test case the rotor and stator meshes are linked together at an intersection plane. After each time step the boundary conditions are updated. The circumferentially averaged velocity components $\bar{c}_r, \bar{c}_\phi, \bar{c}_z$ on each torus surface are passed from the stator to the rotor and the static pressure $\bar{p}$ is transferred from the rotor to the stator. The boundary conditions are then compared and updated using the line method.

3.4. RESULTS

To calibrate the numerical models used, the GAMM-Turbine stage consisting of stay vanes, wicket gates and runner has been analyzed. At the EPFL in Lausanne velocity measurements up-and downstream of the runner have been carried out /6/.
Fig. 3 shows the distribution between stator and rotor of c_r for the four

test cases in comparison with the measurements. The radial velocity component shows up the absolute value and the distribution of the massflux in the measuring plane. The c_r-distributions considering testcases 1 and 2 are approximately uniform due to the nearly two dimesional boundary conditions. Test case 3 and 4 show a much more accurate distribution of c_r especially near the shroud. This improvement results from the pressure field at the leading edges of the rotor which has a strong influence on the meridional flow distribution at the trailing edges of the guide vanes. As a direct consequence similar effects can be found looking at the absolute inflow angle,see Fig. 4. There is a bigger difference between test case 1 and 2 due to the fact that the boundary condition in test case 1 was only a geometrically defined angle. Nevertheless in test cases 1 and 2 the inflow angel α is strongly underpredicted at the shroud by 15 degree. Test cases 3 and 4 coincide much bettter with the experimental data. The results clearly show that the rotor-stator-interaction has to be taken into account during the design process to optimize a rotor blading. The rotor strongly influences the flow within the stator and this effect can be predicted well applying the numerical models used in test case 3 and 4.

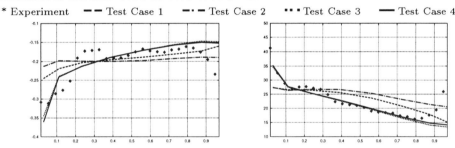

* Experiment   — — Test Case 1   — · — Test Case 2   · · · Test Case 3   —— Test Case 4

Fig. 3: Radial velocity c_r Fig. 4: Absolute inflow angle α

3.5. COMPUTATIONAL EFFICENCY

To analyze the computational effiency the effort to set up the test cases and the CPU time needed for a converged solution has to be considered. Since the meshes are generated using the Integated Flow Analysis System developed at the Institute here the time needed for the mesh generation can be neglected.

A remarkable difference between the four methods can be found looking at the CPU time needed. Since the test case 1 represents the common approach to a 3D runner flow analysis, all CPU times are related to this case, which requires about 70 minutes on a RS6000 workstation having a performance of 50 mflops using a grid of 95 * 30 * 9 mesh point. Table 1 shows a comparison of the CPU times needed for the four test cases.

Due to the slightly faster convergence of the stator computation the effort for test case 2 is a bit less then twice the effort for test case 1. Since the accuracy is approximately the same the approach 2 is not very useful. Test case 3 needs the largest amount of CPU time because only one calculation can be performed at the same time, so the rotor computation has to wait for the stator result and vice versa. Test case 4 shows almost identical results as test case 3 but it needs only 60% of the computational effort. Using two CPUs the numerical effort can be halved for this test case due to the minimum communication effort between the processes compared with the time needed for one time step. Here the most important point is that two domains can be calculated in parallel wheras the calculations in test case 3 have to be performed one after the other, and the CPU time directly depends on the number of iterations needed to satisfy the boundary conditions.

Test Case	1	2	3	4
1 CPU	1	1.8	5.4	3.2
2 CPUs				1.6

TABLE 1. Comparison of the Computational Efficiency

4. Rotor-Side-Space-Interaction (RSSI)

4.1. THE NAVIER-STOKES CODE

Based on the Euler code described a Navier-Stokes code has been developed. A special derivate is in use for rotor side space calculations. The flow field in the rotor side space is assumed to be constant in circumferential direction. Applying this simplification, the three dimensional flow field can be computed in an two and a half dimensional approach. A 2D mesh is set up in the meridional plane of the side space. On this mesh the governing equations for p, c_r, c_z and a circumferentially averaged value $\bar{c}_\phi$ are solved. This approach yields a short CPU-time and a fast convergence as the number of mesh points is relatively small. As the turbulence model of the Navier-Stokes code is still beeing tested the turbulent flow has been computed by the commercial code TASCflow /5/, using the standard $k - \epsilon$ model with log-law or two-layer approach for solid walls.

4.2. ROTOR SIDE SPACE GEOMETRY

As a first step a simplified rotor side space geometry has been defined. Based on this simplified geometry the influence of Reynolds number Re, leakage flow rate q_L and circumferential momentum c_ϕ of the leakage flow on the coefficients of friction torque C_T and axial force C_{Fax} acting on the rotor is studied.

$$C_T = \frac{2T_f}{\rho\omega^2 r_a^5}; \quad F_{ax} = \frac{2F_{ax}}{\rho\omega^2 r_a^4}; \quad q_L = \frac{Q_L}{Q}; \quad Re = \frac{\omega r_a^2}{\nu}$$

The computational domain is characterized by the relative axial gap width of $\lambda_a = 0.0046$ and relative radial gap width $\lambda_r = 0.0232$. The mesh consist of 48300 points so that a sufficiently accurate resolution of the boundary layer and recirculating flow is guaranteed. At the inflow the mass flux and direction of the velocity vector are specified as boundary conditions, at the outflow the averaged static pressure is prescribed.

Fig. 5 and 6 show the influence of Reynolds number on the friction torque and the axial force acting on the rotating disk.

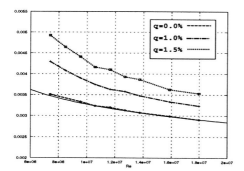
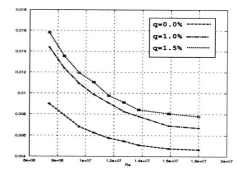

Fig. 5: Torque coefficient C_T *Fig.* 6: Axial force coefficient C_{Fax}

Fig. 7 shows the variation of friction torque and axial force for different leakage flow rates q_L. Due to the higher pressure difference between inflow and outflow at increasing flow rates the axial forces are growing linearly with the flow rate. The momentum coefficient also grows linear with increasing flow rates. Similar effects were presented in /7/.

Aditional calculations were made to show the influence of inflow swirl. The circumferetial velocity profile is changed by the increasing inflow swirl leading to a higher average circumferential velocity and so the friction torque is reduced. The axial force is increased due to a higher pressure gradient.

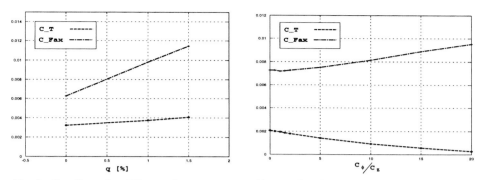

Fig. 7: C_T, C_{Fax} for different flow rates   Fig. 8: C_T, C_{Fax} for different c_ϕ at inlet

4.3. COUPLED IMPELLER SIDE SPACE COMPUTATIONS

A pump turbine having a specific speed $n_q = 35$ was chosen for a coupled calculation of the flow within the rotor and side space. Fig. 9 shows a meridional section of the pump turbine. The radial gap width has been assumed to be 0.1% of the rotor diameter. A pump turbine configuration was chosen to be able to study the flow in both operating modes without changing the geometry. Thus comparison of the occuring effects and of the interaction between runner and side space may be studied.

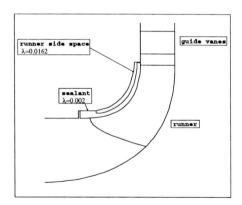

Fig. 9: Meridional section of pump trubine PT35

The flow in the runner is calculated by means of the 3D Euler code described above. As the Reynolds number in the rotor side space is $Re = 1.1 * 10^7$ the flow in the rotor side space is assumed to be tubulent. The coupling of the runner and side space calculation was realized in a similar way as the rotor stator coupling shown in test case 3. The velocities at the inlet and the static pressure at the outlet of the side space are circumferentially averaged and fed into the rotor side space calculation as boundary conditions. The following side space calculation yields new boundary coditions for the rotor at the interfaces. The correction of the boundary values is carried out, as explained above. This iterative process is continued until the exchanged boundary conditions satisfy a prescribed convergence criterion.

4.4. RESULTS

In turbine mode only one calculation has been performed considering the optimum point. As the leakage flow leaves the main flow passage between stator and rotor and is fed back on the suction side of the runner the flow rate in the runner is smaller then the design flow rate. Fig. 10 shows the pressure distributions of the runner near the shroud comparing the flow with and without the RSSI. The reduced flow rate leads to a smaller relative flow angle β and consequently to a reduced loading at the leading edge, see Fig. 10. Beside this local effect on the pressure distribution at the runner inlet near the shroud the rotor side space interaction has no significant influence on the runner flow. The leakage flow rate has been computed to be $q_L = 0.0225$, i.e. 2.25% of the discharge.

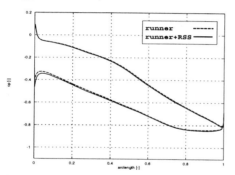
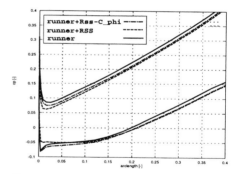

Fig. 10: Turbine pressure distribution   *Fig.* 11: Pump pressure distribution

In pumping mode three different operating points have been considered, 100%, 85% and 75% of the optimum flow rate. Fig. 12 shows ψ_t versus the flow rate φ_r for the three operating points. The Euler calculation of the runner yields a linear increase of ψ_t with decresing flow rates. The leakage flow which is pumped through the rotor side space, is added to the overall mass flow in the runner. Fig. 11 shows the pressure distribution on the blade near the shroud. Similar to the effects in turbine mode the loading at the leading edge is reduced due to a higher relative flow angle β. An aditional runner calculation was performed setting the inflow swirl of the the leakage flow to zero. As shown in Fig. 11 the effect of the swirl brought into the runner flow is only minor. The leakage flow rate has been calculated to be $q_L = 0.0249$ at optimum point and increasing with reduced discharge. In Fig. 13 Torque and axial Force are shown. They both show a similar behaviour as in the simplyfied geometry, s. Fig. 8. The torque is reduced due to the increasing circumferential velocity component entering the rotor side space.

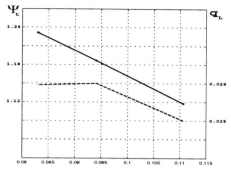

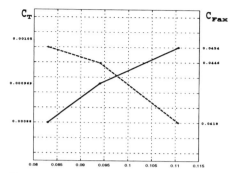

Fig. 12: Ψ_t, q_L versus φ_r *Fig.* 13: C_T, C_{Fax} versus φ_r

5. Conclusions

In this study different approaches to predict the fluid flow interactions in hydraulic machiney are presented. Numerical tools for calculation of RSI and RSSI have been developed. It could be shown that the fluid flow interactions have sigificant influence on performance and efficiency of hydraulic machinery.

6. References

1. W.N. Dawes: Torward Improved Throughflow Capability: The Use of 3D Viscous Flow Solvers in Multistage Enviroment, ASME Paper 90-GT-18, 1990.

2. H. Haas, N. Riedel, A. Fernandez, ,Ch Watzelt, R. Schilling: An Integrated Flow Analysis System for the Recalculation of Hydraulic Bladings, Preprints; 5th International Symposium of Transport Phenomena and Dynamic of Rotating Machinery ISROMAC-5 Maui USA 1994.

3. N. Riedel: Entwicklung eines 3D Euler Codes mit Riemann-Solver, Int. Bericht des Lehrstuhls für Hydraulische Maschinen und Anlagen, 1995.

4. J. Fritz: Numerische Berechnung der Strömung im Radseitenraum hydraulischer Maschinen, Diplomarbeit Lehrstuhls für Hydraulische Maschinen und Anlagen, 1995.

5. TASCflow User Documentation, Theory Documentation, 1995.

6. E. Parkinson: Turbomachinery Workshop ERCOFTAC II, Test Case 8: Francis Turbine, EVA94-PAE, IMHEF Ecole Polytechnique Federal de Lausanne

7. R. Schilling: Strömungen in Radseitenräumen von Kreiselpumpen, Habilitationsschrift Universität Karlsruhe, 1980

NUMERICAL OPTIMIZATION OF HIGH HEAD PUMP-TURBINES
A Challange for Mechanical and Hydraulic Design

H. BUCHMAIER, B. QUASCHNOWITZ,
W. MOSER, D. KLEMM
Voith Hydro GmbH
D-89509 Heidenheim, Germany

1. Summery

In order to cope with the technical challenges of high head pump-turbines, new methods like simultaneous engineering and numerical studies are required. In this paper the application of simultaneous engineering is shown with an example, as well as typical pump-turbine concepts and design features.

2. Introduction

The design process of the High Head Pump-Turbine is an increasing challenge to the development teams of today:
Operational behavior and efficiency trends have to be predicted exactly, requiring an optimized hydraulic layout. At the same time, the development time between the placement of the order and the start of manufacturing is decreasing drastically. This forces a trend toward numerical studies. In addition, the strong competition among the hydro turbine companies can only be met with cost optimized designs.

The process of numerical design and optimization will be presented from both a hydraulic and a mechanical standpoint, as they are realized using simultaneous engineering.

3. Pump-Turbine Design Optimization (typical example: Spiral Case)

Because of increased evaluation of efficiency, more importance has to be attached to the optimization of (pump-)turbines. The optimization should not only be conducive to the reduction of losses, but should improve the relation of costs to benefit as well.

E. Cabrera et al. (eds.), Hydraulic Machinery and Cavitation, 160–169.

Optimizing an assembly group as complex as a pump-turbine spiral case means that single parameters are varied while the degree of target-fulfillment has to be checked with every step.

Figure 1 shows the results of the optimization steps of one parameter for a high head pump-turbine. Civil construction limits are not considered.

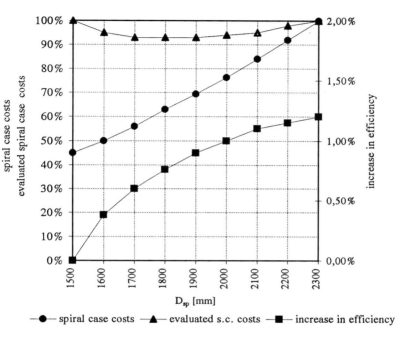

—●— spiral case costs   —▲— evaluated s.c. costs   —■— increase in efficiency

Figure 1. Sample of design optimization of a high head pump-turbine: throat diameter D_{sp} versus increase in efficiency and costs.

The diagram shows clearly that the evaluated costs of the spiral case do have their minimum between minimum throat diameter D_{sp} on the one hand and maximum efficiency on the other hand.

To reliably define all effects of the design variations within a short period of time, and to estimate the quality in addition, a fast tool with sufficient accuracy is needed.

The use of simultaneous engineering proved to be very efficient for this task (see figure 2). The backbone is formed by a program system called KE that permits the generation of the most important hydro-mechanical (pump-)turbine assembly groups as 3-D computer internal models in the shortest time possible.

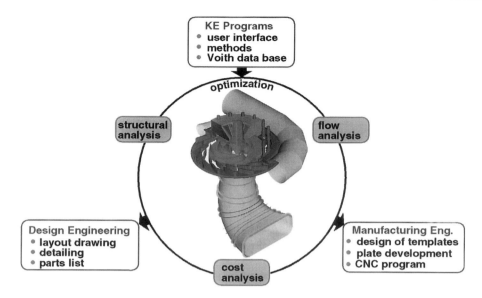

Figure 2. The concept of Simultaneous Engineering at VOITH.

With the aid of attached programs for flow-, structural- and cost analysis, the design variants can be discussed with regard to losses, stability and costs. This allows the run through of several optimization cycles in a fairly short time period and is therefore a guarantor for a good layout.

Besides design optimization, the 3-D model is used as a base for detailing and to derive manufacturing aids in the next project phase.

With our example pump-turbine spiral case, the enlargement of the throat diameter causes some changes in the conditions: e.g. the pressure area is increased due to the larger spiral case sections, while the velocity of the flow is decreased due to the larger cross section. All changes have to be examined carefully.

Since the loads on the stay vanes are increased, a repeated structural analysis is definitely necessary. A large number of load cycles with a considerable portion of dynamic stress have to be expected during the life cycle of pump-turbines. Therefore a load universe as input for the fatigue analysis is generated. In doing so, the flow-induced vibrations of the stay vanes must not be ignored. They are determined through experiments and calculations.

The resulting loads are input into the Finite Element Method calculation. Stress and strain at the most critical spots can be extracted from the results of the computation (see Figure 3).

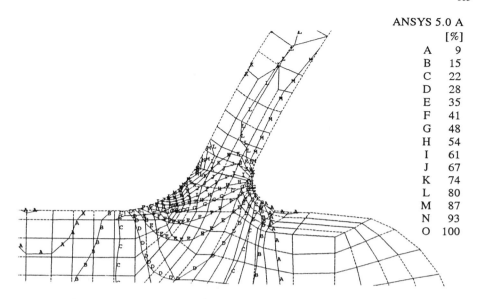

| ANSYS 5.0 A |
| [%] |

	[%]
A	9
B	15
C	22
D	28
E	35
F	41
G	48
H	54
I	61
J	67
K	74
L	80
M	87
N	93
O	100

Figure 3. Equivalence-stress distribution, shown as iso stress lines
at the transition of spiral case section to upper stay ring deck.
Sample of a design variant of GUANGZHOU II at load rejection (H = 775 m).

Locations of max. stress (see fig. 3 e.g.), are examined with methods of fracture mechanics. To prove that certain defects which are smaller than the detectible size do not become critical during the life cycle, crack-propagation is applied. It is desirable to select materials and working stresses such that a defect will grow through the wall of the component before it will fracture due to brittle failure.

The spiral case losses are relatively small in many cases. Nevertheless, it is important to find the best individual design, and for that the losses have to be determined.

During the simultaneous engineering, different flow calculation methods are used. In order to compare different designs or to optimize some dimensions regarding losses, fast and rather simple algorithms are necessary. For a deeper understanding of the loss mechanism, more time consuming numerical tools which account for the three dimensional and turbulent behaviour of the flow are applied. Due to the basic equations which describe this kind of flow, they are called Navier-Stokes-methods. These sophisticated methods are also used to calibrate the above mentioned simple approaches. For more detailed information see [1].

Fig. 4 exemplarily shows the distribution of the total pressure in the symmetry plane of the spiral case of a pump turbine (turbine mode).

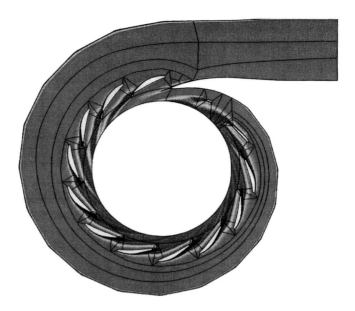

Figure 4. Distribution of the total pressure in a pump-turbine spiral case.

With such detailed results and the total component losses, the designer has the appropriate hydraulic information. Together with the stress analysis results and the manufacturing costs, an optimization of the spiral case according to the special wishes of the customer is easily possible. With similar methods all important components can be designed/optimized separately or together with neighboring components.

4.   General Pump-Turbine Concepts and Design Features [2]

During the last decades the application of reversible pump-turbines has been extended to high heads above 300 .. 400 m. Due to the lack of knowledge about the dynamic forces occurring primarily in transient operating conditions, some premature component failures occurred in the beginning, and reduced the availability of reversible pump-turbines significantly compared to separate pump and turbine units as used for many such high head applications.

In the meantime, the know-how progressed due to extensive theoretical and practical studies. The combination of experience gained on prototypes together with simultaneous engineering including FE-computations enables the building of pump-turbines with heads of more than 600 m, with an expected availability similar to that of other types of turbines. In order to meet this high demand, special design concepts have been developed. Figure 5 shows a typical concept for a high head pump-turbine.

This paper highlights typical design features of these concepts and gives some recommendations based on VOITH´s experience for the optimized design of high head pump-turbines. These are in detail:
- Shaft systems and dismantling methods.
- Embedment of stay ring/bottom ring assembly.

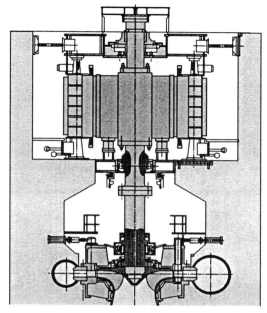

Figure 5.
Typical concept for a high head pump-turbine.

4.1. SHAFT SYSTEMS AND DISMANTLING

The following three different concepts for the arrangement of the shaft system of pump-turbines are discussed (figure 6):
1- Intermediate shaft with bracket supported thrust bearing on top of the motor-generator.
2- Separate turbine/generator shafts with bracket supported thrust bearing below the motor-generator.
3- Single shaft with head cover supported thrust bearing.

For the first concept, the components are to be removed laterally (side removal). For the concepts no. 2 and 3, they are to be removed vertically (top removal).

4.1.1. *Top Removal of the Pump-Turbine Parts*
Up to approx. 300 m head, the motor-generator stator is larger than the pump-turbine head cover, therefore it is possible to remove a one-piece head cover vertically. This is an advantage, if the transportation limits do not prevent it. With regard to powerhouse volume considerations, VOITH's recommendation for top removal is for runner diameters larger than 4 m and heads smaller than 600 m.

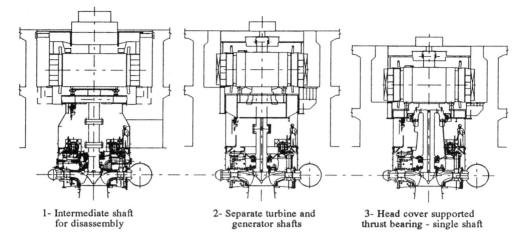

| 1- Intermediate shaft | 2- Separate turbine and | 3- Head cover supported |
| for disassembly | generator shafts | thrust bearing - single shaft |

Figure 6. Different concepts for shaft systems.

4.1.2. *Side Removal of the Pump-Turbine Parts (Figure 7)*

From approx. 300 m head, the motor-generator stator diameter becomes smaller than the pump-turbine head cover diameter; thus for vertical removal it is necessary to split the head cover in 2 or 4 sections. This is a disadvantage, if the diameter of a one-piece head cover is below the transportation limits. For plants with a runner diameter smaller than 4 m, all components are generally shipped in one piece.

Plants with lateral removal normally need an intermediate shaft, thus an extension of the shafting is required. For a surface power plant with the spherical valve directly ahead of the spiral case, lateral removal can be performed in the upstream direction without extending the powerhouse dimensions.

For plants with a runner diameter larger than 3 m the vertical clearance between the motor-generator and the pump-turbine due to the required vertical height in the turbine pit, is such that for lateral removal no significant increase of the shafting length is required.

One of the most important arguments for lateral removal however is the reduction of time for the dismantling of a pump-turbine by approx. 2-3 weeks, which is one reason why this solution was finally adopted for the GUANGZHOU II power plant.

In general, the VOITH recommendations for lateral removal consist of the following:
- Lateral removal is advisable for plants up to runner diameter of approx. 4 m.
- Lateral removal is possibly advisable for heads higher than 600 m.

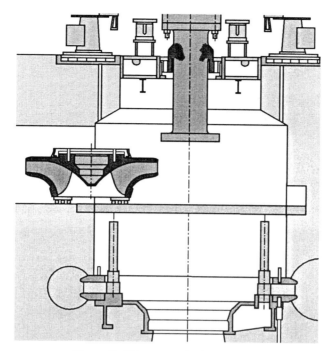

Figure 7. Side removal of the runner

4.1.3. *Bottom Removal of the Runner*

Bottom removal of the runner, while possible, has been adopted very rarely, since it requires an expensive design of the foundations and the removable draft tube section. Furthermore this design can lead to detrimental vibrations and a high level of noise.

4.2. EMBEDDING OF THE STAY RING/BOTTOM RING ASSEMBLY

The design criteria for the stay ring/bottom ring assembly of a pump-turbine are of much more importance than for a Francis turbine.

A comparison of many pump-turbines shows that, in general, the following two design concepts are applied:
- Non-embedded removable bottom ring, freely accessible from below.
- Embedded stay ring/bottom ring assembly without access to the lower wicket gate bearings.

Non-Embedded Bottom Ring (Fig. 8). The stay ring is partially embedded and fixed in the foundation. The bottom ring is not embedded but, like the head cover, bolted to the stay ring and accessible and removable. The static vertical hydraulic forces acting on the head cover (upward) and on the bottom ring (downward) are balanced and result in small foundation loads only. For the absorbtion of the dynamic forces of the bottom ring and discharge ring however, the stay ring requires heavy anchorage in the foundation.

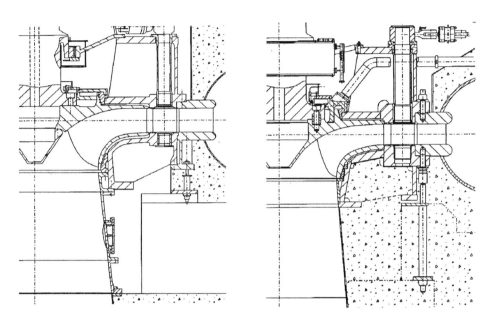

Figure 8.   *Figure 9.*
Non-embedded stay ring/ bottom ring assembly.   Embedded stay ring/ bottom ring assembly.

Embedded Stay Ring/Bottom Ring (Fig. 9). The stay ring/bottom ring assembly is completly embedded and fixed in the foundation. Due to a special embedment procedure, the vertical hydraulic forces from the head cover and bottom ring are nearly balanced and result in reduced foundation forces. The different forces acting on the stay ring/bottom ring assembly are determined by finite element computation (fig. 10). The results are given to the civil engineer (fig. 11).

VOITH´s recommendation. If sand erosion and early wear of the wicket gate bearings are expected, an accessible and removable bottom ring is more favourable. In pumped storage plants, however, water normally does not carry abrasive materials. Therefore erosion does not play an important part.

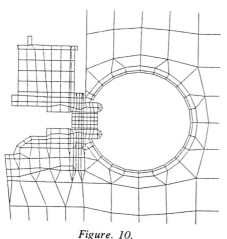

Figure. 10.
FE computation of foundation forces

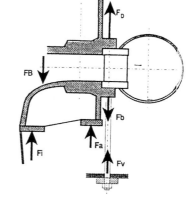

Figure. 11.
Forces on stay ring/bottom ring assembly

Generally a number of important considerations speak in favour of embedding the bottom ring:
- The dynamic forces are transmitted to the foundation directly.
- No axial movement of the discharge ring and draft tube occurs in operation.
- The rigid embedding reduces deflection and vibrations of the bottom ring, thus also reducing the wear of the lower wicket gate bushings.
- No vibrations and therefore low noise level in the lower distributor.
- The smaller deflection reduces the clearance and therefore minimizes the cross flow between the wicket gate ends and the bottom ring. This has a positive effect on part load efficiency and reduces the risk of leakage erosion.

Cosidering all these aspects, this solution was finally adopted for the GUANGZHOU II power plant.

References

1. Riedelbauch, S., Klemm, D. and Hauff, C. (in press) Importance of interaction between turbine components in flow field simulation, IAHR 1996, Valencia.

2. Heine, W., Meazza, G. and Wüst, M. (1995) Comparison of mechanical design concepts for high head pump-turbines.

NUMERICAL HILL CHART PREDICTION BY MEANS OF CFD STAGE SIMULATION FOR A COMPLETE FRANCIS TURBINE

HELMUT KECK
Sulzer Hydro AG
CH-8023 Zürich, Switzerland

AND

PETER DRTINA AND MIRJAM SICK
Sulzer Innotec AG
CH-8401 Winterthur, Switzerland

Abstract. A full stage simulation method is used to calculate the first numerically predicted hill chart of a high specific speed Francis turbine. The numerical method simulates rotating and non-rotating components and their mutual interactions within a single computation. An efficiency hill chart is determined numerically and compared to experimental results from a model test whereby good agreement is obtained.

1. Introduction

Three-dimensional Computational Fluid Dynamics (CFD) is a key technology without which a modern design of a hydraulic machine could not be imagined. Sulzer Hydro has now applied a 3D-Euler code for advanced runner design for 10 years. This was developed together with the EPF-Lausanne (Goede and Rhyming, 1987; Keck *et al.*, 1990) and has been used for over 70 different contracts, allowing wide experience to be gained.

Parallel to the 3D-Euler code, Sulzer Hydro also applies 3D-Navier-Stokes codes for the design of components with adverse pressure gradients, i.e. impellers in pump mode (Goede *et al.*, 1992), for the analysis of the losses in turbine components (Drtina *et al.*, 1992), and for the prediction of turbine hill charts by calculating the flow in a complete turbine from the spiral casing inlet to the draft tube outlet.

E. Cabrera et al. (eds.), Hydraulic Machinery and Cavitation, 170–179.
© *1996 Kluwer Academic Publishers. Printed in the Netherlands.*

The calculation of a complete turbine has been made possible by the development and validation of stage capability in a Navier-Stokes code (Sick *et al.*, 1996). The Reynolds-averaged Navier Stokes equations are solved using the TASCflow-code applying the k-ε-turbulence model and with a new interface condition to extend the code to include stage capability (see (Galpin *et al.*, 1995)). The new stage capability is a major innovation compared to conventional CFD-calculations of individual components linked together in a stacking technique ((Shyy and Vu, 1994)).

With the ability to carry out a CFD calculation of a complete turbine an old dream has come true. But what do we really gain from it? Simply adding the calculations of individual components together is not very challenging except for the memory and CPU requirements of the computer. The real challenge is the prediction of the hill chart and for this not only the interactions between the components are important, but also the increase of losses due to off-design conditions of adjacent components.

Experiences will be presented from a CFD-simulation of a Francis turbine of very high specific speed ($n_q = 116$). We have selected a high specific speed turbine for this work because:

- For high specific speeds the streamline curvature at wicket gate outlet (runner inlet) is more pronounced than for low specific speeds. Hence, specifying boundary conditions in this inter-cascade passage is more difficult and the advantage of the stage capability can be fully exploited.
- For high specific speeds the losses of the draft tube have a higher influence on the total turbine efficiency than for low specific speeds. Therefore, a stage simulation where runner and draft tube are calculated simultaneously is of high significance.
- High specific speed turbines have strongly "distorted" hill charts which represent a larger challenge for numerical predictions than the more regular hill charts of low specific speed machines.

The numerical analysis has been performed for a representative number of operating points within the hill chart. Of special interest is the question whether the shape of the efficiency curves $\eta = f(ku)$ for different guide vane openings can be predicted correctly.

2. Computational Method

2.1. DISCRETISATION METHOD AND TURBULENCE MODEL

The calculations were carried out using TASCflow. This CFD code solves the 3D Reynolds-averaged Navier-Stokes equations in strong conservative form for structured multi-block grids. The system of transport equations is discretized using a conservative finite element based finite volume method and is solved for the primitive flow variables (pressure and cartesian velocity components) using a coupled algebraic multigrid method. The discretization scheme is second order accurate. Turbulence effects are modeled using the standard k-ε model.

2.2. STAGE CAPABILITY

The steady-state interaction between stationary and rotating components in a turbomachine is simulated by a mixing plane between the components. Each component is calculated in its own frame of reference and the blade rows can be reduced to single blade channels with periodic boundaries. The method is based on stage simulation ideas of Denton (1992) which were extended by Galpin (1995) and installed into TASCflow under the partnership of ASC, Sulzer Hydro and Sulzer Innotec. Details of the validation of the stage capability are described in the companion paper (Sick et al., 1996) and by (Galpin et al., 1995).

2.3. GEOMETRY AND GRID GENERATION

The geometry data were directly transfered from the CAD system to the CAD based grid generating software ICEM-CFD. This software enables complex multi-block grids in complicated geometries to be generated with a high degree of freedom.

The computational grids are generated for each component of the machine separately. The spiral casing grid also includes the stay vanes. The grid of the distributor consists of stay vanes (2 passages) and wicket gate (3 passages). The grids were overlapped so that the inlet conditions for the distributor could be obtained from the calculation of the spiral casing flow.

The spiral casing and the draft tube are discretised by a butterfly grid. The computational grids for the blade rows (stay vanes, wicket gate and runner) consist of a combination of O-grids around the blades and H-grids. This grid topology resolves steep flow gradients near the blades accurately and is flexible enough for different blade rows. The combination of all grids for the entire Francis turbine is shown in figure 1.

2.4. BOUNDARY CONDITIONS

The flow in the spiral casing is a function of the Reynolds number only, just as the flow in the penstock or the branch pipe. To determine the losses in the spiral casing over the whole hill chart it is only necessary to calculate the viscous flow in the spiral casing at one operating point. The losses are then a parabolic function of the volume flow (Q^2), see Drtina and Sebestyen (1996). For the stage simulation it is therefore sensible to calculate the flow

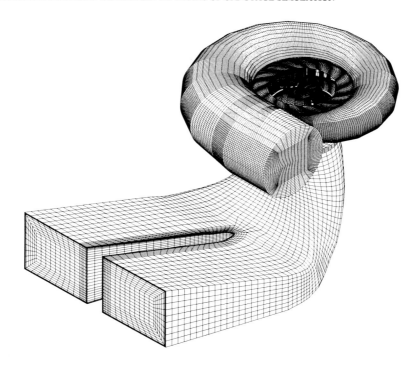

Figure 1.   Grid for entire Francis turbine.

in the spiral casing separately (and in one half of the symmetric casing only) to avoid unnecessary large grids for each operating point.

The flow field depends on the operating point between the inlet of the distributor and the outlet of the draft tube. This whole flow regime, including two stationary cascades (stay vanes and guide vanes), one rotating cascade (runner) and one stationary outlet section (draft tube) was computed simultaneously in a single stage simulation. No intermediate boundary conditions are needed and this minimizes the amount of user intervention during the whole analysis. The boundary conditions for the spiral casing are given as follows:

Inlet: mass flow and turbulence intensity
Outlet: constant pressure

The outlet of the spiral casing simulation is chosen to be downstream of the stay vanes such that the flow at the inlet to the stay vane cascade is correctly modelled. The boundary conditions for the stage including distributor, runner and draft tube are:

Inlet: swirl in a reference plane upstream of the distributor as
                calculated in the spiral casing analysis
Outlet: constant pressure in an outlet plane downstream of
                the draft tube exit (this allows for the typical non-
                uniformities of the flow at draft tube exit)

2.5. COMPUTATIONAL DATA

The calculations presented were carried out on a SGI workstation (Power
Challenge XL with six R8000 processors) on a computational grid consisting
of 279, 300 nodes for half of the spiral casing including the stay vanes and
265, 795 nodes for distributor, runner and draft tube.

The solutions were estimated to be sufficiently accurate when the maximum residual was below 10^{-4}. The flow simulation in the spiral casing
took 170 iteration steps and required 22 hours CPU, that is 1.7 msec cpu
per node. The first stage simulation of the combined distributor, runner
and draft tube needed 600 iterations whereby each iteration took 700 cpu
sec, giving a total time of 5 days and a computational time per node of 2.6
msec. The solution of the first simulation is taken as initial condition for
other operating points. This reduces the number of subsequent iteration
steps to 100-300 (1 to 3 days) depending on the operating point.

3. Hill chart prediction

3.1. EXPERIMENTAL DATA BASE

The hill chart for a high specific speed Francis turbine has been determined
experimentally on a model test rig. The turbine efficiency has been evaluated for more than 200 operating points which yields a good approximation
of the turbine performance for the operating range of interest. Figure 2
shows the resulting non-dimensionalized η/η_{opt} hill chart.

3.2. EVALUATION OF EFFICIENCY

To calculate the turbine efficiency from the numerical simulation data the
spiral casing (casing and stay vane ring) and stage (distributor, runner and
draft tube) calculation results are examined together in order to evaluate
the total pressure loss over the entire machine.

Efficiencies are determined for each component (i) by applying

$$\eta_i = \frac{\overline{p}_{tot,i,in} - \overline{p}_{tot,i,out}}{\Delta \overline{p}_{tot,em}} \tag{1}$$

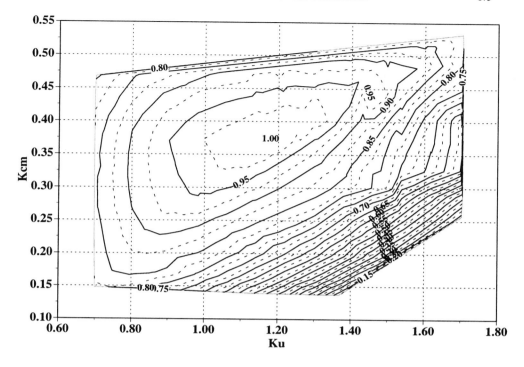

Figure 2. Hill chart based on all experimental data points.

with $\Delta\bar{p}_{tot,em}$ being the difference in total pressure over the entire machine. Note that for the runner one has to take into account the work delivered by the runner.

3.3. NUMERICAL HILL CHART

Due to the high demand in computer memory and CPU-time the number of operating points which could be simulated was restricted. In the present case 14 operating points have been chosen, whereby three different guide vane openings (volume flow rates) and six different heads are investigated.

A comparison between experimentally and numerically evaluated hill charts only makes sense if the number and distribution of data points is identical. For this purpose, the experimental hill chart shown in figure 3 is based only on the experimental data corresponding to the operating points that were also calculated. Figure 4 shows the resulting hill chart based on the 14 operating points from the stage calculations.

The qualitative agreement of both hill charts is impressive. All general features are captured by the simulations. The best efficiency point is iden-

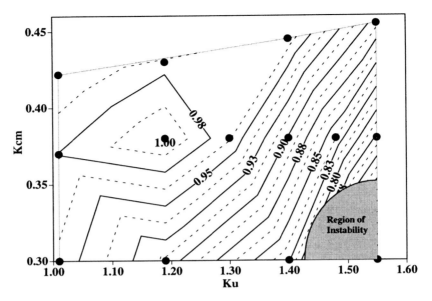

Figure 3. Hill chart based on 14 experimentally obtained efficiency values. Data points are marked by dots.

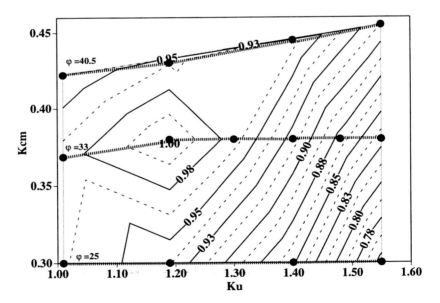

Figure 4. Hill chart based on 14 numerically obtained efficiency values. Data points are marked by dots. Constant guide vane opening is indicated by dashed lines.

tical in both hill charts and all gradients show a very similar behaviour. Please note that the two scales Ku and Kcm are in absolute, not in relative terms, to be able to check the correct location of the best efficiency point. Quantitatively there is some discrepancy in the steepness of the gradients, in particular for severe off-design points with high values of Ku.

A detailed comparison for the three different guide vane openings under investigation can be obtained from figures 5 - 7. For each guide vane opening the normalized efficiency is plotted as a function of Ku.

The shape of the efficiency curves for all three guide vane openings is predicted correctly. For small openings the best efficiency point occurs at low Ku-values (high head). At large openings the high efficiency region occurs at high Ku-values (low head). This typical characteristic of a high specific speed turbine is perfectly represented in the analysis.

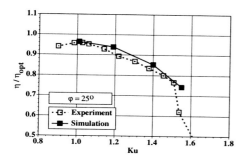

Figure 5. Turbine efficiency for $\varphi = 25°$.

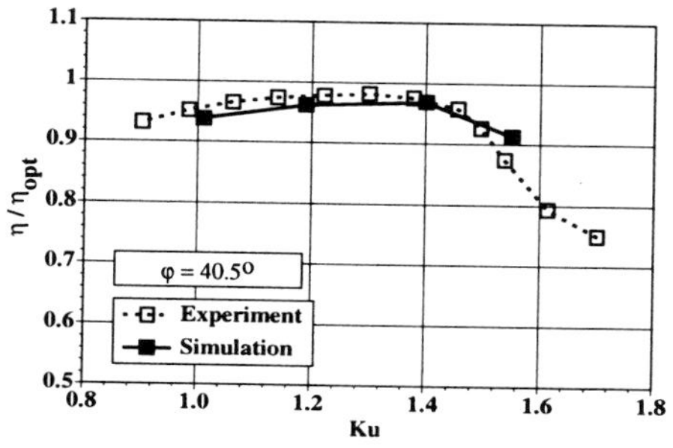

Figure 6. Turbine efficiency for $\varphi = 40.5°$.

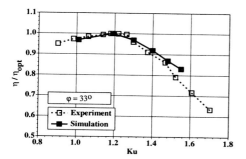

Figure 7. Turbine efficiency for $\varphi = 33°$ (design mass flow).

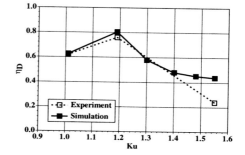

Figure 8. Draft tube efficiency.

Two weaknesses of the present analysis should be mentioned. Firstly, whereas the relative shape of the hill chart is perfectly simulated, the absolute level of the calculated efficiency is about 3% lower than that measured.

This is attributed to the use of a coarse grid. A check of one operating point calculated with a finer mesh produces a reduction of friction losses.

Further improvements are expected if the near wall regions are treated by advanced turbulence formulations, such as two-layer models. Calculations in individual components have shown that the application of two-layer models increase the accuracy of loss prediction, but also increase the computational time.

Secondly, the rate of convergence is rather poor for high Ku-values at small guide vane opening. Large separation zones occur in the draft tube in this case. This is probably the reason why the steep gradients of the measured efficiency curves could not be reproduced in the numerical analysis. However, it is known from the model test that the turbine exhibits a very rough and unstable operational behaviour at Ku $\geq$ 1.5 especially for small openings. The unstable conditions make it difficult to compare the time averaged data of strongly unsteady experimental signals with numerical data of a basically steady analysis.

This is confirmed by the comparison of measured and predicted diffusor efficiency of the draft tube, see figure 8. At high values of Ku, where the flow separates, the numerical simulation currently overestimates the off-design performance of the draft tube.

4. Conclusion

The 3D-turbulent flow analysis in a complete Francis turbine for 14 operating points over a full hill chart has been presented. Of special interest is the fact that the whole assembly of distributor, runner and draft tube has been calculated in a single computation using the newly developed stage capability of the TASCflow code. The advantages of this approach are:

- no need for the user to specify interface boundary conditions between stationary and rotating parts of the machine, thereby eliminating possible errors and time-consuming interventions of the CFD-engineer.
- more accurate simulation of interaction phenomena between adjacent turbine components, especially at off-design conditions. This enables the optimisation of the turbine as a whole and not only of individual components.
- prediction of the shape of a complete hill chart. This is of high practical interest if by modification of certain turbine components special shapes of the hill chart can be generated or if for upgrading projects the impact of existing poorly designed components on the performance of the unit equipped with a new runner has to be predicted.

Future work will concentrate on improvements of the absolute accuracy without excessively increasing the CPU-time and on further automation of the chain of calculations from pre- to post-processing.

5. Acknowledgements

The authors gratefully acknowledge the helpful contributions of Dr. M.V. Casey and D. Wiss, Sulzer Innotec, to the preparation of this paper.

6. List of symbols

$Ku = \omega R(2E)^{-\frac{1}{2}}$	$[-]$	speed coefficient
$Kcm = Q(R^2\pi)^{-1}(2E)^{-\frac{1}{2}}$	$[-]$	flow coefficient
φ	$[-]$	guide vane angle
R	$[m]$	radius at runner blade outlet
ω	$[s^{-1}]$	rotational frequency
E	$[Jkg^{-1}]$	specific hydraulic energy
Q	$[m^3 s^{-1}]$	flow rate
$n_q = 157.8 \cdot Ku \cdot Kcm^{\frac{1}{2}}$	$[-]$	specific speed

References

Drtina,P., Goede,E. and Schachenmann,A. (1992) Three-dimensional turbulent flow simulation for two different hydraulic turbine draft tubes, Proceedings of the First European Computational Fluid Dynamics Conference, September 7-11, Brussels, Belgium

Drtina, P., Sebestyen, A. (1996) Numerical Prediction of Hydraulic Losses in the Spiral Casing of a Francis Turbine, *Proc. XVIII IAHR Symposium*

Galpin,P.F., Broberg,R.B. and Hutchinson,B.R.(1995) Three-dimensional Navier-Stokes predictions of steady state rotor/stator interaction with pitch change, Third Annual Conference of the CFD Society of Canada, June 25-27, 1995, Banff, Alberta, Canada

Goede,E. and Rhyming,I.L.(1987) 3-D computation of the flow in a Francis runner, Sulzer Technical Review 4/87

Goede,E., Sebestyen,A. and Schachenmann,A. (1992) Navier-Stokes Flow Analysis for a Pump Impeller, 16th Symposium of the IAHR, september 14-19 1992, Sao Paulo, Brazil

Keck,H., Goede,E. and Pestalozzi,J. (1990) Experience with 3D Euler flow analysis as a practical design tool, IAHR Symposium, Belgrade, September 1990

Shyy, W., Vu, T.C. (1994) A CFD-Based Computer Aided Engineering System for Hydraulic Turbines, *Proc. XVII IAHR Symposium*

Sick, M., Casey, M.V., Galpin, P. (1996) Validation of a Stage Calculation in a Francis Turbine, *Proc. XVIII IAHR Symposium*

DEVELOPMENT OF A NEW GENERATION OF HIGH HEAD PUMP-TURBINES, GUANGZHOU II

D. KLEMM, E.-U. JAEGER, C. HAUFF
Voith Hydro GmbH
D-89509 Heidenheim, Germany

1.   Introduction

For the Guangzhou Pumped Storage Plant in the People's Republic of China we had to develop the Pump-Turbines of the second phase considering the very tight delivery and erection schedule. The hydraulic, as well as the mechanical design, the manufacture of a new complete homologous model machine and the detailed model tests had to be performed within a very short time.

In order to achieve all the goals and requirements the application of most advanced methods of numerical   flow calculation was necessary (CFD = Computational Fluid Dynamics). Furthermore new ideas for the mechanical design of the model were necessary for a time saving, low cost and very accurate manufacture of the model machine. A powerful test rig with automatic data acquisition and processing of the static and dynamic measurements made possible a very fast and accurate evaluation of the test results.

According to the customer's requirements after the preliminary model tests in our own laboratory in Germany model acceptance tests in an independent laboratory had to be performed. These tests confirmed all our results. The extremely high number of the individual guarantees was met for all performance characteristics and also for the extremely low pressure fluctuations.

2.   Guangzhou II

2.1   DESCRIPTION OF THE PLANT

The Guangzhou Pumped Storage Power Station is located at Lutian Town,

180

E. Cabrera et al. (eds.), Hydraulic Machinery and Cavitation, 180–189.
© 1996 *Kluwer Academic Publishers. Printed in the Netherlands.*

reduced grid frequency of 49 Hz at highest pump head can be operated without any problems due to the very stable Q-H characteristic curve.

5.1.2 *Cavitation*

The model test visual observations of the cavitation phenomena at the leading edges of the blades confirmed the CFD-results. In the normal operating ranges the blades are absolutely free of cavitation.
The range of the suction heads at site is between -70 m and -82.4 m.

5.1.3 *Pressure Fluctuations*

According to the contract for different operating conditions maximum pressure fluctuations at several locations of the water passage way were guaranteed. For the measurement of these fluctuations pressure transducers were used in the spiral, in the vaneless space between runner and guide vanes and in the draft elbow at three different positions, one at the outer side, one at the inner side and one at the draft cone.
The fluctuations at the different locations were measured simultaneously. The analysis regarding characteristic amplitudes of the fluctuations in the time domain as well as dominating amplitudes and their frequencies in the frequency domain was established with the help of the computerised data acquisition and processing system. The following tables No. 2 and 3 show the most interesting results.

TABLE 2.   Amplitudes of pressure fluctuations in the draft elbow

Draft Tube	Measured amplitudes in % of H		
	Cone	Outer side	Inner side
Turbine rated load	0.3	0.3	0.3
Turbine bestpoint	0.18	0.16	0.15
Turbine partial load	1.7	1.5	2.1
Turbine without load	1.1	1.4	1.2
Pump mode	0.25	0.21	0.42
Runaway	5.5	4.2	4.3
Pump no discharge	2.1	2.1	2.1

TABLE 3. Amplitudes of Pressure Fluctuations between Runner and Guide Vanes

Between runner and and guide vanes	Measured amplitudes in % of H
Pump best point	1.68
Normal pump worst case	1.73
Turbine 50 % load	3.9
Turbine rated load	2.9
Pump no discharge	17.8
Runaway	33

These pressure fluctuations are very low. They represent the high hydraulic quality of this Pump-Turbine and they meet the guarantees. However it should be remarked that older Pump-Turbines are operated with higher amplitudes very smoothly and without any damages caused by pressure fluctuations.
The three locations at the draft elbow show nearly the same amplitudes of the pressure fluctuations. Normally it seems to be sufficient to make measurements at only one position of the draft elbow..

5.1   MODEL TESTS AT A THIRD PARTY'S TEST RIG

The results of the internal development tests and the official model acceptance tests in an independent laboratory in the presence of the customer and his consultant showed in all cases a very good agreement. All guarantees for the model as well as for the prototype could be met.
For example the best efficiency of the turbine mode was measured 0.071 %-points higher at the acceptance test than at VOITH test; the weighted average turbine efficiency was only 0.007 %-points lower. In pump mode the differences of the weighted average efficiency was 0.173 %-points and 0.183 %-points at the best point. All these differences are within the systematic uncertainty of the measurements which is +/- 0.218 % for both test rigs. The pump Q-H-curve was shifted by about 1 %. This may be caused by different inflow conditions to the draft tube at the different test rigs.
With respect to the measurement of the pressure fluctuations at the acceptance tests, most of the piezoelectric transducers were those of the Voith tests. However some transducers and the total measuring technique such as amplifiers, the data acquisition and processing as well as the evaluation method were the normal standards of the other company.
Amplitudes very similar to our preliminary tests have been measured. In the normal operating modes the differences of amplitudes between the two

measurements were in the range of +/- 0.5 %-points.   Only at off-design operating conditions, such as runaway, higher differences occurred. Here for comparison the dynamic behaviour of the whole test rig should be taken into account.

6.   Conclusion

☐ Modern design methods such as CFD and CAD are necessary to get high efficiencies and specially defined characteristics precisely within a relative short time and with low costs.

☐ The number of modifications of a model configuration and the number of model tests can be minimised essentially applying CFD and CAD.

☐ Model tests at two different test rigs, with an independent staff of the third party, confirmed the calculated results.

☐ Model tests still remain essentially for the final verification of a hydraulic design.

References

[1]   Buchmaier, H., Quaschnowitz, B., Moser, W., Klemm, D. , (1996) Numerical Design and Optimization of High Head Pump-Turbines, *Proceedings of the XVIII IAHR Symposium, Valencia., Spain*, Kluwer Academic Publisher, to appear

[2]   Riedelbauch, S., Klemm, D., Hauff, C. (1996) Importance of Interaction between Turbine Components in Flow Field Simulation, *Proceedings of the XVIII IAHR Symposium, Valencia., Spain*, Kluwer Academic Publisher, to appear

DEVELOPMENT OF INTEGRATED CAE TOOLS FOR DESIGN ASSESSMENT AND ANALYSIS OF HYDRAULIC TURBINES

BERNARD MASSÉ, HENRI PASTOREL AND ROBERT MAGNAN
Hydro-Québec (IREQ)
1800 boul. Lionel-Boulet
Varennes, Québec, Canada, J3X 1S1.

1. Abstract

Recent advances in numerical analysis can lead to substantial improvements in turbine design. In light of the present tendency to increase the unit power and operating speed of these machines at better cost, electric utilities have to develop their expertise in the field of numerical simulation with a view to improve technical decisions, validate conceptions and, in the long term, enhance plant reliability. Applications of numerical simulation range from analysis of problems specific to hydraulic machines such as manoeuvre margin estimation of turbines in operation, correction of design flaws, design validation and technical assistance for planners, designers and operators. This paper describes an integrated system of Computer Aided Engineering (CAE) tools developed to assess designs, optimize retrofits and provide insight for problem analysis.

This system is described using examples to illustrate the use and limitations of the different tools. Selected applications are presented such as the flow simulation in a water intake, the geometric measurements of Francis runners, computed and measured pressure and stress distributions on runner blades, vibration analysis of a Francis turbine, draft tubes flow validation, spiral casing flow and friction losses due to the roughness in a hydraulic tunnel.

2. Introduction

Manufacturers have improved their design methods to use numerical simulations more intensively. Public utilities have several advantages to gain from this technology to improve the reliability, safety and life expectancy of their equipment, especially in view of today's high capital costs and mounting environmental pressures. Numerous challenges are involved in improving plant design and in rehabilitating or making better use of existing plants.

With almost 96% of its production based on hydroelectric resources, Hydro-Québec has a vested interest in increasing its hydraulic turbine expertise in order to design and maintain generating systems in top operating condition. An effort is ongoing within the framework of a research project called MATH, directed toward simulating the fluid flow in hydraulic turbines and estimating the stress level of the mechanical structure for various operating conditions. The MATH project makes use of several numerical techniques to analyze the

190

E. Cabrera et al. (eds.), Hydraulic Machinery and Cavitation, 190–199.
© 1996 *Kluwer Academic Publishers. Printed in the Netherlands.*

behavior of water turbines components. Various software tools were developed and integrated in a CAE environment focused on the needs of electric utilities.

Utilities do not always share the same viewpoint as manufacturers. Performance and high-quality products are important for both, of course, but the owner of a hydraulic machine is particularly interested in its good dynamic behavior, high life expectancy, and low cavitation damage. Numerical simulation tools developed so far do not allow easy computation of unsteady conditions, forced vibration analysis and life prediction. With the rapid growth of computational capabilities, such computations will be available in the near future. The CAE system is designed to allow the use of this numerical simulation technology.

3.   CAE tools for turbine analysis.

The CAE system is a flexible network of commercial software, custom tools and data translators set up to insure the flow of information from the geometric and operational parameters to the required engineering results. This environment integrates geometric modeling and meshing tools, fluid-flow computation codes, stress and vibration software, performance analysis codes, all linked to a common data base.

The tools are designed to allow the analysis from the water intake to the draft tube exit. For this purpose, various simulations are required in the different hydraulic components, and interpolation of the boundary conditions insures flow continuity from one component to the next. This approach allows specific modelling of each hydraulic passage taking advantage of possible optimization. Numerical tools capable of simulating the whole range of turbine operation do not yet exist and the development of relevant algorithms can be a very difficult task.

Despite all the progress made, numerical simulation methods approximate the physical phenomena more or less accurately. The modelling of turbulence, unsteady flows, cavitation and hydroelastic effects, as encountered in hydraulic turbines, are good examples of such limitations. Consequently, the confirmation of numerical models by experimental testing is very important and takes time and resources. The CAE system is still under development to include unsteady fluid flow, hydroelastic effects, forced response and fatigue analysis.

Figure 1 shows the overall structure of the various tools in use. Geometries of the hydraulic turbine components, collected from drawings and measurements, are stored in a database which is the input for geometric modeling and meshing. The numerical solvers use the data produced in the preceding steps to perform different numerical simulations. The last step consists of viewing and analyzing the solutions.

To exchange data between the different software tools, a common data format and program translators are needed. Although this approach is not general, it works well and requires a minimum of conversion programs to maintain. Conversion of geometric data is among the most involved and a specific program is endowed with this responsibility: VEDGE. It handles a variety of geometric format among which IGES is perhaps the most important since I/O with CAD packages is done using that standard exchange format.

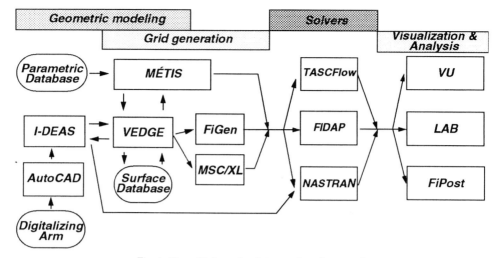

Fig. 1. Flow of information between the software tools

4. Modeling and Meshing of Geometries

Important efforts were spent on geometric modelling and mesh generation. A custom made tool (MÉTIS) can automatically construct a geometric model and meshes for spiral casings, runners and draft tubes. MÉTIS starts with parametric data describing the main features of the components. For other geometries or to model fine details that cannot be described through parametric data, commercial CAD packages are used to hand-build a geometric model.

MÉTIS can automatically produce meshes for most geometries. The geometric model computed by MÉTIS can also be used by itself as an input to FiGen (a mesh generator included in FIDAP) to allow a better control of the meshing. The geometry constructed with CAD packages follows a similar path. In particular, FiGen makes it possible to use unstructured meshes in part of the domain. This can prove useful in many instances by greatly reducing the effort required to mesh complex geometries and sometimes considerably reducing the number of nodes. The flexibility of the tool network allows us to combine coarse geometry coming from MÉTIS with precise surfaces measured in the field and modeled with CAD software. This can greatly simplify the task of analyzing complex, measured or detailed geometries.

Figure 2, produced using VU, shows a spiral casing modelled from drawing informations and on-site measurement for details near the baffle vane region. This example illustrates the use of several tools to obtain the complete mesh. Geometric sections were generated using MÉTIS. FiGen was used to create the mesh and measured curves were used where needed.

Figure 3 shows an example of complex modelling and meshing for structural analysis of a Kaplan blade. In this case, most of the work has been done using I-DEAS, the blade profiles coming from the database.

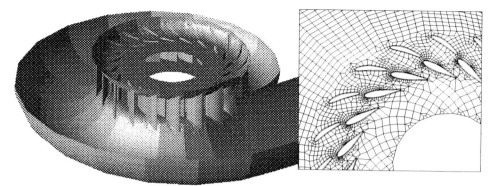

Fig. 2. Geometric model and meshing of a spiral casing

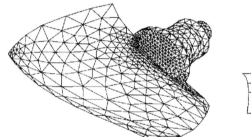

Fig. 3. Structural mesh of a Kaplan runner blade   *Fig. 4. Measure of two hydraulic passages at LG3*

4.1.   GEOMETRIC MEASUREMENT

On-site experimental measurements are required for several parts such as runners, stay and guide vanes and spiral casing baffle vane. Fine fluid-flow analysis asks for improved measurement methods and geometric modelling.

Measurement can be used as a quality control method and as a method to obtain surface data of critical parts such as fillets, leading and trailing edges and complex surfaces. These methods can be useful for retrofitting turbine units, allowing comparison between old and new designs. Measured data are stored in the geometric database for fluid-flow and structural analysis and could be used for rebuilding cavitated areas.

In the past, a laser interferometer and a mechanical digitizing arm were used. The former is expensive, accurate (in the order of 0.01 mm) but not well coupled to CAD systems. The latter is cheaper, less accurate (0.1mm) but has a direct interface with a CAD system, which allows on-site generation and validation of measured surfaces.

Figure 4 shows the reconstructed geometry of a runner at LG-3. Measurement were taken in the field using the mechanical digitizing arm.

5.   Solving the equations

The equations to calculate fluid flow, stress distributions, frequencies and modes of the structural components are well known. Commercial tools such as FIDAP and TASCFlow for fluid flow and NASTRAN for structures are generally used. In fluid-flow simulations, the convergence of the numerical solution depends on several parameters such as the upwinding scheme, relaxation factors, and boundary and initial conditions. An inappropriate choice of solution parameters can lead to divergence of the solution. A cyclic symmetry formulation has been applied for calculating the modes and frequencies of the turbine runner. Solving the various equations using numerical tools calls for optimized parameters. However, the process demands a significant amount of experience.

5.1.   FLUID-FLOW

In fluid-flow simulations, the 3D Reynolds-averaged Navier-Stokes equations are solved. The flow is assumed steady and incompressible. However, the first assumption is not always valid at partial load and full power, where a rope forms in the draft tube and oscillations occur, but it is imposed by the available computer resources. The Reynolds number is very high ($>10^6$) and the flow can be considered fully turbulent. The momentum equations to be solved with these assumptions are the Reynolds-averaged Navier-Stokes equations including Coriolis, centrifugal and gravitational components. The Reynolds stress tensor is calculated using the k-ε model of turbulence. In this model, the characteristic turbulent velocity and length scales are related to the kinetic energy of the turbulence and its rate of viscous dissipation. It adds two semi-empirical transport equations to be solved with the momentum equations. The k-ε model has been calibrated for several turbulent fluid flow problems but validity of this model in hydraulic turbine is still under investigation. The standard k-ε turbulence model is based on the Boussinesq eddy-viscosity model which assumes a linear function between the Reynolds stress and the strain rate tensor. Anisotropic eddy-viscosity models which renders Reynolds stress tensor a higher order function of the strain rate tensor are now being considered.

5.2.   STRESSES AND VIBRATIONS

For mechanical structures, in addition to the well known stress equations from a pressure loading, the development effort is oriented toward solving for different modes and frequencies of rotating runners. Boundary conditions are then applied, including cyclic symmetry, together with loading forces such as centripetal acceleration and flow pressure.

The whole runner (not only a sector), the presence of the fluid and rotating inertia terms must be considered to compute vibration characteristics of turbines. To take the whole runner into account, classical theory of cyclic symmetric structures is applied [7], computing what are called rotating mode shapes. Each family is characterized by a phase angle relating the deflection of one blade to its neighbors. Table 1 shows typical results for a Francis runner[5]. Addition of Coriolis terms is straightforward but work is ongoing for

Numerical method

Standard Galerkin finite element method is used for space discretisation .P1isoP2 tetrahedrons with 10 nodes are also used. Linear interpolation is used for the pressure and linear interpolation on sub-elements for the velocity ,k and ε.

The characteristics method with a Runge Kutta integration is used to solve the advection step.

The diffusion step is solved using an Uzawa or a Chorin algorithm

3-DESIGN AN ANALYSIS APPLICATION TO THE TWO STAGE PUMP TURBINE

3-1 DESIGN OF A RETURN CHANNEL BLADE   USING THE INVERSE APPROACH

The aim of the return channel is to supply   the second runner in pump and turbine mode. This return channel included two kinds of blades: one is on the elbow and the second on the flat part. We present here the design of the second one. With the usual direct method, we could not find a shape satisfying at no flow separation and a good velocity profile for the second runner inlet in pump mode or for the other blade inlet in turbine mode. The inverse method permitted to avoid these problems.

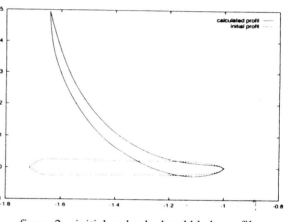

figure 2 : initial and calculated blade profil

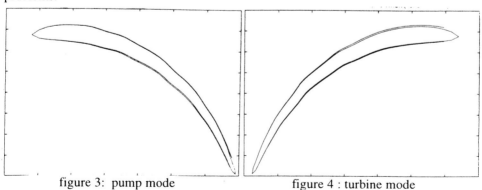

figure 3: pump mode   figure 4 : turbine mode
Return channel : boundary layer comutation

The data were the velocities at the inlet and outlet, and the pressure difference between the pressure and the suction side of the profile .The profile azimuth thickness was also given.The obtained blade satisfied all the design criteria such as the boundary layer no separation and the velocity field at the outlet.

A boundary layer computation in pump mode and in turbine mode is shown on figure 3 and 4.

3-2 ANALYSIS OF THE FLOW IN THE GUIDE-VANES, STAY-VANES AND THE SPIRAL CASE IN PUMP RUNNING USING N3S

In this paragraph we present the results of an N3S computation in the domain constituted by the guide vanes , stay vanes and the spiral case of the two stage pump turbine.

The calculation was performed at the best delivery point Q=168l/s on a mesh of about 110 000 nodes .At the inlet (of guide vanes) we have imposed the axisymetrical velocity field delivered by a meridian calculation in an S1S2 computation of the runner (figure 5). The k-ε values were approximated to $V^2/100$ for k and $V^3/1000$ for ε .we have noted that the calculation was not sensible to these values . At the outlet we have imposed a zero normal stresses condition .

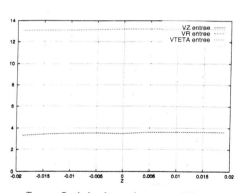

figure 5 : inlet boundary conditions

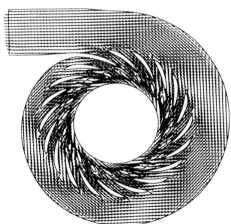

figure 6

On figure 6 we present the velocity field in the median plan. The computation shows that the delivery and the pressure are equi-distributed in the all the inter stay and guide vanes channels . And no recirculation have been detected .

The radial velocity field, presented on two plans (figure 7 and 8) (versus the azimuth θ= 180°, 360°) shows that :

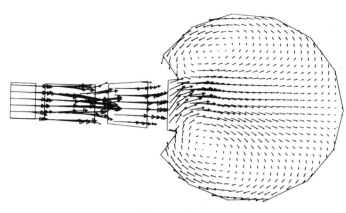

figure 7: θ = 180°

- issued from a quasi uniform radial velocity at the inlet , the radial flow in the stay vanes and in the guide vanes present an acceleration in the median region and a deceleration near the walls .
- the flow issued from the stay vanes generates two contra-rotative eddies in the spiral case

Besides the calculation of the total pressure at the inlet and outlet of the guide vanes stay vanes and the hole domain let us guess that the major part of the losses have been between the domain inlet and the stay vanes inlet .

To confirm this analysis we have performed a computation with the same code in only one inter stay and guide

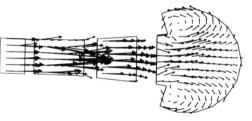

figure 8: θ = 360°

vanes channel. The mesh was refined(60 000 nodes) and the same boundary condition were taken .

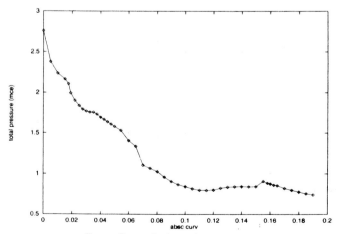

figure 9 :evolution of total pressure along stay vane and wicket gate

Figure 9 shows the total pressure evolution in the channel and confirm the assumption we have made while the hole domain computation. Besides, this run, with a refined mesh, shows not only a flow deceleration near the walls but also a recirculation zone in the channel .

3-3 ANALYSIS OF A DRAFT TUBE IN TURBINE RUNNING USING N3S

In a two-stages Pump Turbine, the draft
tube is characterised by the presence of
the machine shaft which crosses it, due to
the necessary bearing under the lower
runner.
The calculation of this draft tube is a
Navier-Stokes calculation in turbine
running. The geometry and the mesh are
shown on figures 1 and 10.
The running conditions for calculation are
corresponding at a partial load:
    Q= 119.1 l/s H= 30 mwc (reduced
                    model)

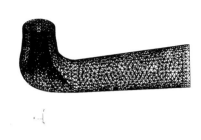

figure 10

Two calculations have been realised with two sets of boundary conditions at the inlet
(figure 11), to study the influence of the velocities distribution under the runner.
- the first distribution is the same as that measured in the laboratory on an old reduced
model (figure 12)
- the second is a theoretical ideal velocities distribution, which is aimed during the
design of the runner (figure 13)
The boundary conditions at the outlet are a zero normal stresses conditions.

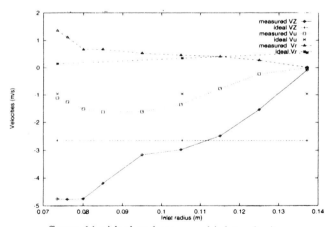

figure 11 : ideal and measured inlet velocity

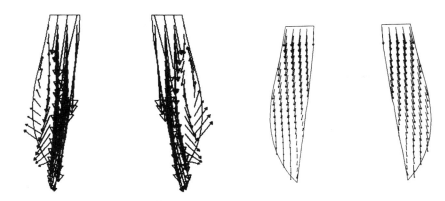

figure 12: Measured boundary conditions   figure 13 : Ideal boundary conditions

Figures 12 and 13 are a comparison of velocities distribution in a vertical cut of the draft tube for the two inlet boundary conditions.

Figure 15 is a comparison of the losses curves for the two inlet boundary conditions.

When the velocities distribution at the inlet is that measured, the calculation head losses in the draft tube are $\Delta H = 2\%$ .The same measured head losses are $\Delta H = 2.21\%$. When the velocities distribution at the inlet is ideal, the head losses are $\Delta H = 0.57\%$.

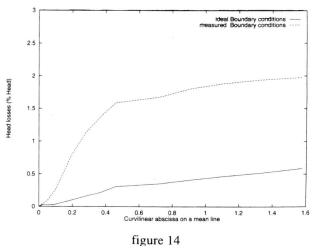

We observe that the losses are concentrated in the elbow of the draft tube, due to the recirculation (figure 13), and calculation shows clearly the relation between the reduction of the losses and the velocities distribution along the elbow.

Thus, the runner design will respect at best the ideal velocity distribution at the inlet of the draft tube.

figure 14

3.4 ANALYSIS OF RUNNER CALCULATION IN PUMP MODE

This calculation was realised with an unstructured mesh of about 40 000 nodes of speed.

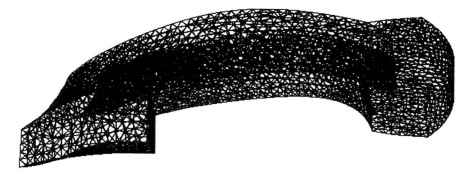

figure 15

The used boundary conditions are:
* At the inlet , an homogenous velocity field without rotational speed.
* At the oulet , a Neuman condition has been used
* In the rotating frame, an wall condition is imposed on blade, shadow and crown.
* Out of the runner , a wall condition is imposed in the fixed frame.

All the calculation is done with a speed of 1200 rpm for a runner diameter of 400 mm. Two discharges are calculated: Q= 168 l/s (nearly the maximum of the hill chart) and Q= 154 l/s.The figure 17 shows the comparison between the calculated values and the experimental' ones.

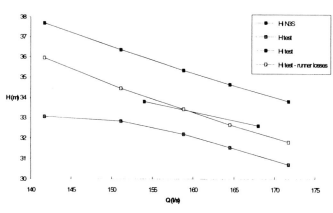

Hi test represents a measured value:

$$H_i test = \frac{N_a - N_{fr}}{\omega(Q + q)}$$

Na output measured on shaft
Nfr losses by friction
q : clearance discharge
Q : discharge of runner
ω : rotational speed
Hi N3S represents the intern head calculated of the runner:

$$H_i = \omega \frac{\Delta rVu}{g},$$

The value of Hi test - runner losses (measured) are to be compared with the calculated value of Hi . We find a good correlation between the two values.

The figure 18 shows the distribution of meridian velocity at the runner outlet   which allow to better the flow in guide vane.

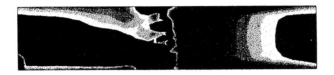

figure 18 : velocity at the runner outlet

CONCLUSION:

This paper deals with a design methodology in the specific case of a two-stage pump-turbine. The type of calculations used has been described for each component of the machine. Among the series of codes available, it is essential to choose the best suited tool for each shape to be generated. The model tests are still required as final confirmation of the design and the calculation / measure comparisons permitted to validate our approach.

REFERENCES

(1) Eremeef L.R, Philibert R :(1985) :*Modélisation quasi-tridimensionnelle des écoulements dans les turbomachines hydrauliques* ,La Houille Blanche n°7/8
(2) Francois M, Philibert R (1988) :*Ecoulements dans les baches spirales*, Symposium AIRH, TRONDHEIM
(3) Soares Gomes F, Kueny JL, Reynaud G, Combes JF (1992) :*Numerical analysis of three-dimensional flow in a pump turbine spiral casing*, Symposium AIRH, SAO POLO
(4) Caudiu E, Grimbert I, ElGhazzani EM, Verry A, Philibert R (1988) : *3D flow computation in turbomachinery* ,

STUDY ON HIGH SPEED AND HIGH HEAD REVERSIBLE PUMP –TURBINE

T.NAKAMURA   *Toshiba Corporation*
              20-1,Kansei-cho,Tsurumi-ku,
              Yokohama 230   Japan
H.NISHIZAWA   *Electric Power Development*
              Co.,Ltd.   Japan
M.YASUDA   *Electric Power Development*
              Co.,Ltd.   Japan
T.SUZUKI   *Toshiba Corporation*   Japan
H.TANAKA   *Toshiba Corporation*   Japan

ABSTRACT

In this paper, improvement of hydraulic performance, particularly pressure fluctuation and cavitation, in order to speed up the runner revolution for a pump-turbine of 500m head class, is numerically and experimentally studied. High speed design of a pump-turbine offers an economical merit, because the size of the pump-turbine and generator-motor can be made smaller and the efficiency is improved. However, there are some problemes to be solved in order to develop a high speed pump-turbine. The first one is the increase of pressure fluctuation of runner outlet at partial load turbine operation. The second one is the deterioration of cavitation performance at pump operation. The third one is the increase of stress and vibration for runner. In order to overcome these problemes, the optimization of runner shape is carried out at design stage by using numerical simulation and its results are experimentally verified.

1 Introduction

Recently, the demand for larger capacity of pumped-storage power station is increasing because of expansion of power system and for inprovement of the plant economy. In the planning of a pumped-storage power station, higher head and larger unit capacity are selected usually to reduce the cost per capacity (cost/kW). But these involve some problems, for example, size limitation of

E. Cabrera et al. (eds.), Hydraulic Machinery and Cavitation, 210–219.

transportation, efficiency decline due to using low specific speed machines, etc. In Japan, several large scale pumped-storage projects are now being planned. Most of the projects are over 500m head and unit capacity is more than 300MW to reduce the size of reservoirs and machine cost. Advanced type high specific speed pump-turbines (Nsq= 40~45 m-m3/s) are envisaged for a pumped-strage project, now Electric Power Development Co. is planning to make the machine size smaller and to achieve higher efficiency(Figure 1). There are some problems to be solved in order to adopt high specific speed machines. The first one is the increase of pressure fluctuation of runner outlet at partial load of turbine operation. The second one is the deterioration of cavitation performance in pump operation. These are attributed to the faster peripheral velocity of runner outlet and the larger absolute swirl velocity at runner outlet thereby in turbine operation, and to the lower static pressure of runner outlet in pump operation. The third one is the increase of stress and vibration of runner due to higher speed.

In order to overcome the above problems and to adopt higher rated speed, the optimization of runner shape is carried out at design stage by using numerical simulation and its results are experimentally verified. Further, for the strength of runner, FEM analysis and real head model test are carried out, then the stress level and resonance characteristic of the runner are verified.

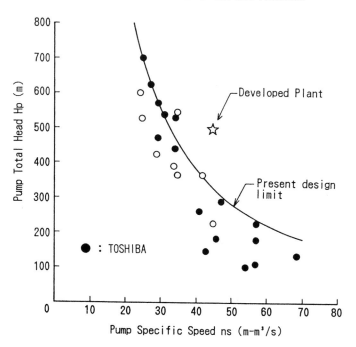

Fig. 1 Pump head versus spacific speed

2 Computational Method

2.1 Computational Grid

The three-dimensional Euler code with the pseudo-compressibility[1] is applied to flow simulation in pump-turbine runner. Figure 2 shows the solid-model and the grid in one runner blade channel of pump-turbine. In this analysis, in order to obtain flow distribution at runner outlet related to pressure fluctuation in the draft tube at partial load turbine operation, the simulation is carried out for the flow domain between runner inlet and conical part of draft tube. For the boundary surfaces other than runner blade surfaces, velocity and pressure are added periodically. The number of grid points is about 44,000.

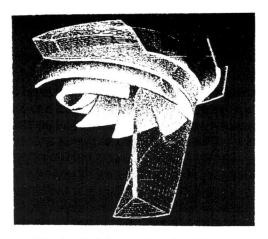

Fig. 2   Solid-model and Grid

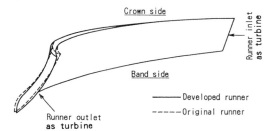

Fig. 3 Comparison of runner profiles

2.2 Calculated Condition

Flow analysis is carried out for model pump-turbine. In the case of turbine flow, calculated conditions are three cases about design point, maximum power point and partial load point. In the case of pump flow, calculated conditions are two cases about maximum and minimum head.

2.3 Geometry

Figure 3 shows the comparison of the profiles for the original runner(A) and the newly developed one(B). The modifications are limited to the area near the outlet of blades because the head-discharge curve and the optimum point must be kept as those of the original runner as possible. The purpose of the development of the new runner is improvement of cavitation characteristics in pump operation and that of pressure fluctuation in partial load turbine operation. In the course of this development, attention is paid so that no significant change in principal turbine and pump performance is involved to retain excellent efficiency characteristics of the original runner.

3 Computational Results

3.1 Cavitation Performance

Figures 4 and 5 show the pressure contours on the surface of blade at runner inlet for pump flow. The static pressure on blade surfaces which is obtained from the flow analysis is normalized by the pressure coefficient(Cp). Since it is thought that the cavitation inception occurs when the minimum static pressure (Hst) on blade surfaces becomes lower than the vapour pressure(Hv), the following relation is led :

$$\mathrm{Cpmin} = \frac{(\mathrm{Hv} + \mathrm{Hsi} - \mathrm{Ha})}{\mathrm{H}} = -\sigma_i \qquad (1)$$

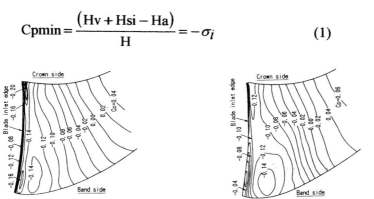

(a) Original runner A   (b) Developed runner B

Fig. 4 Pressure coefficient contours on suction surface of blade
at maximum head operation

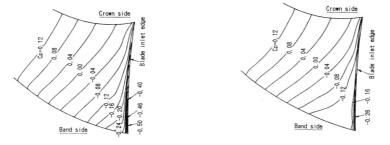

(a) Original runner A   (b) Developed runner B

Fig. 5 Pressure coefficient contours on pressure surface of blade
at minimum head operation

Therefore, the minimum pressure coefficient which is obtained from the flow analysis is compared by using eq.(1) with the inception sigma of cavitation obtained from the experiment.

For the maximum head operation, as shown in Figure 4, the pressure on the suction surface at runnner inlet edge becomes lower. As observed in these figures, the low pressure region of the developed runner(B) becomes smaller than that of the original runner(A). It can be seen that the developed runner(B) shows remarkable improvement in the pressure distribution at the crown side by the optimization of the runner profile. Also, in the minimum head operation, the improvement on pressure side surface is remarkable. In this case(Figure 5), the extre-mly low pressure region exists on pressure side surface of blades in the vicinity of the inlet near band, especialy for runner(A). It is recognized that the minimum pressure region of the modified runner(B) is reduced to narrow limited area. In addition, the minimum value of pressure coefficient Cp is considerably improved. As shown in Figure 6, the num-erical results show very good agreement with experiments. It indicates that the numerical simulation technique presented in this paper can predict the cavitation characteristics of pump-turbines satisfactorily.

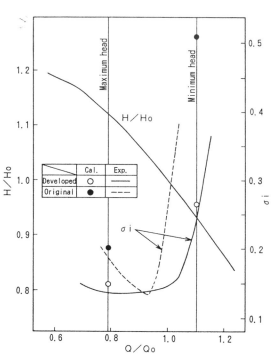

Fig.6 Comparison of cavitation performance

3.2 Flow Patterns and Velocity Distributions

When the runner revolution speed for a pump-turbine is increased, the increase of pressure fluctuation due to whirl in the draft tube at the partial load is concerned because of the faster peripheral velocity at runner outlet. Since this problem is attributed to the runner outlet flow and blade-to-blade flow, numerical simulation to investigate them is conducted. As an example, Figure 7 and Figure 8 show the numerical results of the flow patterns and velocity distributions at the partial load for the original runner(A) and the newly developed one(B) respectively. It is very different from each other. The flow in runner(A) is distorted, particularly the partial flow from crown side to band side is observed in the result. Then the streamlines near band is twisted. On the other hand, in case of the runner(B), the profile of runner outlet passage is rectified properly and the flow in it becomes much smoother than runner(A). The effect of this rectification is also recognized clearly in the velocity distributions at the runner outlet shown in Figure 8. Here, Ca and Cu mean the axial and absolute swirl velocity normalized by the mean axial velocity of runner outlet V0. This figure indicates that the swirl velocity coefficient Cu becomes smaller in case of the newly developed runner. It indicates that the intensity of the swirl is reduced. The decrease of swirl velocity is related to the decrease of rotating energy. Namely it suggests that the pressure fluctuation becomes smaller.

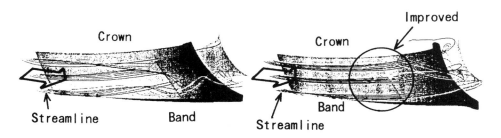

(a) Original runner A   (b) Developed runner B
Fig. 7 Flow patterns in runner at 65% partial load

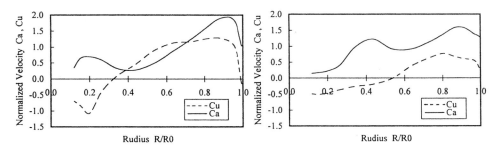

(a) Original runner A   (b) Developed runner B
Fig. 8 Velocity distributions of runner outlet at 65% partial load

4 Pressure Fluctuation of Runner Outlet

Pressure fluctuation of the newly developed runner was measured by the model test. Figure 9 shows the comparison of the experimental results of pressure fluctuation for the pump-turbine of conventional design (specific speed Ns=34 min^{-1},m^3/s,m) and for the newly developed one (Ns=41). It is demonstrated from the figure that the pressure fluctuation of the developed runner is less than that of the conventional one.

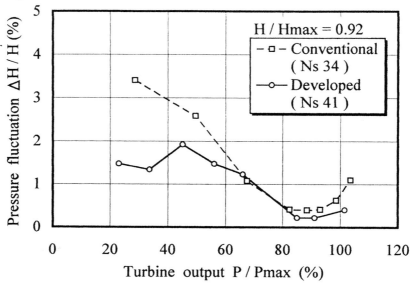

Fig. 9 Pressure fluctuation of draft tube in turbine operation

5 Stress and Vibration Analysis

When high specific speed design is adopted to increase the rotational speed of the runner, the width of a channel becomes relatively larger, so that the rigidity of runner and guide vanes may deteriorate. For static strength, FEM analysis was carried out and it was verified that the stress of the runner can be retained below the conventional stress level.

Regarding the dynamic strength, there has been no experimental result available for high head and high specific speed runner with larger height of inlet channel. Therefore, the experiments by means of real head model testing which conformed to hydro-elastic similitude were conducted to simulate the dynamic behaviour of the runner and to make clear the magnitude of the dynamic stress fluctuation.

5.1 Measurements

The real head model test arrangement consists of two model pump-turbines. The effective head for the model pump-turbine to be tested as turbine is provided by the other model pump-turbine operated as pump which is coupled to the end of a common shaft. The model has homologous passages and a runner made of the same material with prototype. Verious quantities(for example,stress and vibration,pressure fluctuation etc.)in model are related to those of the prototype by the hydro-elastic similitude law. The dynamic stress is measured by strain gauges stuck on the runner. The model speed is changed from 3,000min^{-1} to 6,600min^{-1}.

5.2 Results

During the 1980s, experimetal studies, as well as field measurements of the dynamic stress of runners, were conducted by Toshiba to explore the vibrational behaviour of high head pump-turbine runners. Detailed results of the studies were published in 1990 [2]. Similar to the past studies, investigation on vibrational behaviour of the high specific speed runner developed this time was conducted. An example of the mesurement results is shown in Figure 10 and Figure 11. Figure 10 shows the trend diagrams of frequency spectra with the change of test speed. It shows a dominant frequency corresponding to (NZg)Hz. Figure 11 shows a resonance curve obtained by the model test. From this figure, no significant resonance peak is observed in the characteristics of the developed runner. This means the runner vibration stress is sufficiently low, as long as it is used within the tested speed range. In order to confirm the strength against fatigue failure, the modified Goodman's diagram is applied to the assessment of the fatigue strength (Figure 12). It is indicated that the stress amplitude is sufficiently low, which reveals that the safety factor against fatigue strength is 9.2 .

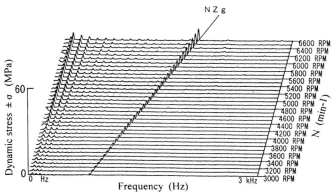

Fig. 10 Diagrams of frequency spectra

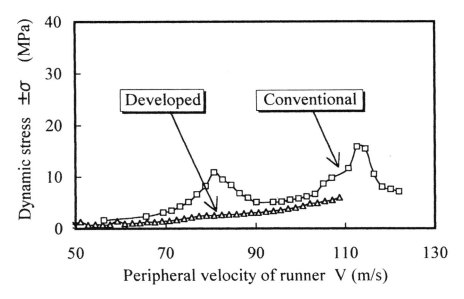

Fig. 11 Resonance characteristics of the runner

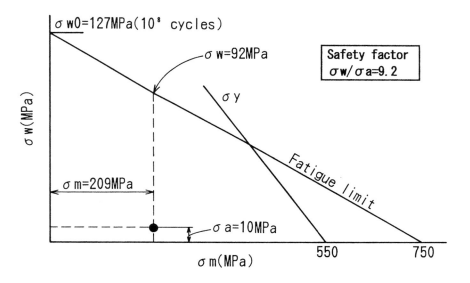

Fig. 12 Assessment of the dynamic stress of the runner
using the modified Goodman's diagram

6 Conclusion

In order to accomplish the speeding up the runner revolution and reducing the size of pump-turbines of 500m head class, there are three subjects to be explored. The first one is the decraese of pressure fluctuation of runner outlet at partial load turbine operation. The second one is the improvement of cavitation performance at pump operation. The third one is the decrease of stress and vibration for runner.

This paper describes the summary of the studies conducted on the above subjects. The optimization of the design of the blade profile was carried out at the design stage by using flow analysis and FEM analysis and its effect was verified by model test. Regarding the strength of runner, a real head model test was conducted to verify the stress level. Finally, it was confirmed that the newly developed high speed pump-turbine of specific speed 41 can be applied to 500m class high head pumped storage project.

References

[1] Nagafuji.T., Uchida.K., Nakamura.T. 1995, "Numerical and experimental studies of three-demensional flow fields in hydraulic turbine runners", proc. of the 1995 ASME/JSME Fluids Engng. Conf., FED-Vol. 227, pp.37-44.
[2] Tanaka.H., "Vibration Behaviour and Dynamic Stress of Runners of Very High Head Reversible Pump-turbines", *Proceedings,* IAHR Belgrade, Yugoslavia; 1990.

ANALYSIS OF THE PERFORMANCE OF A BULB TURBINE USING 3-D VISCOUS NUMERICAL TECHNIQUES

Y. QIAN   *Fuji Electric Co., Ltd.*   *Japan*
R. SUZUKI   *Fuji Electric Co., Ltd.*   *Japan*
C. ARAKAWA   *The University of Tokyo* *Japan*

Abstract

In this paper, a high-accuracy 3-D viscous numerical technique, which has been developed previously by the authors, is introduced to analyze the flows through a bulb turbine passing through the guide vanes, runner and draft tube. The techniques of predicting the interaction between the guide vanes(stationary) and runner vanes(rotating), the tip clearance, and a new turbulent model which was previously developed by the authors, are introduced to this analysis. In order to check the shortcomings of the k- ϵ two equations turbulent model, the limitation of the k- ϵ turbulent model in the application to vortex flows are discussed by various swirling rates in this paper. The purpose of this paper is to attempt to show that 3-D numerical techniques have been developed to analyze the flows through the all flow channels, guide vanes, runner, and draft tube. The flow patterns, separation phenomenon and performance of bulb turbine can be simulated depending on the characteristics of discharge rate and head only. The simulations are carried out under a series of discharge rate and rotating speed conditions. That is, when the runner opening is fixed and when the guide vane openings are changed to several opening angles to determine the On-cam operating conditions following the discharge rate and rotational rate.

1. Introduction

The authors have previously published several papers in analyzing the flow phenomenon through hydraulic turbine. Arakawa and Qian(1991) successfully developed a 3-D numerical code to simulate the flow through a hydraulic turbine. Qian(1994) proposed an interface method to deal with the interaction between the rotating block(runner) and stationary block (stay and guide vanes) in Francis type turbines. This code considers the influence of viscosity in some turbulent models and can be successfully applied to simulate the phenomenon of channel vortexes in Francis turbines. This interface method provides very good agreement in comparison with the experimental results. This interface technique is introduced in this paper to solve the flow associated with the rotor-stator vane configurations. Furthermore, Qian(1995) developed the code to analyze the flow in a bulb turbine runner. An embedded H-type mesh topology was successfully utilized to resolve the gap at the blade tip region. In order to simulate the turbulent flows in the gap, a new wall condition were proposed

220

E. Cabrera et al. (eds.), Hydraulic Machinery and Cavitation, 220–229.

to provide the turbulent parameters near the wall.

It is well known that the k- ε turbulent model is not suitable for predicting strong vortex flows as for example, discussed for basic confined swirling flows by Hogg and Leschziner(1989). The swirling flow in the draft tube is always one of the most formidable problems in hydraulic turbine design. Clearly, rationally controlling the swirling flow will greatly improve the loss creation and pressure recovery in the draft tube. Several research papers concerning draft tubes have been reported. One of the researches of this topic is Vu (1990), who comparatively provide complete numerical researches of hydraulic turbine draft tubes with viscous numerical techniques. However, these previous researches were limited in individual parts. In this paper, a highly-accurate 3-D numerical code is introduced to analyze the flow patterns in bulb turbines including all the flow channels of the guide vane, runner, and draft tube. The boundary flow conditions are obtained by exchanging the information substituting the results from each flow parts into the iterations of the others.   The authors check the limitations of the k- ε turbulent model for draft tubes under several uniform swirling rates. The shortcomings of k- ε the turbulent model are found under strong swirling flows. This phenomena shows that the k- ε turbulent model has a limitations in the case of strong vortex flows.

2. Major symbols of the bulb turbine model

The major parameters of bulb turbines are specified by the following symbols.

n_{11}	: Unit rotating speed
Q_{11}	: Unit discharge rate
H	: Effective head
D_1	: Runner diameter
U_1	: Runner peripheral speed
V	: Flow velocity
C_i	: Flow velocity coefficient
P	: Pressure
Pref	: Reference pressure
Cp	: Pressure coefficient
Ci	: Velocity coefficient
m	: Swirling rate

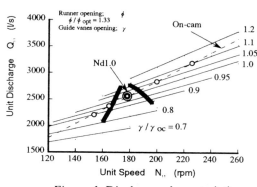

Figure 1. Discharge characteristics

3 . The geometry of the bulb turbine

The specifications of this bulb turbine under design operating conditions is presented in table 1. The computations are carried out under several operating condition shown in figure 1. The outline of this model turbine is shown in figure 2. The computational domain consists of the guide vanes, runner, and draft tube.

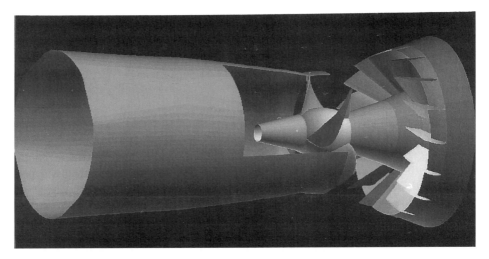

Figure 2.   Outline of the bulb turbine

TABLE 1. Specification of the bulb turbine under design operating conditions

H(m)	D1(mm)	Z_R(blade)	Z_g(blade)	n_{11}(l/s)	Q_{11}(l/s)
12	500	4	16	150	1800

4. Block structured grid geometry and boundary conditions

The guide vanes consist of 16 blades and the runner consists of 4 blades. The guide vanes are set in an absolute coordinate system and the runner is set in a relative coordinate system under the constant rotational speed. The boundary conditions between the runner inlet and stator outlet are in the unsteady flow conditions. However, to overcome these difficulties, a mesh is used which is divided into three blocks, one that includes the runner, one that includes the guide vanes and another that includes the draft tube. The flow region associated with the runner and guide vanes can be described by patched grid blocks as proposed by Qian(1994). The inlet boundary conditions of the draft tube are obtained from the exit of the runner. The meshes chosen for the calculation are described with and H-type grid. The size of guide vanes grid is 70x21x21, runner is 70x28x26, and the draft tube is 41x21x21. All of the computations are carried out on an EWS computer.

5.   Numerical method and turbulent model

The pseudo-compressibility idea is introduced into this research, which employs the numerical technique of compressible flows to solve hyperbolic-type partial differential equations. This means that the scheme conveniently covers both in incompressible flows and air turbines using the same algorithm. We take the so-called Beam-Warming scheme(1976) proposed for compressible flows, which is based on an implicit algorithm, such as an approximately factored method, to obtain a large time step. The upwind based highly-accurate TVD scheme(1985) is used without any artificial dissipation terms. The diagonalized ADI method is used to improve computational efficiency, and the operator is inverted by the forward and backward sweep techniques.

The standard k- ε turbulent model is employed to simulate the flow of a high Reynolds number. The wall function is used to decrease the number of grid points near the wall. A modified wall function is introduced to describe the turbulent boundary conditions on the points very close to the wall, as in the gap region of the runner.

6. Results

6.1 Guide vanes

To confirm the numerical techniques, the flows through the guide vanes (eliminate the runner) are computed independently. Figure 3 shows the velocity and pressure coefficients on the exit of the guide vanes. These results are in very good agreement with the experimental results. After this confirmation, the computations are carried out including the influence of the runner. Figure 4 shows the tangential velocity (V_θ)under two types of channel shapes from the guide vanes outlet to the runner inlet under the same operating parameters and same guide vanes and runner. The discrepancy between the two types of channels can be found by analysis of the results in figure 4. It is well known that the V_θ volumes in the span direction are one of the important conditions for the runner design. Hence, optimally designing the V_θ at the leading edge will not only approve the maximum efficiency, but will also extend the operating regions by suppressing cavitation. After optimally considering the attack angle of the runner, operating region, and cavitation, the best channel form is selected in this paper.

Figure 5 shows the guide vane pressure coefficients on the three sections from the boss side to the discharge ring side under one of the On-cam operating conditions, where sections 2, 5, 9 represent the section near the discharge, middle, and boss ring respectively. Figure 6 shows one of the On-cam velocity and pressure coefficients on the exit (as the inlet boundary condition of runner).

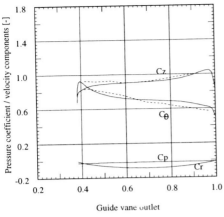

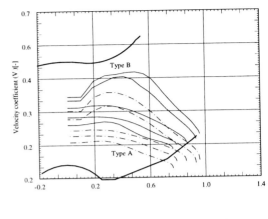

Figure 3. Velocity and pressure coefficient on the exit guide vanes
——— : Computation, ········ : Experiment

Figure 4. V $_\theta$ along the two types shape of channels

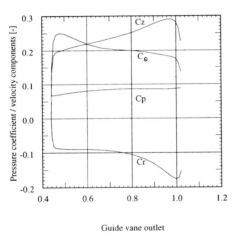

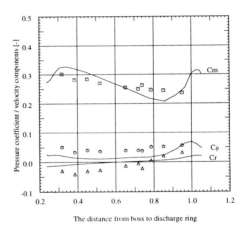

Figure 6. Velocity and pressure coefficient on the exit guide vanes

Figure 8. Velocity and pressure coefficient on the the runner exit. Symbols: Experiment

6.2 Runner

In the computation of the runner, the boundary conditions are obtained from the exit of the guide vanes at the initial iteration. After performing a back iteration and exchange on the overlapping region, final boundary conditions are computed in figure 6. A series of computations are carried out under various rotating speeds and discharge rates for each guide vane opening. The operating condition at the maximum efficiency is defined as the On-cam operating condition. Figure7 shows the pressure coefficients on

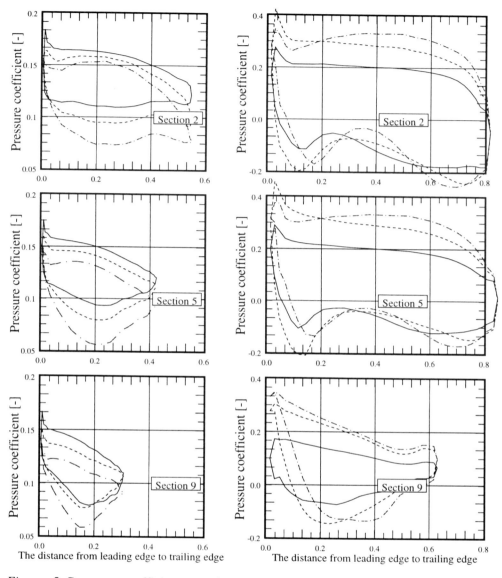

Figure 5 Pressure coefficient on the guide vanes ——— $\gamma/\gamma_{opt}=1$
-- -- -- $\gamma/\gamma_{opt}=1.1$ ······· $\gamma/\gamma_{opt}=1.2$

Figure 7 Pressure coefficient on the runner ——— $\gamma/\gamma_{opt}=1$
-- -- -- $\gamma/\gamma_{opt}=1.1$ ······· $\gamma/\gamma_{opt}=1.2$

the sections along the streamwise direction under three On-cam operating conditions. The velocity and pressure coefficients on the exit of the runner are shown in figure 8.

These exit results are selected as the inlet boundary conditions for the computation of the draft tube. A peak of C_m can be found near the discharge ring, This phenomena may be considered that the boundary layer becomes thicker along the blade tip from the leading edge to the trailing edge, so that the flow can easily pass through the gap near the exit of the runner as a wall jet, as observed in the pressure dumping near the trailing edge in the figure 7.

6.3 Draft tube

In order to check the limitations of k- ϵ turbulent model application for swirling flows, a series of uniform rotation velocities(ω x R) are introduced to the test. The swirling rates(= $\int \rho V_z V_u dA$ / R$\int \rho V_z^2 dA$) are displayed in figure 9 under these 5 kinds of flow

patterns. The pressure recovery characteristics Cp_q and pressure coefficients P_q are shown in figure 10 and figure 11. Cp_q and P_q are defined as follows;

$$Cp_q = \frac{Po - Pq}{Ps} , \qquad P_q = \frac{Pqi - Po}{\rho \overline{u} / 2} \qquad (1)$$

where P_o is the static pressure of the exit, u is the axial velocity, and P_d ($= \dfrac{\int (\rho V^2) V_z dA}{Q}$) is the dynamic pressure coefficient. The parameter of Cp_q is the average pressure-rise and P_q represents the relation of the transport energy from draft tube inlet to the outlet. When the swirling rate is larger than m=0.5, the computational iteration processing appears to not converge in the turbulent dissipation equation. However, the swirling rate under the On-cam operating conditions are almost always less than 0.3 as shown by the dotted line on the figure 9. Figure 12 shows the velocity vectors from inlet to outlet under one of the On-cam inlet boundary conditions obtained from the computation of the runner.

Following the successfully development of the above techniques for the bulb turbine design, the total performance of the bulb turbine can be analyzed the using the numerical tools. The flow patterns and performance of all features of the bulb turbine, such as the blade attack angles, loss phenomena, and cavitation, can be provided by computation before the experiment. One set of the analysis results is provided in this paper. Figure 13 shows the pressure contours and velocity vectors from the guide vanes to the runner under the design operating conditions, and figure 14 displays the cavitation coefficient at the inlet of the bulb turbine runner. A more detailed report will be presented in later research paper.

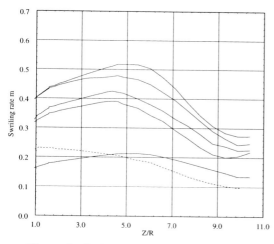

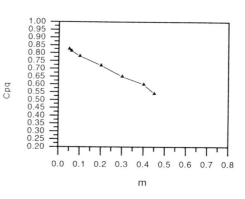

Figure 9 . Swirling rate in the draft tube

Figure 10. Pressure recovery under various swirling rate

7. Conclusion

High-accuracy 3-D viscous numerical techniques have been successfully introduced to analyze the performance of bulb turbines including the guide vanes, runner, and draft tube. In the simulation of guide vanes and runner, the authors select a series of different channel shapes to the optimum shape. The computations provide very good cavitation results at the leading edge of the runner because the influence of tip clearance is accurately considered in the numerical tools. These results show the operating regions of the bulb turbine and provide guidance to improve the cavitation phenomenon and extend the operating regions. The computations also present several flow patterns in the draft tube. The pressure recovery and pressure coefficient will be important data for draft tube design.

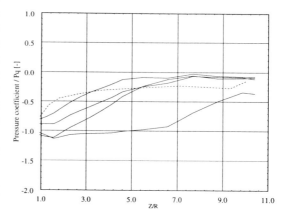

Figure 11. P_q under various swirling rate

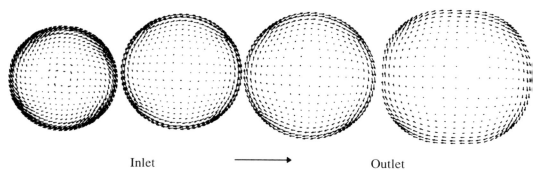

Inlet   ⟶   Outlet

Figure 12. Velocity vectors on the sections from the inlet to the outlet of draft tube

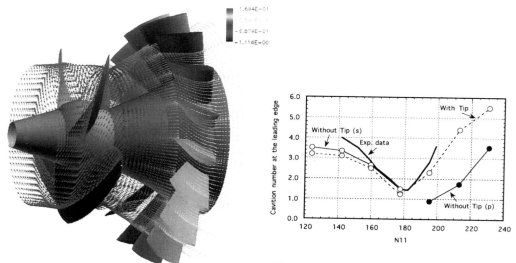

Figure 13. Pressure and velocity vectors

Figure 14 Cavitation parameters at
the runner leading edge

As a consequence, the computational results can be obtained depending only   on the characteristics of the head and discharge rate. The computational results provide a lot of important information to correct the initial design. These results have already been adopted as a part of our design routine. However, in order to verify the applicable regions of the computations, extended studies will be required. to improve the shortcomings of the turbulent model and the develop highly-accurate turbulent models. The rotor-stator interaction under Off-cam operating conditions will also have to be addressed in future work.

Reference

Arakawa, C., Qian, Y., and, Samejima, M.(1991) Turbulent flow Simulation of Francis Water Runner With Pseudo-Compressibility, *GAMM-Conference on Numerical Methods in Fluid Mechanics, 259-268*

Beam,R.F. and Warming, R.M.,(1976), An Implicit Finite-Difference Algorithm for Implicit Finite-Difference Hyperbolic System in Conservation Laws, Hyperbolic System in Conservation Laws, *Journal of Computational Physics, Vol.22,*

Chakravarthy,S.R. and Oshers,S.,(1985), An new Class of High Accuracy TVD Schemes for Hyperbolic Equations for Steady-State and Time-Dependent Program, *AIAA Paper 89-0463.*

Chorin,A.J., (1967) A Numerical Method for Solving Incompressible Viscous Flow Problems, *Journal of Computational Physics, Vol.2,pp.12-26*

Hogg, S. and Leschziner,M.(1989), Computation of Highly Swirling confined Performance Prediction of Mixed-Flow with a Reynolds Stress Turbulence Model, *Vol.27 No.1, pp.57-63*

Obayashi,S. and Fujii,K.,(1985), Computation of Three-Dimensional Viscous Transonic Flows with the LU Factored Scheme, *AIAA Paper 85-1510*

Qian,Y.,Arakawa,C.,and Kubota,T.,(1994), Numerical Flow Simulation on Channel Vortex in Francis Runner, *IAHR Symposium ,1994*

Qian, Y.,Arakawa,C.(1995), 3-D Numerical Analysis of Bulb Turbine Runner Performance with and Without Tip Clearance, FED-Vol.227, Numerical Simulation in Turbomachinery ASME 1995.

Vu, T.C., Shyy, W. (1990), Navier-Stokes Flow Analysis for Hydraulic Turbine Draft Tube, *ASME Journal of Fluid Engineering, pp. 199-204*

ANALYSIS OF THE INLET REVERSE FLOWS IN A PUMP TURBINE USING 3-D VISCOUS NUMERICAL TECHNIQUES

Y. QIAN   *Fuji Electric Co., Ltd.*   *Japan*
R. SUZUKI   *Fuji Electric Co., Ltd.*   *Japan*
C. ARAKAWA   *The University of Tokyo*   *Japan*

Abstract

The purpose of this paper is to apply advanced technology to predict the phenomenon of reverse flows created at the inlet during pump operation of Francis type pump-turbines. The authors introduce 3-D viscous numerical techniques, which were developed for hydro-turbine runners several years previous, to predict and analyze the recirculation flows at the inlet during pump operation. The motions of the separation flows are predicted by reducing the discharge rate from design to off-design operation conditions. In order to accurately capture the strong vortex, a new boundary condition of turbulent parameters on the wall is introduced in this simulation. The computational results are found to be in reasonable agreement with the experimental results. The limitations of the k- ε two equations turbulent model in the application of vortex flows are discussed in this paper. Finally the authors present a basic scale for use with Navier-Stokes simulation techniques with k- ε two equations model when using the model as design tool in the case of reverse flows.

1. Introduction

Many researchers have published numerous papers on the reverse flow phenomena using measurement techniques. In a series of combined analytical and experimental researches, Kurokawa(1994) provided experimental results of reverse flow patterns over a wide discharge range. These experimental results provided a detailed description and became well established as research and design tools in industry. On the other hand, following the development of computational hardware and Computational Fluid Dynamics(CFD), CFD techniques have been widely used to simulate the flow patterns and predict the performance of hydraulic turbines and are known as "numerical experiment". Qian and Arakawa(1991) successfully developed a 3-D numerical code to simulate the flow through a hydraulic turbine. The code considers the influence of viscosity with some turbulent models and can be successfully applied to simulate the phenomenon of the channel vortex in Francis turbines(1994). This work is a attempt to apply this code to predict the inlet reverse flows of pump turbines during pump operation.

E. Cabrera et al. (eds.), Hydraulic Machinery and Cavitation, 230–237.

It has been made known that the k- ϵ turbulent model is not suitable for predicting strong vortex flows, for example, discuss for basic confined swirling flows by Hogg and Leschziner(1989). Qian(1996) checked the difference vortex flows through the draft tube against the k- ϵ turbulent model simulation. The shortcomings of the k- ϵ turbulent model are found in strong swirling flow. When the swirl rate 'm' is increased to greater than 0.5, the computational iterations can be diverged to relational results comparable with the experiments. This phenomena confirms the fact that the k- ϵ turbulent model have a limitation in the case of strong vortex flows. However, it is well known that if higher level turbulent models are selected, such as Rynolds stress model and LES model, longer computational time are needed. At present, it is very difficult for industrial design to simulate the flow patterns, due to the huge CPU time required. Therefore, in this paper, the k- ϵ model is still selected to model the eddy viscosity in the Reynolds average Navier-Stokes equations, even though its shortcomings have been made known. According to the above reason, Qian(1995) proposed a new ideal model to modify the turbulent parameters near the wall in the k- ϵ model. This new method is also introduced in this paper. The simulation results can be used with confidence as a design tool, and the process of the iterations converge smoothly despite the instability of the vortex progress during the iteration.

2. Major symbols of the pump turbine model

The major parameters of pump-turbines are specified by the following symbols.

n   : Rotating speed
Q   : Discharge rate
H   : Effective head
D_1 : Runner diameter
U_1 : Runner peripheral speed = $\pi\, D_1 n/60$
C_i : Flow velocity coefficient = V_i/U_1
P   : Pressure
Pref   : Reference pressure
Cp   : Pressure coefficient = $2(P-Pref)/\rho\, U_1^2$

3 . The geometry of the pump turbine

The pump-turbine model runner analyzed is one of the test models being researched and developed by the authors. The specifications of this pump-turbine are presented in table 1, and the outline of this model runner is shown in figure 1. Several new hydraulic design ideas are introduced in this series pump turbines, such as have 9 blades. As one of the most important tools, numerical analysis techniques predict the performance before the model test and support the new model design.

TABLE 1 Specifications of the pump-turbine

H(m)	D1(mm)	Z_R(blade)	n(r/min)	Qd(l/s)
60	500	9	1200	400

Figure 1. Outline of pump-turbine runner

4. Numerical method and boundary conditions

Pseudo-compressibility was adopted as in previous papers (Arakawa, 1991,1994). The Navier-Stokes equation is transformed into the linear Delta form using the Beam-Warming method(1976), and moreover the TVD scheme (Chakravarth, 1985) based on the upwind and approximately factored method (Obayashi, 1985) was used. The turbulent model that the authors use for the simulation is the standard k- ε 2 equations turbulent model with wall function condition and a modified new wall boundary condition proposed by Qian(1995).

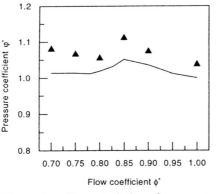

Figure 2. Characteristics of pump

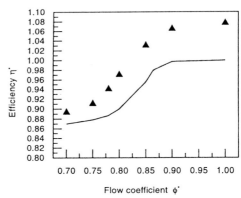

Figure 3. The efficiency of pump

The computational domain includes the runner and the prolonged parts at the inlet and outlet of the runner. The inlet boundary conditions are given by the discharge rates. For the boundary at the exit, the differentials of velocity components in the streamwise direction are set at 0 for the initial iteration, and then a non-reflective boundary condition proposed by Dulikravich(1993) is introduced and the velocities are corrected in order to conserve the mass flow rate to be the same as the inlet discharge. On the crown, band and blade surfaces, impermeable conditions are specifying. The tangency condition is enforced by specified the contravariant velocity W=0 on the crown and band surfaces, and V=0 on the blade surfaces. The contravariant velocities and their pressures are obtained by linear extrapolation from the interior points when required.

5.   Analysis results and performance characteristics of the pump turbine

The analyses are carried out under the design discharge(Qd) and partial discharge at 90%(0.9Qd), 85%(0.85Qd), 80%(0.8Qd), 75%(0.75Qd) and 70%(0.7Qd) of the design discharge. The simulation results are presented in the related forms of the flow coefficient ($\phi = \dfrac{Q}{nD_1}$) and pressure coefficient ($\psi = \dfrac{P - \mathrm{P}ref}{\dfrac{1}{2}\rho U_1^2}$) as shown in equation 1.

$$\phi * = \phi / \phi_{opt}$$
$$\psi * = \psi / \psi_{opt} \tag{1}$$

ϕ_{opt} , ψ_{opt} are the experimental flow coefficient and pressure coefficient under the highest efficiency. The numerical results(symbols) of $\phi *, \psi *$ are shown comparison with experimental results in figure 2. When $\psi *$ rises to the peak of $\phi *, \psi *$ distribution, the $\psi *$ tends to become less against the decrease of $\phi *$. The decrease in experimental $\psi *$ starts after $\phi * < 0.85$ and that of the computational $\psi *$ starts after $\phi * < 0.80$. According to the visualization of the experiment, the separation flows near the inlet of the pump are first appear close to this operation condition. The efficiency presented in this paper is defined by the authors as equation 2.

$$\eta_c^* = \eta_c / \eta_{opt}$$
$$\eta_e^* = \eta_e / \eta_{opt} \tag{2}$$

Where, η_c, η_e, η_{opt} are the computational efficiency, experimental efficiency, and highest experimental efficiency in the experiment respectively. Figure 3 shows the results of the efficiency. The efficiency reduction is observed against $\phi *$. The numerical solutions predict a larger efficiency when compared with the experiments. The computational efficiency is higher than the experimental results, because the loss given by computation are only taken from the runner, whereas the measurements not only cover the rotating part but also include the losses created in the stator such as guide vanes and casing.   However, the discrepancy between the two results are almost constant following the different $\phi *$.

Figure 4 shows the results of the absolute velocity and pressure distributions at the inlet of the pump under design operation conditions. C $_\theta$, C$_r$, C$_z$ are respectively the tangential, centrifugal, and axial velocities averaged between the rotating blades in the circumferential direction, and C$_p$ is the pressure coefficient, all of which are made non-dimensional using the spouting velocity corresponding to the net head $\sqrt{2gH}$.

Figure 5 shows the results of the velocity and pressure coefficients, which are defined in figure 4 under the partial discharge Q=0.7Qd. The axial velocity created a negative volume near the band side and the tangential velocity rise at the same region. The results show that the reverse flows are created near the band side of the pump inlet when the discharge is reduced to less than 75% of the design discharge. The reverse flow separation point can be accurately obtained by data analysis. On the other hand, the reverse flow patterns may be observed by computational graphic visualization on the computer display. Some of the results are shown in figures 6,7. Figure 6 shows the pressure contours of Q=Qd, where figure 7 displays the results under Q=0.7Qd. Figures 8, and 9 show the velocity volume along the streamline from the inlet to the outlet of the runner at the Qd and 0.7Qd operation conditions, and the curve numbers represent the sections from the crown side to the band side. This result clearly shows that a vortex exists at the inlet near the band and extends a long way in the downstream direction along the band surface.

In order to observation the reverse flow pattern in more detail, several velocity vectors on side sections are shown in figure 10. When the discharge is reduced to less than 75%Qd, a reverse flow is created at the inlet and the reverse region enlarges from the suction surface to the pressure surface of runner blades.

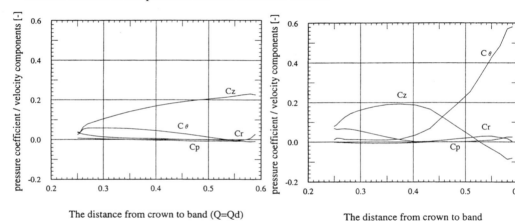

The distance from crown to band (Q=Qd)

Figure 4. Velocity and pressure coefficient   under Q=Qd

The distance from crown to band

Figure 5. Velocity and pressure coefficient under Q=0.7Qd

Figure 6. Pressure contours under Q=Qd

Figure 7. Pressure contours under Q=0.7Qd

Figure 11 shows the pressure coefficient distributions along three chord-wise sections of a runner at several discharge rates, while the 2,5,9 sections present the sections near the crown, in the middle and the band respectively. A negative pressure peak can be found at the leading edge on the suction surface near the band side when the discharge rate is reduced to less than 0.75Qd.

6. Conclusion

This paper reports an attempt to analyze and observe the pump reverse flow patterns using numerical techniques. The recriculation flow phenomena, reverse flow starting conditions, and the characteristics of the pump-turbine have been predicted in the paper. These results have already been adopted as a part of the design routine for daily use after introduction connection scales.

The authors believe this paper is only a first report for the analysis of pump-turbines. In the near future, a further series of analyses will be reported. Several points will be studied, for example, introducing a conservative algorithm to handle the unstable flows, computing the flow including the guide vanes, and developing highly-accurate turbulent models which can be computed on small scale computers such as desk-top workstations.

Reference

Arakawa, C., Qian, Y., and, Samejima, M.(1991) Turbulent flow Simulation of Francis Water Runner With Pseudo-Compressibility, *GAMM-Conference on Numerical Methods in Fluid Mechanics, 259-268*

Beam,R.F. and Warming, R.M.,(1976), An Implicit Finite-Difference Algorithm for Implicit Finite-Difference Hyperbolic System in Conservation Laws, Hyperbolic System in Conservation Laws, *Journal of Computational Physics, Vol.22,*

Chakravarthy,S.R. and Oshers,S.,(1985), An new Class of High Accuracy TVD Schemes for Hyperbolic Equations for Steady-State and Time-Dependent Program, *AIAA Paper 89-0463.*

Chorin,A.J., (1967) A Numerical Method for Solving Incompressible Viscous Flow Problems, *Journal of Computational Physics, Vol.2,pp.12-26*

Dulikravich, G. Ahuja, V. and Lee, S. (1993), Three-Dimensional Solidfication With Magnetic Fields and Reduced Gravity, *31st Aerospace Sciences Meeting & Exhibt, AIAA 93-0912.*

Hogg, S. and Leschziner,M.(1989), Computation of Highly Swirling confined Performance Prediction of Mixed-Flow with a Reynolds Stress Turbulence Model, *Vol.27 No.1, pp.57-63*

Kurokawa, J., Kitahora,T. and Jiang, J.(1994), Pumps Using Inlet Reverse Flow Model, *IAHR Symposium, No. A13,1994*

Obayashi,S. and Fujii,K.,(1985), Computation of Three-Dimensional Viscous Transonic Flows with the LU Factored Scheme, *AIAA Paper 85-1510*

Qian,Y.,Arakawa,C.,and Kubota,T.,(1994), Numerical Flow Simulation on Channel Vortex in Francis Runner, IAHR Symposium ,1994

Qian, Y.,Arakawa,C.(1995), 3-D Numerical Analysis of Bulb Turbine Runner Performance with and Without Tip Clearance, FED-Vol.227, Numerical Simulation in Turbomachinery ASME 1995.

Qian, Y., Suzuki, R., Arakawa, C., (1996) Analysis of the Performance of Bulb Turbine Using 3-D Viscous Numerical Techniques, *IAHR, Symposium, 1996.*

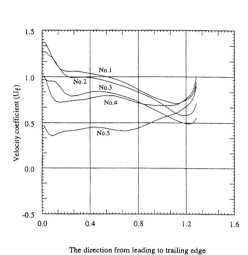

Figure 8. Axial velocity from inlet to outlet under Q=Qd

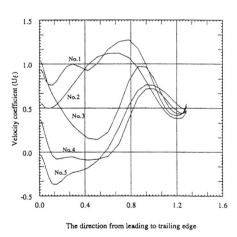

Figure 9. Axial velocity from inlet to outlet under Q=0.7Qd

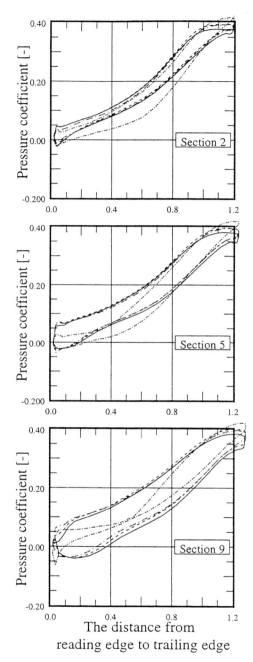

Figure 11. Pressure coefficient along the streamwise direction. ———— :Qd, ········· :0.9Qd, –·–·– :0.8Qd, –··–··–: 0.7Qd

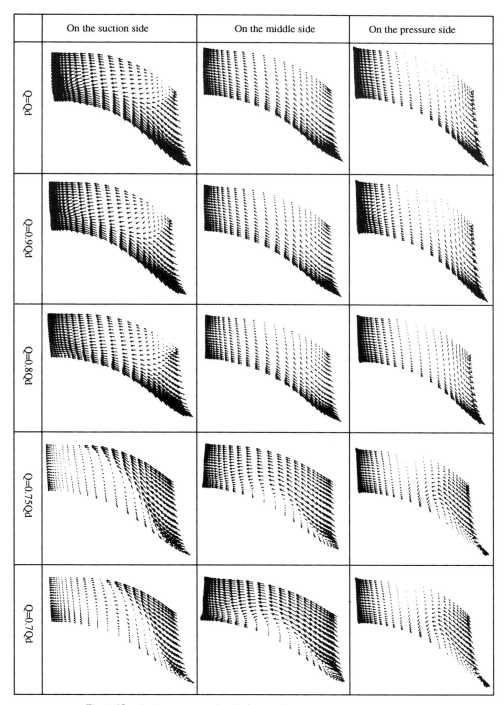

Figure 10. velocity vectors on the side from suction to pressure side

IMPORTANCE OF INTERACTION BETWEEN TURBINE COMPONENTS IN FLOW FIELD SIMULATION

S. RIEDELBAUCH, D. KLEMM, C. HAUFF
Voith Hydro GmbH
D-89509 Heidenheim, Germany

1.   Introduction

The application of viscous flow prediction methods, i.e. Navier - Stokes methods, has drastically increased during the development of hydraulic machines in the course of the last several years. This proceeding supports the design process through a detailed knowledge of the viscous flow behaviour in all components of the hydraulic machine. Usually, a broad operation range is investigated, covering the optimum conditions, too.

Another reason for the large increase of theoretical flow prediction applications is the increased reliability of the corresponding results. In most cases, these predictions show a flow behaviour being comparable to that observed during model test investigations.

The main reason for this agreement is the following: The numerical flow field simulation also takes into consideration the influence of the neighbouring hydraulic components. Thus, the flow is approximated in a much more realistic way.

This paper will focus on the used numerical method including coupling procedures. Some selected results will demonstrate the influence of component interaction and the related success during the product development phase.

2.   Numerical method and component coupling

The numerical method used here is based on the incompressible Reynolds averaged Navier - Stokes equations and employes the well-known finite volume concept. The influence of turbulence is considered with the two - equation k-ε turbulence model.

E. Cabrera et al. (eds.), Hydraulic Machinery and Cavitation, 238–247.
© *1996 Kluwer Academic Publishers. Printed in the Netherlands.*

Owing to restrictions in computer capacity, a complete hydraulic machine is usually subdivided into several components:

- semispiral or full spiral including stay vanes
- tandem cascade of stay vane and wicket gates
- runner
- draft tube

The modelling of individual turbine components is more simple than that of a complete turbine in a product development environment, even if the turbine model is available in a CAD system (Buchmaier *et al.*, 1996).

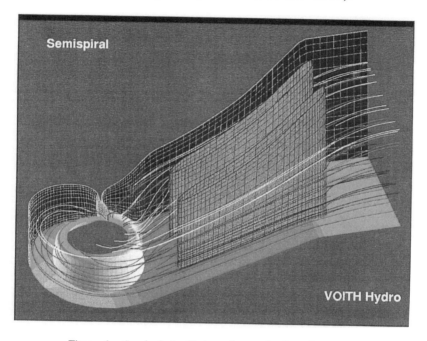

Figure 1. Semispiral with two piers and selected streamlines.

The viscous flow through these components has been analyzed for many years (Riedelbauch *et al.*, 1994 and Ruprecht *et al.*, 1994). The simulation of the flow through a semispiral is one example (fig. 1). This picture presents a semispiral with two piers and a set of selected streamlines coloured with the height level at the entrance. The nose vane is also well seen. The streamlines show a very smooth flow behaviour. Flow separation is only present in the wakes of the piers. Here, constant inflow conditions were chosen. However, the result of this flow case is strongly dependent on the boundary conditions - formulated overrefined: Any boundary condition will give any result possibly being far away from the real flow situation. This is especially true for the tandem cascade, the runner and the draft tube.

Of course, the main goal is to predict the flow in each of these components as realisticly as possible for steady state flow conditions. Here, the word prediction is emphasized in contrast to analysis. E.g. for turbine operation, necessary input parameters for prediction are only discharge, wicket gate opening, angular velocity of the runner and, of course, the entire geometry. All other magnitudes are results of the flow prediction, e.g. operation point and efficiency. This minimum set of parameters is sufficient for the simulation of the complete machine. However, additional assumptions have to be introduced, if individual components are considered alone. These assumptions are basically boundary conditions obtained through simple physical considerations. In most cases, however, these are constant conditions and stiff in this sense. Appropriate coupling procedures permit the adjustment of those boundary conditions in a non-stiff automatic way resulting in more realistic flow approximations. This is particularly important for the stay vane / wicket gate - runner interaction and the runner - draft tube interaction.

In principle, three different types of coupling are possible:
- three dimensional unsteady simulation
- frozen rotor, i.e. fixed relative location of runner and wicket gate and/or runner and draft tube
- circumferential flow averaging

Of course, each option may vary due to implementation details.

The three dimensional unsteady simulation is supposed to be the most realistic approach, provided that the corresponding flow code is designed for time accurate flow simulation. Real machines possess a non-integer ratio of wicket gates and runner blades. As a consequence, all wicket gates combined with the entire runner have to be considered. The number of nodes of these computational grids is huge - and so is the needed simulation time. Today, three dimensional unsteady simulation of viscous flow is considered to be of academic interest only. Effective support of product development needs shorter turn-around times.

The second possible approach is the so-called frozen rotor approach. This terminology means that the relative location between rotating parts and non-rotating parts remains fixed. At a certain surface of revolution, the flow components have to be transformed from a fixed frame of reference to a rotating frame of reference. Due to the non-integer ratio of wicket gates and runner blades, the complete geometry has to be considered being very similar to the unsteady approach. Consequently, no problem reduction is possible through the use of cyclic periodic boundary conditions.

Another coupling approach is to average the flow variables in circumferential direction at the boundary between rotating and non-rotating parts (sliding boundary). This proceeding allows to reduce the size of the problem to one periodic flow channel for axisymmetric geometries, i.e. runner and tandem cascade. Of course, the size of the draft tube grid and the spiral grid is not influenced. The choice of the location of the sliding boundary is dictated by the flow physics. The runner - draft tube sliding boundary has to be defined within the axisymmetric part of the geometry. The wicket gate - runner sliding boundary has to be positioned between the trailing edge of the wicket gate and the leading edge of the runner. From a physical point of view, it should be as far upstream as possible for the runner to account for the upstream influence of the runner blades.

For the tandem cascade, however, it should be as far downstream as possible. For this reason, we are working with overlapping grids and not only with a sliding boundary, i.e. the runner grid extends very close to the trailing edge of the wicket gate, while the tandem cascade grid extends to the leading edge of the runner. The runner - draft tube boundary is treated in a similar way.

3. Selected results

This chapter is devoted to present some selected results, considering the flow influence of neighbouring turbine components. Here, the flow averaging in circumferential direction has been employed.

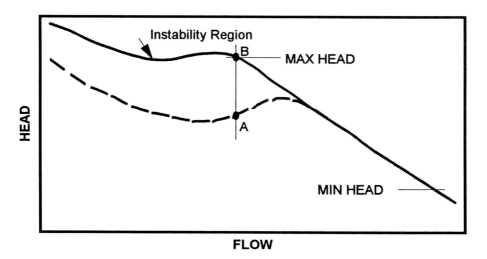

Figure 2. Sketch of the pump characteristic of a Pump-Turbine.

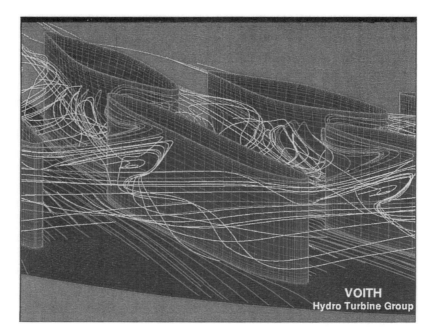

Figure 3. Selected streamlines through tandem cascade <u>with</u> runner interaction.

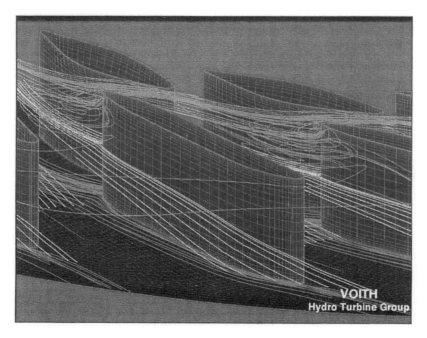

Figure 4. Selected streamlines through tandem cascade <u>without</u> runner interaction.

3.1   TANDEM CASCADE - RUNNER FLOW INTERACTION

Pump storage plants are often operated by pump-turbines. This type of machine is capable of pumping fluid into the upper reservoir as well as running in turbine mode generating energy. One runner handles both, pump and turbine operation. Thus, the development of high performance pump-turbines is a challenging task. The efficiency level has to be very high for pump as well as for turbine mode. See Klemm *et al.*, (1996) for a recent development.

Typically, the pump operation is limited by an instability region at high heads (Fig. 2). This instability region is related to massive flow separation. It may occur within the stay vane/wicket gate component and/or the runner (Fisher *et al.*, 1978). The separation of flow and the formation of appreciable regions of unsteady recirculating flow is characterized as stall. Rotating stall may occur if stall appears only in one part of the flow channels, while it periodically moves around in a certain rotating sense. Pump-turbines should not operate within this instability region because considerable pressure fluctuations are induced which, finally, lead to component damage.

The qualitative characteristic of a Pump-Turbine at pump operation is displayed in figure 2. The plant head range is indicated as well as the instability region. Each of the pump characteristics belongs to a specific geometry, denoted with A and B at a considered flow rate.

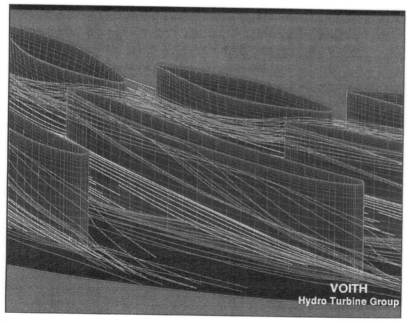

Figure 5. Selected streamlines through modified tandem cascade <u>with</u> runner interaction.

At first, consider operation point A indicating that severe flow separation phenomena occur. The visualization of the flow through the runner shows a very smooth behaviour (not shown here). The flow through the tandem cascade is presented in figure 3. The streamlines close to the crown meridional contour clearly emphasize a large region of vortical flow between the stay vanes creating high flow losses. Close to the band meridional contour, the flow moves smoothly towards the spiral. Note, the runner influence on the stay vane/wicket gate component has been considered for this prediction.

If runner interaction is not taken into account, Euler's angular momentum equation has to be applied to determine the inflow conditions at the tandem cascade for pump operation. The resulting streamlines are presented in figure 4. Flow separation is also observed close to the crown meridional contour, while everywhere else smooth flow prevails. The comparison of figure 3 with figure 4 reveals a considerable difference of the flow field.

Owing to the project requirements, the pump-turbine has to operate at conditions B, too (cf. fig. 2). Thus, no large regions with flow separation are permitted. A modification of the tandem cascade geometry meets this requirement (fig. 5). The flow moves fairly smooth towards the spiral over the entire pad height. Note, this result has been obtained with consideration of runner interaction. In this case, the runner flow also behaves very smooth.

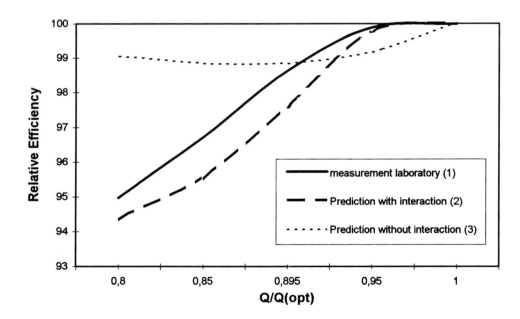

Figure 6. Relative efficiency versus non-dimensional flow rate.

These described results are summarized in figure 6. It presents the relative efficiency, i.e. η/η_{max}, versus a non-dimensional flow rate, i.e. Q/Q_{opt}. Curve one through curve three belong to geometry 1. The first curve represents measured data from the laboratory, while curve two presents the flow prediction with flow interaction of runner and tandem cascade (cf. fig.3). Curve three shows the flow result without the consideration of runner and tandem cascade interaction (cf. fig.4). There is a good agreement between the measurements in the laboratory and the flow prediction, if component interaction is considered. The neglect of flow interaction between neighbouring turbine components shows a wrong flow behaviour under high head operation conditions as well as near optimum operation. The flow characteristics of the modified geometry (geometry 2) are fairly different. It produces nearly constant flow losses over a large portion of the operation range (cf. fig. 5).

3.2 RUNNER - DRAFT TUBE FLOW INTERACTION

A high pressure recovery in the draft tube is a necessary prerequisite for high turbine efficiencies. The kinetic energy remaining in the flow downstream of a runner has to be transformed into static pressure. This process should occur with minimum losses to maintain high efficiencies.

The flow through the draft tube is strongly dependent on the flow distribution and the flow properties downstream of the runner. This automatically implies the necessity to take the runner - draft tube flow interaction into consideration. A corresponding flow field simulation result of a Francis turbine draft tube is presented in figures 7 and 8. Both flow fields differ in their operation point. Figure 7 corresponds to near machine optimum conditions (OP1), while figure 8 represents a larger flow rate and gate opening at the same head (OP2).

Operation point OP1 shows a counter-clockwise inlet swirl distribution (runner rotation is counter-clockwise) across the entire draft tube entrance. The streamlines reveal a vortex originating at the rotation axis. Their colour represent the absolute speed. This vortex moves into the right half of the draft tube. Near the end of the draft tube, the transport velocity (main flow component) displays a low mass flow rate through the right half of the draft tube, while it is considerable higher on the left half. The wake downstream of the pier is clearly visible, too.

Operation point OP2 produces a different inlet swirl distribution. It is counter-clockwise at large radii, but clockwise near the rotation axis. The center vortex is less marked. This flow field exhibits a very similar transport velocity on both sides of the pier. The flow losses of the draft tube at operation

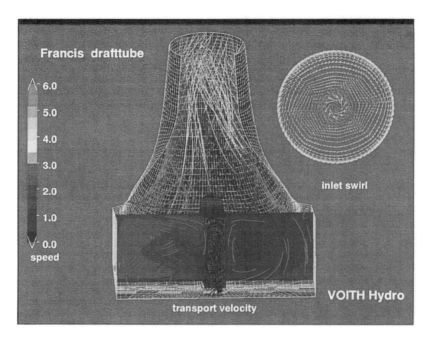

Figure 7. Inlet swirl, streamlines and transport velocity of a Francis turbine draft tube (OP1).

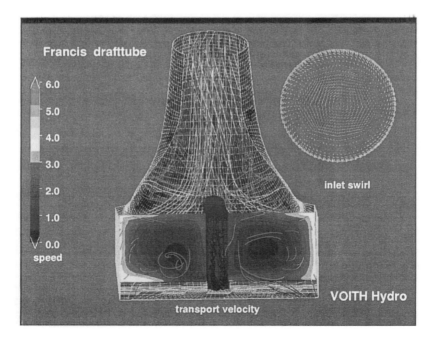

Figure 8. Inlet swirl, streamlines and transport velocity of a Francis turbine draft tube (OP2).

point 1 are less than one percent, while those of operation point 2 amount to about two percent. This difference can not be neglected for turbines with high efficiencies.

4. Concluding remarks

Numerical simulation of viscous flow fields has proven to be a very powerful tool for the design and development of hydraulic machinery. It gives the designer detailed insight into the flow field and leads to an improved understanding of the flow physics.

Different methods of considering flow interaction between turbine components are shortly described. For the present investigation, a method has been chosen which allows an effective support during the turbine development phase. Results being obtained through the flow simulation without component interaction will be very likely subject to misinterpretation of the flow behaviour and, consequently, of the turbine design.

The presented results strongly emphasize the necessity to consider flow interaction between turbine components. During the development of our hydraulic machines, these advanced methods are applied for the benefit of power generation.

5. References

Buchmaier, H., Quaschnowitz, B., Moser, W. and Klemm, D.(1996) Numerical Design and Optimization of High Head Pump-Turbines, *Proceedings of the XVIII IAHR Symposium, Valencia, Spain,* Kluwer Academic Publishers, to appear.

Fisher, R.K. and Webb, D.R. (1978) Effect of Cavitation on the Discontinuity Point and On Alternating Pressures and Gate Torques on a Pump/Turbine Model in the Pump Cycle, ASCE-IAHR/AIHR-ASME Joint Symposium, Colorado Springs, Colorado, USA.

Klemm, D., Jaeger, E.-U. and Hauff, C. (1996) Development of a new Generation of High Head Pump-Turbines, GUANGZHOU II, *Proceedings of the XVIII IAHR Symposium, Valencia, Spain,* Kluwer Academic Publishers, to appear.

Riedelbauch, S., Fisher, R.K. and Riva, P. (1994) Utilization of Three Dimensional Viscous Flow Simulation for the Design and Optimization of Hydraulic Machinery, *Proceedings of the XVII IAHR Symposium, Beijing, China 1994* 1, 353-364.

Ruprecht, A., Bauer, C. and Riedelbauch, S. (1994) Numerical Analysis of Three-Dimensional Flow through Turbine Spiral Case, Stay Vanes and Wicket Gates, *Proceedings of the XVII IAHR Symposium, Beijing, China 1994* 1, 71-82.

FROM COMPONENTS TO COMPLETE TURBINE

NUMERICAL SIMULATION

M. SABOURIN   Y. LABRECQUE   V. DE HENAU
Gec Alsthom   *Gec Alsthom*   *Gec Alsthom*
Tracy, Qc, Canada   *Tracy, Qc, Canada*   *Tracy, Qc, Canada*

1. Introduction

In the quest for turbine improvement, computational fluid dynamics (CFD) is the emerging tool. The fast development of computers combined with the new capabilities of turbulent flow codes make possible the calculation of an entire turbine unit. Until recently, fluid simulations for turbines were done considering the various components separately. This convenient way of analysis can however present some major drawbacks. As components can be strongly coupled, the unknown introduced by assumed boundary conditions needs to be clarified. One consequence of a complete turbine numerical simulation is to avoid assumptions of internal boundary conditions.

In this paper, the implementation of a strategy to simulate the interactions between rotating and non-rotating components is presented. The capacity of the method to predict beginning of cavitation, power output, head losses and other hydraulic behaviour makes this tool very useful for the turbine designer.

2. Historic

Since the beginning, the design of hydraulic turbines has been an experimental science. However, step by step CFD has emerged as a cost effective complement to model testing. Commercial codes for three dimensional turbulent flow simulation, which are now available, yield significant improvements in accuracy of simulation due to their capacity to solve real fluid flowing in real geometry.

E. Cabrera et al. (eds.), Hydraulic Machinery and Cavitation, 248–256.
© 1996 *Kluwer Academic Publishers. Printed in the Netherlands.*

The calculation of losses in turbine components treated as separate entities has been the subject of numerous papers [1,5,6]. This approach brings to the designer interesting information on the behaviour of the turbine. It presents however some difficulties:

- the specification of boundary conditions stands on experimental data or assumptions based on designer experience,

- the analysis of results must segregate local improvement from overall benefit,

- comparison with experimental results is partial as only the performance of the complete turbine is measured with accuracy.

Additional information concerning the strength of the hydraulic coupling between component can be evaluated.

A publication from De Henau [1] demonstrated the importance of interaction between static components like the casing and the distributor. The full understanding of this interaction can be accomplished only through a complete 3D model including the casing and the complete distributor. The pressure condition at the distributor outlet coming from the runner modifies the flow distribution with more or less influence, depending on the turbine type and the hydraulic conditions.

This interaction between runner and distributor has been integrated for 20 years in our flow simulation software used in the design of GEC ALSTHOM's turbines [4]. Although this software uses ideal fluid concept the prediction of inlet cavitation is more accurate with this method. The interaction stator-rotor has a major influence in turbine behaviour.

Similarly, the draft tube, responding to the flow field delivered by the runner provides a back pressure resulting from its kinetic energy recuperation capacity. During the optimization process of the draft tube, a modification of the geometry can have significant influence on the runner back pressure and therefore, on the velocity field at draft tube inlet. This feed back loop escapes to the simulation of the isolated draft tube. A precise evaluation of losses can therefore be achieved only through precise boundary conditions.

For all these reasons, the adequate simulation of the coupling between turbine components is highly desirable. Moreover, for turbulent flow codes, the passage from the status of analysis tool to the one of design tool necessitates the simulation of interaction between components. The stator-rotor interaction takes more importance in this perspective.

3. Implementation of the stator-rotor interaction

3.1 NUMERICAL CONCEPTS

The basic constraint to the implementation of coupling between components is the limit of computer resources. Experience with calculation of isolated components demonstrated that in many cases the solution is mesh dependent. This makes the implementation strategy more delicate as it must embed two aspects in the numerical process: the numerical zoom and the circumferential averaging.

The numerical zoom is based on the assumption that velocity field is more accurate than the pressure field for a given grid, creating a lever effect that permits to transfer velocity field from a coarse mesh to a fine mesh in order to obtain a more accurate pressure field solution. Thus, the solution is found after two sequences of calculations. The first one has the goal to determine velocity boundary conditions and the second step must refine the losses calculation. Figure 1 shows a local concentration of nodes in the distributor in order to calculate the losses.

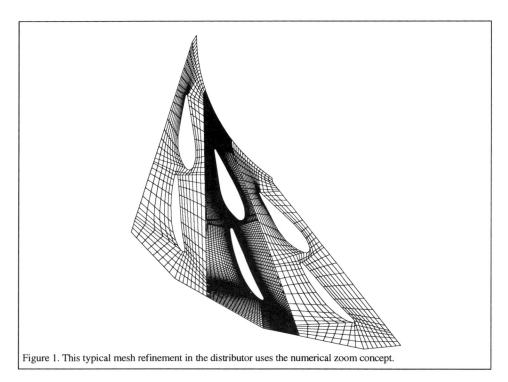

Figure 1. This typical mesh refinement in the distributor uses the numerical zoom concept.

The circumferential averaging is the method chosen to simulate the stator-rotor interaction. Galpin et al. [2] reviewed various methods. He concluded that a practical way of simulation is to use a steady state method using a conservative averaging in the circumferential direction at the sliding interface. A sliding interface is a surface of revolution and most commonly a conical surface. There are two sliding interfaces. A first one is located at the runner inlet between the trailing edge of guide vane and the blade leading edge of the runner blades. A second sliding interface is below the trailing edge of the runner blades. These interfaces constitute internal boundaries where node variables are treated and communicated upstream and downstream.

Two approaches implement circumferential averaging.

The first one, the *simulation by linked components*, manages three numerical models: the casing and distributor model, the runner in a rotational frame of reference and the draft tube. This method, described in a recent paper by Labrecque et al. [3], permits to use natural boundary conditions at the ends of the complete turbine, see Figure 2. At inlet, the total pressure level is specified. Static pressure is specified at the outlet of the draft tube. These are the only boundary conditions needed for the entire simulation. Between the components, at interfaces, boundary conditions are initially estimated and after few iterations, calculated boundary conditions can be obtained. On Figure 2, the descending arrows indicate the communication of velocity and turbulence fields from

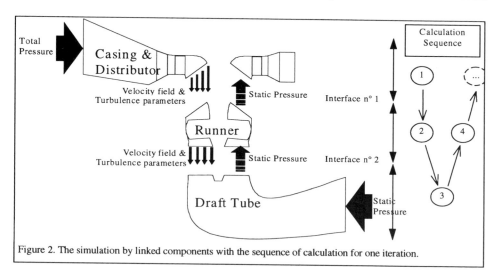

Figure 2. The simulation by linked components with the sequence of calculation for one iteration.

the upstream components to the downstream ones. The ascending arrows communicate the static pressure from the downstream components to the upstream ones. One iteration over the complete turbine necessitates 4 calculations of isolated models. Depending on the initial conditions and on the level of convergence desired, a solution ca be reached within about 10 iterations.

The second approach needs the new software package *stage* now available within TASCflow [2,7]. This package handles multiple frame of reference (MFR) giving the possibility to calculate the entire turbine in one single calculation automatically. However, this type of calculation over an entire turbine may require millions of nodes to obtain reasonable precision.

3.2 THE COMPUTING PROCESS

The calculation of an entire turbine unit is made in an industrial environment: a result must be obtained on a low price computer within a reasonable time. The software

TASCflow can run on various computers. A rule of thumb for memory requirement is 1 Meg for each thousand nodes Although the speed and memory capacity of today's computers have increased significantly in recent years, there is a limit to the largest calculation possible and this limit is a lot smaller than required.

The *linked components* method permits to work with different mesh sizes and topology for each model as no mesh connection is necessary at sliding interface. The sequence of

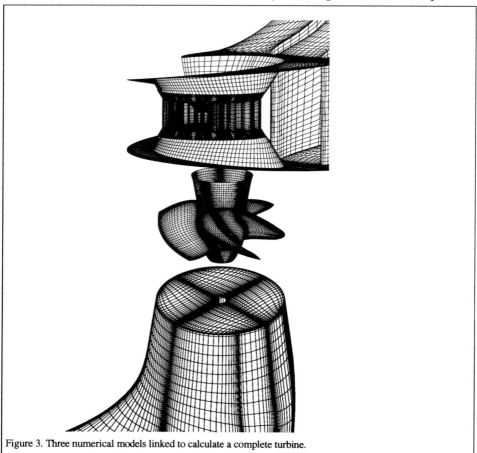

Figure 3. Three numerical models linked to calculate a complete turbine.

calculation is not automated. Depending on the hydraulic behaviour to evaluate, the mesh of each component can be refined at any step, concentrating all the computer resources over the area of interest. Also, at the beginning of the problem, the convergence criteria do not need to be stringent as boundary conditions are approximate at this time. During processing, the draft tube calculation may be skipped a while, as the coupling between it and the runner is weaker than the one between the distributor and the runner. This strategy can contribute to reduce the machine calculation time.

The *stage* method is theoretically able to solve the entire turbine in a single calculation. However, the computer resources limit the accuracy of the solution and a marriage with the linked components method is necessary. The calculation of the casing-distributor model gives the boundary condition to a distributor model limited to one or two guide vane passages. This distributor model is connected through the stage interface with the runner. Therefore, the distributor and the runner are calculated in a single calculation, see Figure 4. The draft tube is calculated separately and the pressure condition at the runner outlet is adjusted when necessary.

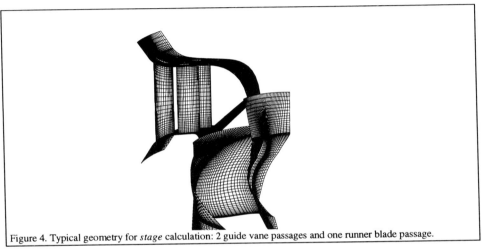

Figure 4. Typical geometry for *stage* calculation: 2 guide vane passages and one runner blade passage.

The optimal strategy of the computing process is to embed the *stage* method in the *linked components* method benefiting the efficiency of *stage* to simulate coupling and the accuracy of *linked components* to obtain more reliable results.

4. Application

In particular, two aspects of the turbine design and analysis are improved with the calculation of a complete turbine. They are cavitation and losses evaluation.

4.1 CAVITATION ANALYSIS

The cavitation in the runner can be found at inlet close to the leading edge or on the profile.

The inlet cavitation originates from the mis-adaptation of the blade angle. Therefore, the coupling between distributor and runner demands the simulation of the stator-rotor interaction. On Figure 5, the result of two simulations is presented. The first one involves the distributor alone where the outlet pressure condition comes from an ideal fluid simulation of stator-rotor interaction. The second one is a calculation of the complete turbine with a turbulent flow code. At the sliding interface between the runner and the distributor corresponding to the outlet condition on the distributor, a comparison is made on velocity and pressure. It is observed that the two methods yield different incidence angles near the shroud. The complete turbine calculation indicates a greater risk of inlet cavitation as observed during model test.

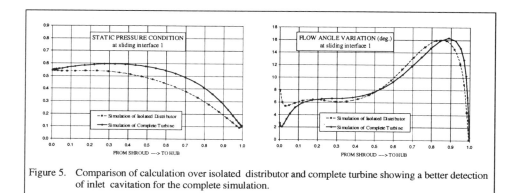

Figure 5.   Comparison of calculation over isolated distributor and complete turbine showing a better detection of inlet cavitation for the complete simulation.

In the case of the profile cavitation, the tailwater level and losses between it and the runner must be calculated to obtain the absolute static pressure. On Figure 6, the absolute static pressure field on suction side of blade is compared to the observation. It can be observed that the area where the pressure is below the vapour pressure is in agreement with the location of bubbles conveyed by the flow.

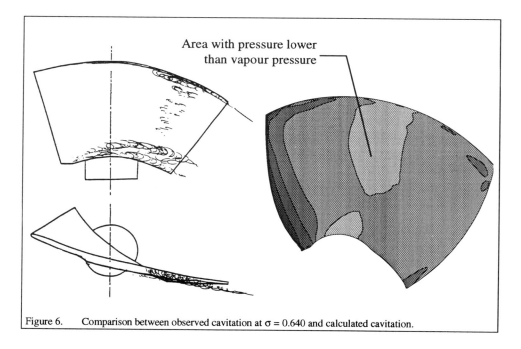

Figure 6.   Comparison between observed cavitation at $\sigma = 0.640$ and calculated cavitation.

4.2 LOSSES ANALYSIS

As seen, the other aspect improved with entire turbine simulation is the evaluation of losses. Losses calculations are linked to the other hydraulic parameters. They are speed, flow and torque coefficient.

Experience shows that torque and power are calculated precisely either by the pressure field or the kinetic momentum. However, the numerical solution is mesh dependent regarding losses calculation. The fact that the mesh is coarse introduces a drift in the static pressure. This drift increases from the outlet to the inlet in the flow direction. As the drift of the static pressure is similar on both sides of the blade, the pressure differential on the blade is precise. The evolution of losses with mesh concentration indicates a convergence to the experimental result as presented on Figure 7.

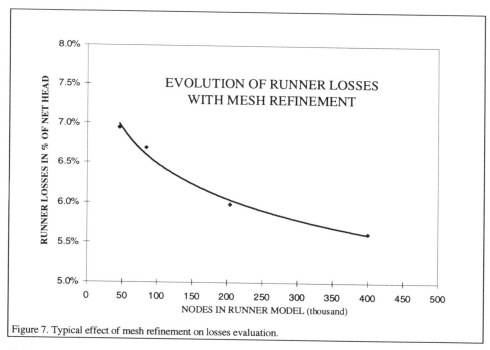

Figure 7. Typical effect of mesh refinement on losses evaluation.

Using similar meshes on various geometries leads to a precise evaluation of losses difference. The calculation of an entire turbine eliminates the need for assumptions on boundary conditions and therefore gives a more precise evaluation of losses. Moreover, this method can evaluate if a local improvement will produce the expected effect on the complete turbine. This makes the calculation of the entire turbine an excellent design tool.

5. Conclusion

The calculation of an entire turbine needs numerical simulation of the interaction between components. This simulation involves numerical zoom and circumferential averaging. The combination of the *linked component* with the *stage* methods permits to optimize the calculation process.

The complete turbine numerical simulation eliminates the need for experimental data to specify boundary conditions. Although this method is less accurate than model testing, it is faster and more economical. However, with continuous development of computers, the actual level of precision is improving.

The results, compared to separated components simulation, provide better information for the turbine designer and permit to see the effect of a local modification on the entire turbine behaviour.

The CFD involving three dimensional turbulent flow codes and stator-rotor interaction can now be considered as the best tool for turbine design.

References.

[1] De Henau V., 1995, "Turbine Rehabilitation: CFD analysis of distributors", Water Power Symposium, San Francisco, USA

[2] Galpin P.F., Broberg R.B., Hutchinson B.R., 1995, "Three-Dimensional Navier Stokes Predictions of Steady State Rotor / Stator Interaction with Pitch Change", CFD Conference, Banf, Alberta, Canada.

[3] Labrecque Y., Sabourin M., Deschênes C., 1996, "Numerical simulation of a complete turbine and interaction between components", Modelling, Testing & Monitoring for Hydro Powerplants, Lausanne, Switzerland.

[4] Moulin C., Wegner M., Eremeef R., Vinh P., 1977, "Méthode de tracé des turbomachines hydrauliques", La houille blanche/n° 7/8.

[5] Sabourin M., Eremeef R., De Henau V., 1995, "Extensive Use of Computational Fluid Dynamics in the Upgrading of Hydraulic Turbine", Canadian Electrical Association, Vancouver, British Colombia, Canada.

[6] Sabourin M., Couston M., 1994, "Turbine Rehabilitation: Experience Gained through Systematic Draft Tube Evaluation", IAHR Symposium, Beijing, China.

[7] TASCflow Users Manual, 1995, Version 2.4, Advanced Scientific Computing Ltd., Waterloo, Ontario, Canada.

VALIDATION OF A STAGE CALCULATION IN A FRANCIS TURBINE

MIRJAM SICK AND MICHAEL V. CASEY
Sulzer Innotec Ltd
CH-8401 Winterthur, Switzerland

AND

PAUL F. GALPIN
Advanced Scientific Computing Ltd.
Waterloo, Ontario, Canada

Abstract. This paper describes the verification of a simulation method for complete hydraulic turbine systems, from spiral casing through distributor and runner to the outlet draft tube. The method solves the steady 3D Reynolds-averaged Navier Stokes equations with a mixing plane at the interfaces between the components, so that the flow in each frame of reference is steady. The steady-state interactions between the components are taken into account but unsteady interactions are neglected. The paper verificates the method by comparison of computations with detailed flow measurements from a high specific speed Francis turbine model. Computations in which each component is computed in isolation are compared to computations using the mixing interface between components to demonstrate the advantages of the new technique.

1. Introduction

The calculation of flows in hydraulic turbines has undergone a rapid transformation during the last ten years. The first major breakthrough in the use of CFD methods for hydraulic turbine design allowed Euler methods to be used for design of water turbine components and, combined with suitable design rules, drastically reduced the number of model tests needed for the achievement of a satisfactory design (Keck *et al.*, 1990). The relatively simple Euler methods predict the important features of the flow, such as

E. Cabrera et al. (eds.), Hydraulic Machinery and Cavitation, 257–266.
© *1996 Kluwer Academic Publishers. Printed in the Netherlands.*

incidence levels at runner inlet, pressure levels at runner outlet (cavitation) and swirl in the draft tube inlet. They are, however, limited in their ability to predict the losses, as viscous forces are neglected.

Recent developments have aimed to solve the Reynolds-averaged Navier-Stokes equations to obtain detail of the loss sources in the individual components and to improve component performance prediction. Mature CFD methods for predicting the flow and losses in the individual components of the turbine stage are now well established features of most hydraulic turbine design systems, e.g.(Keck *et al.*, 1994), and these have made substantial improvements in the cost and speed with which new designs can be assessed.

This paper is concerned with the next major improvement in turbine design methods, that is the ability to calculate the performance of components taking into account their interaction with the upstream and downstream elements. Some steps in this direction have already been reported, for example the Kaplan turbine simulations reported by Goede *et al.* (1990).

The method examined in this paper includes a mixing plane interface between the rotating and stationary components so that a steady solution can be calculated in each domain. The mixing plane interface has been incorporated into a commercially available CFD code (TASCflow from Advanced Scientific Computing Ltd., Waterloo, Canada). The development has been carried out under partnership between Sulzer Hydro, Sulzer Innotec and ASC. This paper describes aspects of the verification of the method by comparison with detailed laser anemometry data from a high specific speed Francis turbine (Sulzer Innotec and Sulzer Hydro).

A companion paper to this describes the use of this newly developed tool for the hill-chart performance prediction of a Francis turbine (Keck, Drtina and Sick, 1996).

2. Theoretical method

2.1. NAVIER-STOKES CODE

The TASCflow commercial software package was used for this work. This software solves the 3D Reynolds-averaged Navier-Stokes equations in strong conservation form. A colocated variable arrangement is employed to solve for the primitive variables (pressure, Cartesian velocity components) in either rotating or stationary coordinate systems. The transport equations are discretized using a conservative, finite element based finite-volume method. Turbulence effects are modeled using the standard k-ε model. A second order accurate skew upwind differencing scheme with physical advection correction is employed. A coupled algebraic multigrid method solves the system of equations (coupled solution of mass and momentum). Liquids, subsonic, transonic and supersonic gas flows can be analyzed. Details regarding the theoretical basis of the software are reported by Raw (1994) and in the ASC Theory Documentation (1995). The software has been previously applied to turbomachinery flows and for flow calculations in a wide variety of other components (Casey, Borth *et al.*, 1995).

2.2. TURBOMACHINERY STAGE CALCULATIONS WITH AN INTERFACE MIXING PLANE

The first three-dimensional flowfield calculations to predict the performance of a whole turbomachinery stage were made by Denton using the steady Euler equations, Denton (1983). The flow in adjacent rotating and stationary blade rows is unsteady, so to carry out steady Euler calculations in each blade row some modelling of the flow processes has to be carried out to remove the unsteadiness. Denton achieved this by simple circumferential averaging of the flow at an intermediate calculation plane (the mixing plane) between the adjacent blade rows. The upstream blade row experienced a circumferentially uniform downstream boundary condition, and the downstream row saw a circumferentially uniform flow approaching it.

Following the first Euler stage computations by Denton, several years elapsed for the development of practical viscous flow solvers for isolated blade rows based on the three-dimensional Navier-Stokes equations. As soon as these techniques were available, it was natural that turbomachinery stage capability with Navier-Stokes codes would be developed, see Denton (1992), Dawes (1992) and more recently, Galpin *et al.* (1995).

A problem with the original circumferential mixing model of Denton was that a circumferentially uniform flow may be forced to exist too close to the leading edge of the downstream blade row. This does not allow the flow to adjust circumferentially to the presence of the blade. As a result the leading edge loading on the blade row may be wrong. The solution of this problem (Denton, 1992) was to allow a circumferential variation of fluxes at the mixing plane (by extrapolation from the upstream and downstream planes) while adjusting the level of the fluxes to satisfy overall conservation.

2.3. GALPIN MODEL FOR SLIDING STAGE INTERFACE PLANE

The frame change and pitch change is accomplished at a sliding grid interface using a control surface approach. A Stage sliding interface is defined when the control surface grid forms bands parallel to the machine motion, forming a conservative "mixing plane".

The interface fluxes are assembled into the volume and surface equations in a fully conservative and implicit manner. The pressure forces at the Stage interface are computed such that the average interface pressure equals the

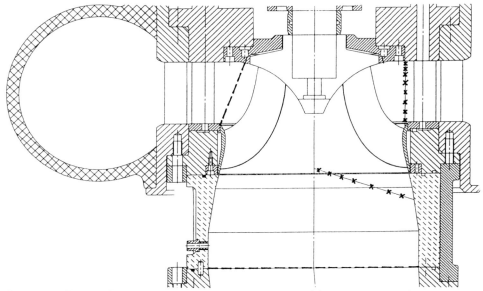

Figure 1. Francis Turbine with measurement positions (x) and interfaces (dashed lines)

control surface pressure, while supporting local nodal pressure variations due to elliptic effects, as shown in fig. 2.

For convenience the control surface equations are placed in the stationary absolute frame of reference, with the pitch-wise extent equal to the pitch of the larger side of the interface. Once all fluxes have been evaluated at all surfaces exposed to the sliding interface, the equation set consists of control surface equations and control volume equations. An algebraic multigrid solver is used to solve this equation set.

3. Test case

3.1. MODEL FRANCIS TURBINE

The test case considered in this paper for the verification of the stage interface method is a model Francis turbine of high specific speed, see fig. 1. The full-scale turbine has a runner diameter of 3.4 m and was designed for a head of 32.7 m at a nominal flow rate of 90 $\frac{m^3}{s}$ with a rotational speed of 166 rpm and a specific speed of 422. The model turbine is a scale model of the original machine with a runner diameter of 0.3 m. The experiments were carried out in the hydraulic turbine test stands of Sulzer Hydro in Zürich. Standard measurement techniques for modern hydraulic practice were used for the derivation of all performance parameters.

The velocity near the draft tube outlet was measured by means of a

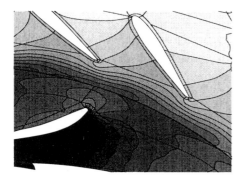

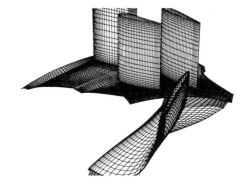

Figure 2.   Pressure contours towards the   *Figure 3.* Computational grid: distributor
interface between wicket gate and runner   and runner

propeller anemometer. Meausurements of the detailed flow velocities were
also carried out in the hydraulic turbine test stands of Sulzer Hydro using
L2F laser velocimetry. Flow traverse planes were established at the runner
inlet and at the runner outlet by insertion of suitable windows in the casing,
see fig. 1. Flow velocities at each of these measurement planes were obtained
at five operating points of the turbine. These operating points were selected
to provide a good overview of the flow at the best operating point and at
a variety of different off-design conditions.

3.2. THE COMPUTATIONAL GRID

Fig. 1 shows a part of the spiral casing, the stay vanes, the wicket gate, the
runner and the beginning of the draft tube and also the locations of the
measurement planes and the position of the grid interfaces between wicket
gate and runner and between runner and draft tube.

Grids were generated with the ICEM-CFD software for the wicket gate,
the runner and the draft tube separately and for combinations of these three
components with interface planes. For simulations of the whole machine the
distributor, i.e. stay vanes and wicket gate, was modelled as a single unit
consisting of two stay vanes and three guide vanes for which different grids
were needed for each wicket gate opening.   Two different types of grid
topology were used for the wicket gate and the runner grid:

– a combination of block structured H-grids only
– a combination of O-grids around the blades inserted within H-grids

The latter more sophisticated grid topology (see fig. 3) was developed for
better accuracy near the blades and to improve the convergence rate with
less skewed grid cells, especially near the trailing edge. The grid for the

whole distributor also consists of a combination of O-grids around the blades with H-grids. The draft tube is discretised by a butterfly grid.

All grids are refined towards the walls in order to get a good prediction of the boundary layer with reasonable values of y^+. In all calculations the values of y^+ were between 10 and 300, and in most cases between 20 and 100. The effect of grid refinement on the computed velocity field has been examined for the runner and the wicket gate and has been found to be small. The results for a runner grid consisting of 27,000 grid points are not largely different from those for a grid of 128,000. The calculated efficiency depends slightly on the grid size within a range of 1 percent.

4. Results

This section presents the results of the CFD simulations and compares these with measurement data. Where possible, useful details of the simulations are highlighted to explain features of the flow pattern of relevance to the simulation and to the performance of the turbine. The simulations are used to highlight the advantages of the full stage simulation over simulations in which the individual components are calculated separately.

4.1. SIMULATION OF INDIVIDUAL COMPONENTS

4.1.1. *Wicket Gate alone up to the Location of the Interface Plane*
The simulation of the flow through the wicket gate alone at design point used boundary conditions of fixed mass flow and flow angle at the inlet and average pressure at the outlet. The grid consisted of 55680 nodes.

The upper part of fig. 4 shows the comparison between experimental data and numerical prediction at the measurement plane close to the wicket gate outlet. The circumferential component of velocity is well predicted. The measurement of the radial velocity component shows a strong increase towards the band because of the strong curvature of the band in the runner region. The numerical simulation predicts a stronger increase of the radial velocity towards the band because the simple outlet boundary condition does not adequately represent the interaction of the flow with the runner. Note that a small part of the measurement plane between the stator and the rotor is not part of the stator grid. Therefore the comparison between experimental data and prediction ends shortly before the band.

This comparison demonstrates the need to account properly for the interaction between the gate and the runner. Without the stage capability the computational domain of the wicket gate needs to be extended artificially into the runner region with "source terms" to model the effect of the runner, or a known pressure profile must be prescribed at the outlet of the gate.

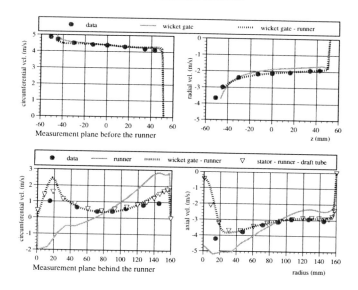

Figure 4. Comparison measurement data and predictions before and behind the runner

4.1.2. *Simulation of the Runner alone*

The runner flow is simulated with a constant energy distribution and a fixed flow angle as inlet condition and average pressure at the outlet. The inlet flow angle results from the wicket gate simulation. The grid consisted of a total of 90390 nodes.

The comparison between experimental data and numerical prediction made at the measurement plane behind the runner can be seen in the lower part of fig. 4. For both the circumferential and the axial component the prediction produces a trend which differs clearly from the measured data. The axial velocity distribution is biased towards the axis of the machine with a corresponding swirl distribution. It can be inferred from the simulations shown below that this is due to the strong influence of the velocity and pressure distribution at the runner inlet (which is very closely spaced to the runner blades) on the flow at the runner outlet.

4.2. SIMULATION OF COMBINED WICKET GATE AND RUNNER

The wicket gate and the runner are joined as one computational domain whereby each part is calculated in its own frame of reference, separated by a sliding interface. The inlet condition at the wicket gate inlet is the same as in Section 4.1.1 (mass flow and flow angle are fixed). The outlet condition below the runner is the same as in Section 4.1.2 (average static pressure). The conditions at the sliding interface are an implicit result of the calculation. The grid consisted of a total of 146,070 nodes.

The comparison of the prediction with experimental data at the measurement plane between wicket gate and runner can be seen in the upper part of fig. 4. The differences between the calculations of single components and the calculation including the interaction of the wicket gate and the runner can be clearly seen. The agreement of both circumferential and radial velocity components with the experimental data is very good. The interaction between wicket gate and runner affects the velocity distribution within the domain of the wicket gate leading to a more accurate prediction.

Fig. 4 also demonstrates the good agreement between experiment and numerical prediction in the measurement plane behind the runner for both the axial and the circumferential component of the velocity. The improvement at the runner outlet results from the better prediction of the flow at the runner inlet by the sliding interface.

4.3. SIMULATION OF THE WHOLE MACHINE

The successful simulation of the combined runner and wicket gate with the new stage capability encouraged us to take a further step, that is the simulation of the whole machine including the interaction between the stay vanes, wicket gate, runner and draft tube.

The inlet conditions to the stay vanes for this simulation result from a circumferential average of a separate spiral casing calculation. Behind the draft tube outlet, the computational domain is extended in order to remove any disturbance of the outlet boundary condition on the draft tube flow. At the draft tube outlet the static pressure is set constant. The grid consisted of a grand total of 265, 795 nodes.

The comparison between test data and prediction behind the runner outlet shows a further slight improvement in the predictions of the circumferential component of the velocity compared to wicket gate and runner computation, see lower part of fig. 4. This demonstrates that there is a small but not insignificant influence of the interaction with the draft tube on the flow in the runner. Fig. 5 shows a comparison of the simulation of the draft tube alone and the simulation of the whole machine with measured data near the draft tube outlet. Both computations show a good agreement between measurement and prediction. The flow through the draft tube alone was simulated by prescribing the circumferentially averaged velocity distribution resulting from the wicket gate - runner simulation at the draft tube inlet. The outlet condition (constant pressure) was the same as for the simulation of the whole machine.

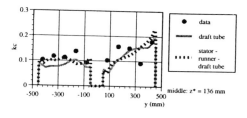

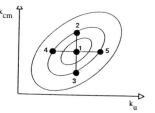

Figure 5. Comparison measurement data and predictions near the draft tube outlet: nondimensionalized velocity

Figure 6. Hill Chart: Situation of the five operating points examined in this paper

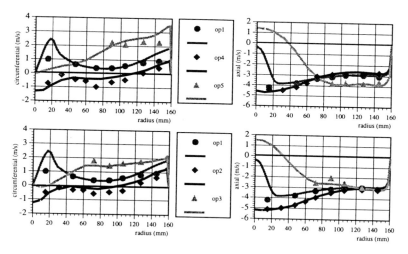

Figure 7. Off-Design: Measurement plane behind the runner. Data shown by symbols, calculated results shown by curves.

4.4. OFF-DESIGN POINTS

Calculations were carried out for the four off-design points shown in the simplified hill chart in fig. 6. The comparison of the measured and calculated circumferential and axial velocity components at the runner outlet (see fig. 7) demonstrates the reliability of the computational method in detail. The good agreement shows that the component interactions are well predicted at both design and off-design points. Further details of the simulations of the whole machine including hill chart prediction are presented in a companion paper (Keck, Drtina and Sick, 1996).

5. Conclusions

The main conclusions of this work are, as follows:

- The stage calculation procedure as incorporated in TASCflow allows for the simultaneous simulation of stationary and rotating frame components by use of a mixing plane at sliding interfaces.
- The mixing plane interface supports circumferential pressure variations due to elliptic effects.
- The stage calculation procedure provides better predictions of the flow-field both for design and off-design operating points as it takes into account the steady-state interactions between all components.
- The most important interaction is seen to be the effect of the runner on the wicket gate and vice versa. Calculations which do not include this interaction may be substantially in error at the runner outlet.

The results show with encouraging clarity that the stage calculation method presented here is a viable method for the hydraulic analysis of a complete turbine. Further work, described in the companion paper (Keck *et al.*, 1996), demonstrates the ability of the method to predict losses and hill charts for a complete turbine.

References

Casey,M.V., Borth,J., Drtina,P., Hirt,F., Lang,E., Metzen,G. and Wiss,D. (1995) The Application of Computational Modeling to the Simulation of Environmental, Medical and Engineering Flows, Speedup Journal, vol 9/2, pages 62-69

Dawes,W.N.(1992) Towards improved throughflow capability: the use of three dimensional viscous solvers in a multistage environment, Trans. ASME Journal of Turbomachinery, vol 114, pages 8-17

Denton,J.D. (1983) An improved time-marching method for turbomachinery flow computation, Trans. ASME Journal of Engineering for Power, vol 105, pages 514-523

Denton,J.D. (1992) The calculation of three-dimensional viscous flow through multistage turbomachines, Trans. ASME Journal of Turbomachinery, vol 114 pages 18-26

Galpin,P.F., Broberg,R.B. and Hutchinson,B.R.(1995) Three-dimensional Navier-Stokes predictions of steady state rotor/stator interaction with pitch change, Third Annual Conference of the CFD Society of Canada, June 25-27, 1995, Banff, Alberta, Canada

Goede,E., Cuenod,R. and Grunder,R.(1990) An advanced flow simulation technique for hydraulic turbomachinery, IAHR Symposium, Belgrade, September 1990

Keck,H., Goede,E. and Pestalozzi,J. (1990) Experience with 3D Euler flow analysis as a practical design tool, IAHR Symposium, Belgrade, September 1990

Keck,H. Goede,E. and Sebestyen,A. (1994) 3-Dimensionale Stroemungsberechnungen in Pumpturbinen, VDI-Kolloquium Wasserkraftanlagen, Dresden, June 28, 1994

Keck,H., Drtina,P. and Sick,M. Numerical Hill Chart Prediction by means of CFD Stage Simulation for a Complete Francis Turbine, 18th IAHR Symposium Valencia, Spain

Raw,M.J. (1994) Coupled Algebraic Multigrid for the Solution of the Discretized 3D Navier-Stokes Equations, CFD 94, CFD Society of Canada, Toronto

TASCflow Theory Documentation, Version 2.4., Advanced Scientific Computing Ltd., Waterloo, Ontario, Canada (1995)

SIMULATION OF FLOW THROUGH FRANCIS TURBINE BY LES METHOD

Charles C. S. Song & Xiangying Chen
Univ. of Minnesota, Mpls, MN, USA
Toshiaki Ikohagi
Tohoku Univ., Sendai, Japan
Johshiro Sato, Katsumasa Shinmei & Kiyohito Tani
Hitachi Ltd., Hitachi-shi, Japan

I. INTRODUCTION

The traditional approach of Francis turbine design which is based on the steady potential flow theory and heavily dependent on model testing and engineering experience, has come a long way in producing efficient and relatively cavitation free turbines. But further improvement of performance for design and off design operating conditions will be extremely difficult with the traditional method because it will depend more on those phenomena such as, boundary layer separation, vortex dynamics, interactions between different components, vibrations, etc., which are not predictable with conventional approach and difficult to measure in physical models.

In recent years the Euler Equation Model has became quite popular, but it still can't deal with the problems listed above. The Reynolds time averaged method such as the k-e model is also being widely tried, but aside from the time averaged boundary layer flow separation effect, most of the problems listed above still remain intractable.

This paper describes recent development and application of the compressible hydrodynamics theory based Large Eddy Simulation model on simulation of flow through spiral case, runner, and draft tube of a Francis turbine system. The results described herein are mostly based on numerical simulations through each of the three basic units separately because the development of a totally unified model has not been completed. But the importance of mutual interactions is suggested in this paper. Some of mutual interaction phenomena are described in a companion paper entitled SIMULATION OF A PUMP-TURBINE.

II. GOVERNING EQUATIONS AND METHOD OF SOLUTION

The governing equations for compressible hydrodynamic flow or weakly compressible flow derived by Song and Yuan (1988), also see Chen (1995), can be written in a conservative form as

$$\frac{\partial G}{\partial t} + \nabla \bullet F = 0 \tag{1}$$

where

E. Cabrera et al. (eds.), Hydraulic Machinery and Cavitation, 267–276.
© 1996 *Kluwer Academic Publishers. Printed in the Netherlands.*

$$G = \begin{bmatrix} p \\ u \\ v \\ w \end{bmatrix} \tag{2}$$

$$F = \begin{bmatrix} Ku & Kv & Kw \\ uu + \dfrac{p}{\rho_o} - \dfrac{\tau_{xx}}{\rho_o} & uv - \dfrac{\tau_{xy}}{\rho_o} & uw - \dfrac{\tau_{xz}}{\rho_o} \\ uv - \dfrac{\tau_{yx}}{\rho_o} & vv + \dfrac{p}{\rho_o} - \dfrac{\tau_{yy}}{\rho_o} & vw - \dfrac{\tau_{yz}}{\rho_o} \\ uw - \dfrac{\tau_{zx}}{\rho_o} & vw - \dfrac{\tau_{zy}}{\rho_o} & ww + \dfrac{p}{\rho_o} - \dfrac{\tau_{zz}}{\rho_o} \end{bmatrix} \tag{3}$$

Song and Chen (unpublished) has shown that this set of equations are the Navier-Stokes equations for small Mach number flows and contains the effect of compressibility inside the time boundary layer (inner solution). For an example, the flow due to impulsively started circular cylinder is highly transient and compressibility dependent within a small time interval, $\delta_t = 5D/a_o$, where D is the diameter and a_o is the speed of sound. Its outer solution is the incompressible flow solution regardless of the value chosen for a_o or the Mach number, M. This set of equations is hyperbolic and can be efficiently solved with a simple explicit method using artificially large value of M.

The idea of LES method is to directly calculate the large scale motions that can be resolved with the finite grid volume averaged quantity but use a subgrid scale turbulence model to approximate the effect of unresolvable small scale eddies. The finite volume averaged equations are identical to Eqs. 1, 2, and 3 except that the shear stress terms should be regarded as the sum of the viscous shear stress and the subgrid scale turbulent shear stress. The subgrid scale turbulent stress can be written as,

$$\tau = \tau_l + \tau_t = \rho(v + v_t) \left\{ \frac{\partial u_i}{\partial x_j} + \frac{\partial u_j}{\partial x_i} \right\} \tag{4}$$

where v_t is the subgrid scale eddy viscosity. We use the simple model due to Smagorinsky (1963) as follows.

$$v_t = (C_s \Delta)^2 (2 \bar{S}_{ij} \bar{S}_{ij})^{0.5} \tag{5}$$

$$\bar{S}_{ij} = \frac{1}{2} \left(\frac{\partial u_i}{\partial x_j} + \frac{\partial u_j}{\partial x_i} \right) \tag{6}$$

In Eq.5 Δ represents the grid size and C_s is an unknown parameter to be determined. For homogeneous isotropic turbulence, Lilly (1966) analytically determined that $C_s \approx 0.23$. Deardorff (1970) found that $C_s = 0.1$ is optimal for turbulent channel flow. For present study we found that using $C_s = 0$ on the wall and increase it to 0.12 outside of the boundary layer makes the numerical results agree with experimental data well.

For the flow through a runner, it is more convenient to use the equations transferred into a rotating coordinate system. Denoting Ω the rotational speed about the z-axis, the governing equations can be transferred to

$$\frac{\partial G}{\partial t} + \nabla \bullet F = T \tag{7}$$

where,

$$T = \begin{bmatrix} 0 \\ 2\Omega v + \Omega^2 x \\ -2\Omega u + \Omega^2 y \\ 0 \end{bmatrix} \tag{8}$$

and G and F are given by Eqs. 2 and 3, respectively. All velocity components now represent the relative velocity with respect to the rotating coordinates.

The governing equations, Eqs.1 & 7, are first averaged over a finite volume $\Delta\forall$ before discritization. During the process of integration, the volume interaction of $\nabla \bullet F$ is converted into a surface integral using the divergence theorem, and the following equation is produced.

$$\frac{\partial \overline{G}}{\partial t} = -\frac{1}{\Delta\forall} \int_s n \bullet F dS + \overline{T} \tag{9}$$

In the above equation bar represents the volume averaged value and n is the unit outward normal vector on the bounding surface S.

MacCormack's (1969) predictor corrector method is used for numerical integration. For detailed description of the method, reference is made to Chen's thesis (1995).

III. GRID SYSTEMS

In earlier days when computers were not so powerful, only very limited space such as one or two flow passages of a runner was isolated from the total system and modeled. In a complex and compact system, such as the spiral case-stay vanes-wicket gates-runner-draft tube of a Francis turbine facility, a small localized model cannot be expected to simulate all the complex flow phenomena. The power of the computer and the computational method can now justify development of a large model containing many related components.

Fig.1 is a typical view of boundary surface grid system of a spiral case containing 20 stay vanes and 20 wicket gates. Note that only small part of the grids are shown in this figure. The flow field is divided into 44 zones, having a total of 78,381 nodes and 58,752 elements. Grids in any of 44 zones can be subdivided into finer grids if so desired. One of the most difficult problem is how to determine the boundary condition on the artificial boundary so that the influence of the unmodeled part can be accurately accounted for. For an example, the most logical boundary condition at the downstream end of the spiral case model is

$$\frac{\partial f}{\partial r} = 0 \tag{10}$$

where f represents p, u, v, and w and r is the radial coordinate. Experience shows, however, that there is a sharp turn of the flow toward the z direction and the boundary condition causes significant amount of error on the velocity and pressure distribution near the spiral case exit or the runner entrance. To improve on this deficiency, an enlarged region as sketched in Fig. 2 was used to model the flow through the spiral case. The crown and shroud of the runner, excluding the runner, plus a fictitious draft tube are attached to the spiral case. The downstream boundary of the model is shifted to the end of the fictitious draft tube. As will be seen later, the velocity profile at the entrance to the runner is significantly modified.

At this stage of the modeling effort, only partial flow passage, has been used to model the runner. It was found by Song and Chen (1991), two flow passage model is actually more efficient than the more commonly used single flow passage model. Two types of draft tubes, one with and one without a divider wall namely a draft tube pier, were modeled.

IV. BOUNDARY CONDITIONS

Expecting upstream effect to be more important than the downstream effect, the simulation process starts with the spiral case, followed with the runner and end with the draft tube. At the entrance to the spiral case, the flow is assumed to be the fully established turbulent flow and the time averaged velocity profile is specified. The pressure is assumed to be constant on the cross-section. No fluctuating components of velocity and pressure were included at the upstream end although it is possible to do so.

As mentioned in the last section, the downstream boundary condition of the spiral case model is applied at the downstream end of the fictitious draft tube. Here the zero gradient condition, Eq. 10 with r replaced with z is imposed. The partial slip condition as described by Chen (1995) is applied to all solid boundary.

The velocity and pressure distribution at the junction between the spiral case and the runner calculated for the spiral case model is used as the upstream boundary condition for the runner model. The zero gradient condition is applied at the downstream end of the runner model.

Because only a partial model is used for the runner, it is not possible to apply a realistic boundary condition at the inlet of the draft tube. For this reason only uniform velocity distributions with and without swirl is applied at the upstream end of the draft tube. As usual, the zero gradient condition is applied at the down stream end of the draft tube.

V. RESULTS

5.1 Calculation Conditions

Fig.3 shows the hill chart determined by the physical model test. The model test was conducted in accordance with International Electro-technical Commission codes. The computed characteristics are compared with measured values at several wicket gate openings such as point A, B, C, D and E in Fig.3.

5.2 Typical Calculated Flow Pattern and Pressure Distribution

The computed time averaged velocity vector field on the middle cross section of the spiral case near some of the stay vanes and wicket gates for 80 % opening condition is shown in Fig.4. The flow appears to behave well and no significant flow separation exists. The corresponding dimensionless pressure distribution is shown in Fig.5. Here, the velocity at the entrance and the pressure at the exit were used as the reference quantities. The time averaged pressure distribution on the solid boundaries of the spiral case at 80 % opening is shown in Fig.6. It appears that the flow in the spiral case is subjected to a slight adverse pressure gradient suggesting that there is a room for some improvement. The computed time averaged velocity distribution at the exit of the wicket gates at three different cross sections are plotted in Fig.7. Note that the angle is measured from the first stay vane and the subscripts t, m, and z, represent circumferential, meridian, and vertical components of the velocity. It is also impotent to note that, without adding the runner casing and the fictitious draft tube, there would be no vertical velocity component. The fact that the velocity on the lower cross section is higher than that of the upper cross section is also due to the flow turning effect which can't be modeled with the zero gradient boundary condition.

Even though only two flow passage model was used to simulate the runner, very fine grids were used and as much computational time was used for the runner as for the spiral case to insure good accuracy. It appears that there is a small flow separation on the suction side of the blade near the leading edge and the hub. Assuming complete symmetry, the pressure distribution on the entire runner can be assembled from the results of the partial model as shown in Fig.8.

Typical time averaged velocity distributions in a draft tube with central divider using the time averaged out flow from the runner for the case of 80% wicket gate opening are shown in Fig.9. The corresponding figure for 60 % wicket gate opening case is shown in Fig.10. Large eddies and flow separation caused by the increased swirl is noteworthy. Although time averaged inflow condition is used, the computed flow in the draft tube is very unsteady. There are unsteady flow separations from the outer walls and horse shoe vortex on the upper part of the central pier. These eddies interact each other and exhibit preferred frequencies of oscillations. Fig.11 shows an instantaneous pressure distribution on the walls of the draft tube with center wall. When the center wall is removed the draft tube becomes a diffuser of large angle and the flow separation becomes much more severe. Flow separates alternatively from two side walls and the jet like flow swings from one wall to the other wall. Fig.12. shows an instantaneous pressure distribution on the wall of the draft tube without divider wall.

All calculated data are normalized by the value of the best efficiency point of the model test. The calculated power outputs of a model turbine at five gate opening conditions, 60%, 70%, 80%, 90%, and 100%, are compared with the model test data in Fig.13. No analysis was made to determine why the agreement is best at the 80% opening case which happens to be the design point. The calculated head loss through each of the three components as functions of the wicket gate opening are shown in Fig.14. The solid line Clearly indicate that most of energy loss in spiral case occurs at wicket gates.

The over all efficiency predicted numerically is compared with the measured value in Fig.15. Here an estimated friction loss and the leakage loss of 1.5% has been added to the computed loss because the numerical simulation doesn't account for this kind of losses.

V. CONCLUSIONS

Although a Large Eddy Simulation approach was used there may still be a significant amount of numerical error because rather coarse grids were used and the coefficient, C_s, need to be selected. As demonstrated in Fig.7 there is a strong downstream effect on the exit velocity distribution of the spiral case. This non uniform exit velocity should produce strong unsteadiness effect on the runner. The calculated flow in the draft tube is not expected to be very realistic because the real outflow from the runner is expected to be much deferent from what was used as the inflow to the draft tube model. The out flow from the runner should be unsteady and much more complex.

The current computation is largely based on three individual models performed in series from spiral case to draft tube. Therefore, much of the detailed interactions between units, especially the unsteady vortex phenomena, are not obtained. The next objective of the development should be an integrated model.

VI. ACKNOWLEDGMENTS

The entire computations were carried out with the super computing facility of the Minnesota Super computer Institute of the University of Minnesota through its Super computer Research Grant program. The authors are very grateful for this support.

VII. REFERENCES

Chen, X.Y. (1995) Multi-Dimensional Finite Volume Simulation of Fluid Flows on Fixed, Moving and Deforming Mesh Systems, Univ. of Minnesota, Ph.D. Thesis.

Deardorff,J.W. (1970) A Numerical Study of Three-Dimensional Turbulent Channel Flow at Large Reynolds Number, *Journal of Fluid Mechanics,* Vol. 41, 492-480.

Lilly, D.K. (1966) On the Application of the Eddy Viscosity Concept in the Inertial Subrenge of Turbulence, *NCAR Manuscript,* No. 123.

MacCormack, R.W. (1969) The Effect of Viscosity in Hyper Velocity Impact Crating, *AIAA Paper,* No. 69-354.

Song, C.C.S. and Yuan, M. (1988) A Weakly Compressible Flow Model and Rapid Convergence Methods, *Journal of Fluid Engineering,* Vol. 110, 441-445.

Song, C.C.S. and Chen, X.Y. (1991) Calculation of Three-Dimensional Turbulent Flow in Francis Turbine Runner, *Waterpower'91,* Denver, July 24-26.

Song, C.C.S. and Chen, X.Y. (unpublished) Compressibility Boundary Layer Theory and Its Significance in Computational Hydrodynamics.

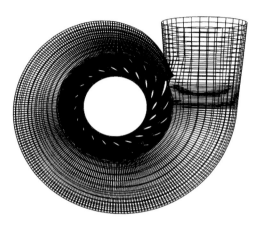

Figure 1 A top view of the mesh system on selected surfaces for spiral case computation.

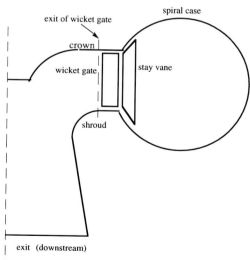

Figure 2 A sketch of the configuration for the computation of spiral case.

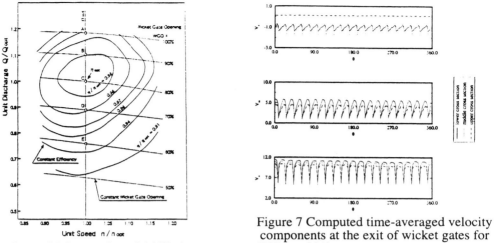

Figure 3 Measured model hill chart.

Figure 7 Computed time-averaged velocity components at the exit of wicket gates for WGO = 80%.

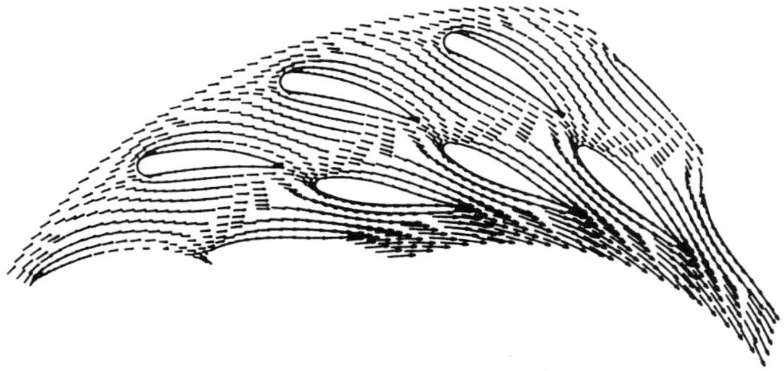

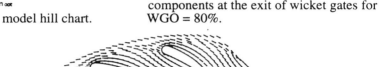

Figure 4 Computed time-averaged velocity field around several stay vanes and wicket gates for WGO = 80 %.

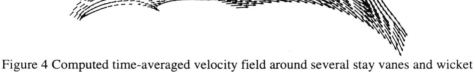

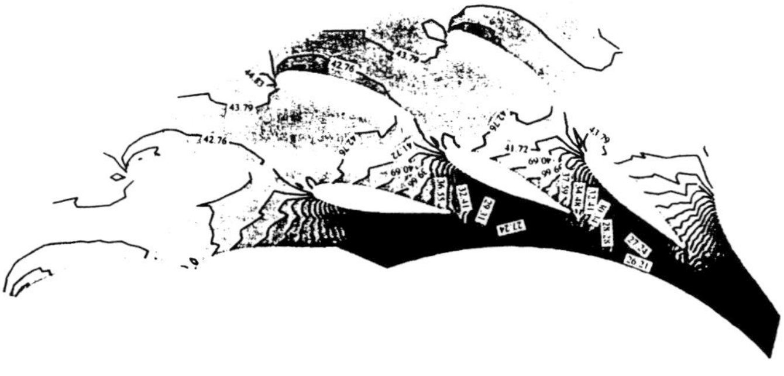

Figure 5 Computed time-averaged pressure field around several stay vanes and wicket gates for WGO = 80 %.

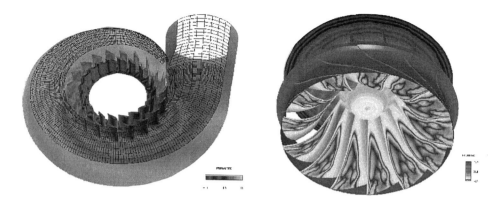

Figure 6 computed time-averaged pressure distribution on the solid surfaces of the entire spiral case for WGO = 80 %.

Figure 8 Computed time-averaged pressure distribution on the solid surfaces of the turbine runner for WGO = 90 %.

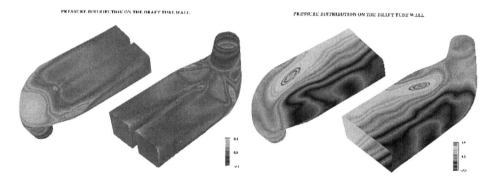

Figure 11 Instantaneous pressure distribution on the solid surfaces of the draft tube with a center pier.

Figure 12 Instantaneous pressure distribution on the solid surfaces of the draft tube without a center pier.

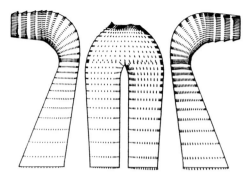

Figure 9 Time-averaged velocity field with inlet swirling flow for WGO = 80 %.

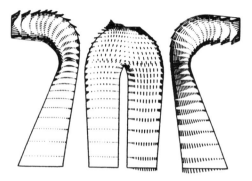

Figure 10 Time-averaged velocity field with inlet swirling flow for WGO = 60 %.

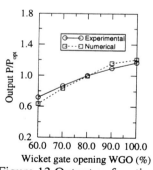

Figure 13 Output as function of the discharge for modeled turbine runner.

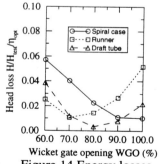

Figure 14 Energy losses in three components of the hydropower system.

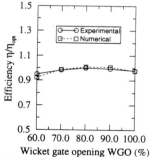

Figure 15 Overall efficiency of entire hydropower system.

SIMULATION OF FLOW THROUGH PUMP-TURBINE

Charles C. S. Song & Changsi Chen
Univ. of Minnesota, Mpls, MN, USA
Toshiaki Ikohagi
Tohoku Univ., Sendai, Japan
Johshiro Sato, Katsumasa Shinmei & Kiyohito Tani
Hitachi Ltd., Hitachi-shi, Japan

1. INTRODUCTION

For years Francis - type reversible pump - turbines have been applied to many pumped storage power plants over the world. So far, they have been used under peak shaving generating operation with AFC (automatic frequency control) function during daytime, and plain pumping - energy consuming - operation at night time. Recently, an adjustable speed pump - turbine was realized in Japan. It begins to add a remarkable AFC function in pumping mode to the conventional pump-turbines.

Figure 1 schematically shows the range of pump input variation of a typical adjustable - speed pump - turbine. The line of rated speed is located usually in the middle of the whole operating area. There is a line of maximum speed on the upper - right side. A line of minimum speed is on the lower - left side. These two lines are determined mainly by the specifications of the frequency converters. On the other hand, there is a line of runner inlet cavitation on the pressure side on the upper - left area of the range. The line of reverse flow at the inlet of the runner is located on the lower - right side. These two lines are determined by the hydraulic performance of the pump - turbine.

In order to make the range of pump input variation larger (to make AFC operation more effective), it is necessary to get those two lines (cavitation and reverse flow) farther from the range. Hitachi has been using various CFD techniques including the industry standard steady - state turbulent flow analysis using $k - \varepsilon$ model to improve those hydraulic performances of the pump - turbine. It was shown that the cavitation inception is predicted very well by industrial standard technique. However, the reverse flow at the runner inlet as well as other flow related to low pump discharge were not simulated very well by the conventional steady - state simulation techniques. In this report the flow in the pump - turbine in pumping mode is simulated by Large Eddy Simulation technique, and it is shown that unsteady flows play very important roles in the hydraulic performance of the pump - turbine.

In turbine mode it is of great importance to further improve the efficiencies not only at an optimum point but at off-design points for both the conventional pump - turbines and the adjustable pump - turbines. A lot of analytical works have been made by using industry standard techniques, however, as we analyze the flow in the turbomachineries more and more in details, we begin to feel that the unsteadiness of the flow should be taken into account. In this report the flow in the pump - turbine in turbine mode is simulated by LES technique, and it is shown that vortex shedding exists even in turbine mode.

E. Cabrera et al. (eds.), Hydraulic Machinery and Cavitation, 277–283.
© *1996 Kluwer Academic Publishers. Printed in the Netherlands.*

2. FLOW IN PUMPING MODE

The flow through runner and spiral case in pumping mode is basically an expending flow with strong downstream effect which is intrinsically unstable. For this reason, it is extremely difficult to model the runner and the spiral case separately. Because there is a strong downstream effect, a proper downstream boundary condition of a runner model cannot be determined a priory. Many such trials failed to produce a convergent solution. Therefore, it was decided to develop a complete model containing the runner and the spiral case with all the runner blades, wicket gates, and stay vanes. An overall view of this model is shown in Fig. 2. The inflow and the outflow boundary conditions are relatively simple to specify. The model turned out to be quite stable but requires large amount of memory and computer time. Some of the more interesting results are described below.

2.1 FLOW SEPARATION ON SHROUD SURFACE

The designer of this machine has been bothered by the existence of reverse flow near the leading edge of the runner blades and sudden drop in efficiency when flow rate is reduced below a critical point. This phenomenon was observed in a physical model but the flow detail and the reason for its occurrence was not clear. The output from the mathematical model with the help of numerical flow visualization technique enabled detailed study of the flow near the junction between the runner blades and the shroud surface. Boundary layer flow separation is observed on the shroud surface slightly upstream of the runner blade when the model flow rate is equal to or slightly less than 0.182 cms. The separation bubble is rather thin but extends beyond the leading edge of the runner blades and affects all blades. As shown in Fig. 3, reverse flow can be observed near part of a runner blade submerged under the separation bubble. Because this separation bubble is not very thick, there is no reverse flow in the mid span part and near the crown of the flow passage. This type of flow separation does not occur for larger flow rate. For an example, no separation can be observed when the discharge is equal to 0.200 cms. The most likely reason for the flow separation on the shroud surface is the existence of large adverse pressure gradient due to small wicket gate opening which constrict the flow downstream of the runner.

2.2 MOVING SEPARATION IN STAY VANE CASCADE

Another unique problem only a comprehensive model can solve is that of unsteady flow separation on stay vanes that was found to occur for relatively small discharge cases. If observation is concentrated on one vane, then the separation bubble will appear to periodically change its size and intensity. But when all vanes are observed simultaneously, there appears to be four wave pattern that rotates slowly, much slower than the rotational speed of the runner. There appears to be four waves of varying separation in the stay vane cascade. An instantaneous velocity field around a few neighboring vanes is shown in Fig. 4. No detailed analysis of the wave length and wave speed as related to other flow conditions has been carried out.

2.3 PRESSURE OSCILLATION DUE TO RUNNER - WICKET GATE INTERACTION

Very strong pressure oscillation generated by the runner blades and the wicket gates interaction is observed every where in the computational domain. Let the rotational frequency of the runner be F_r, the number of the runner blades be N_r, and the number of wicket gates be N_w. Then the pressure on a runner blade will have a strong wicket gates induced pressure oscillation at frequency equal to $F_r N_w$. This is the frequency at which a runner blade passes by wicket gates. A typical time history of pressure at a point on a runner blade is shown in Fig. 5. On the other hand the pressure at a point on a wicket gate will have a strong oscillation at the frequency equal to $F_r N_r$ which is the frequency of a wicket gate passing by the runner blades. A typical time history of pressure at a point on a wicket gate is shown in Fig. 6. These two figures also show a beating phenomenon at the frequency roughly equal to $F_r N_w / N_r$.

3. FLOW IN TURBINE MODE

3.1 STARTING PROBLEM

The numerical model is intended to simulate the actual physical process closely. Therefore, the initial condition as well the boundary conditions should also closely reflect the physical process. A lesson learned is that the runner cannot be given full rotational speed before the flow is nearly established. If the runner is given its full rotational speed too soon, then it will act like a pump generating reverse flow resulting in computational instability. This problem doesn't occur in pumping mode because the runner is an active element in that case, but it is a passive element in turbine mode.

3.2 VORTEX SHEDDING FROM RUNNER BLADES

This particular pump-turbine has six runner blades. Each blade produces a vortex sheet that will detach from the blade and carried by the main stream into the draft tube. Therefore, there are six vortex sheets trailing the runner blade, in this case. However, it is interesting to note that the six vortex sheets quickly merge into three vortex sheets as they move downstream and migrate toward the center line. The detached vortex sheets appear to rotate at slower speed than runner's rotational speed. Fig. 7 shows an instantaneous pressure distribution on a cross - section a short distance downstream of the trailing edges of the runner blades. Fig. 8 shows the corresponding distribution of the vertical velocity component at the same instant and the same cross - section. There are also only three sets of spiral shaped ribbons extending out from the center of cross - section. Three spiral ribbons of relatively high pressure zones and equal number of low pressure zones indicate that there are only three vortex sheets. Notice that the areas of positive velocity correspond to high pressure areas and the areas of negative velocity (reverse flow) correspond to low pressure zones. It would be interesting to extend the present model to include the draft tube and study further evolution of vortex sheets in the draft tube.

4. CONCLUSIONS

In order to meet recent requirement of the customers worldwide, HITACHI has been focusing its effort to develop CFD technique for designing reliable pump-turbines. Among various CFD techniques developed so far LES is the most accurate one and will be the industrial standard in the near future. Present study indicates that flow in each component of a pump-turbine system is so much dependent on other components, a comprehensive model is essential to a good over all design.

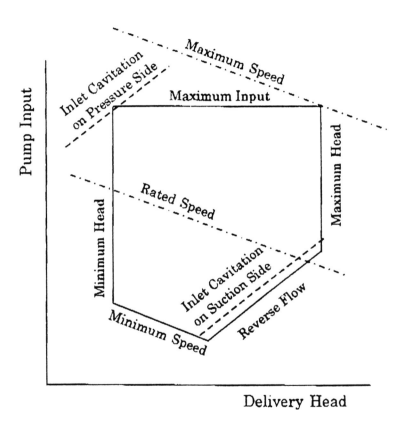

Figure 1 Schematic diagram of pump input variation.

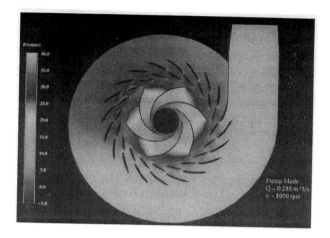

Figure 2 Instantaneous pressure distribution on the middle cross section of pump-turbine system for pump mode (Q=0.285 m^3 / s ; n=1000 rpm).

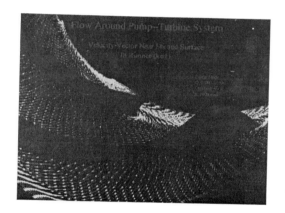

Figure 3 Instantaneous velocity field near shroud surface in runner for pump mode (Q=0.285 m^3 / s ; n=1000 rpm).

Figure 4 Illustration of instantaneous velocity pattern around vanes for pump mode (Q=0.285 m^3 / s ; n=1000 rpm).

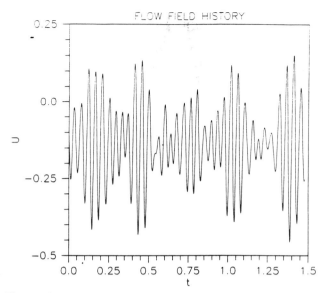

Figure 5 Time history of velocity at a point on a runner blade.

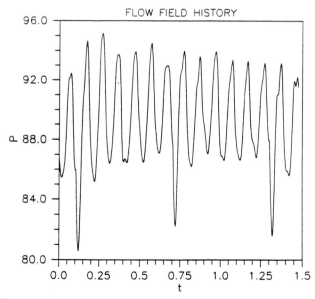

Figure 6 Time history of pressure at a point on a wicket gate.

Pressure

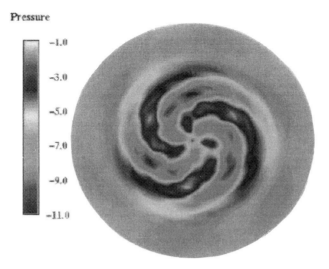

Figure 7 Instantaneous pressure distribution on a cross - section a short distance downstream of the trailing edges of the runner for turbine mode (Q=0.0712 m^3 / s ; n=169.3 rpm).

V(z)

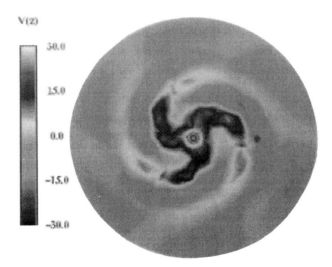

Figure 8 Instantaneous vertical velocity distribution on a cross - section a short distance downstream of the trailing edges of the runner for turbine mode (Q=0.0712 m^3 / s ; n=169.3 rpm).

THE SCALE EFFECT IN KAPLAN TURBINES. NEW RELATIONSHIPS FOR THE CALCULATION OF SCALABLE AND NON-SCALABLE HYDRAULIC LOSSES AND OF THE COEFFICIENT V

I.M. ANTON
Polytechnical University of TIMISOARA
Bd. Mihai Viteazu Nr. 1 TIMISOARA 1900
ROMANIA

Abstract

In this paper new directions are opened concerning the scale effect at Kaplan turbines. The original relationships are developed for the calculation of coefficients δ_M, δ_{ns} and V in the whole domain of operating of Kaplan turbines. Their use needs only the hill diagram of the turbine. Applying a new method there were obtained new relations, for the coefficients δ_{Mo}, δ_{nso} and V_o corresponding to the optimum operating regime of a Kaplan turbine.

1. Introduction

The first paper on this subject was published in 1909 by Cammerer [1]. Since then many researchers worked on the scale effect in Kaplan turbines. S.P. Hutton [2] made a substantial step toward the solution of this problem. The idea of Ackeret [3] to separate hydraulic losses from the turbine in friction and kinetical losses proved to be very useful.

The formula developed by S.P. Hutton [2] was accepted by IEC [4] in 1965 and recommended to predict the efficiency of prototype turbine (P) when one knows the efficiency of the model (M).

J. Osterwalder [5,6] defined the coefficient V of the distribution of hydraulic losses in the turbine and using the results of laboratory measurements established correlations of the type $V = f(Q_x/Q_o)$ for k_u=const. Using such correlations he tried to find the value of V for regimes different from the optimum one.

In order to find out the correct parameters to be used in the formula developed by S.P. Hutton, a lot of research work was necessary. Important contributions are due, among others, to Hutton [7,8], Osterwalder [5,6], Chevalier [9], Fáy [10,11] and Ida [12].

In spite of these the IEC cod from 1992 [13] recommended a unique value for the coefficient V, for all the types of Kaplan turbines and every operating regime.

E. Cabrera et al. (eds.), Hydraulic Machinery and Cavitation, 284–293.
© 1996 *Kluwer Academic Publishers. Printed in the Netherlands.*

In this paper new relationships are presented, which describe the dependence on the efficiency η_M of the friction losses, $\delta_M = f(\eta_M)$, kinetic losses, $\delta_{ns} = f(\eta_M)$ and coefficient V, $V = f(\eta_M)$, for the whole operating domain of the model turbine.

A.   THEORETICAL BASIS

2.   Scalable and nonscalable hydraulic losses in Kaplan turbines

2.1. GENERALITIES

Related to the scale effect formula for turbines, Ackeret [3] suggests the idea to separate the hydraulic losses δ_h into two categories: friction losses, δ and kinetic losses, δ_{ns}. Thus we have

$$\delta_h = \delta + \delta_{ns} \tag{1}$$
$$\eta_h = 1 - \delta_h \tag{2}$$

The hydraulic losses through friction δ depend on the Reynolds number (Re) and on the relative surface roughness (e/D) of the solid flow surfaces, which confer them the property of being scalable losses and thus $\delta = f(Re, e/D)$. On the contrary, the kinetic losses δ_{ns}, which are independent of Re and depend only on the geometry of the flow channel, are nonscalable.

The coefficient V of distribution of hydraulic losses in a turbine, which appears explicitly in the scale effect formula of Hutton and Osterwalder, it is defined in the following way:

$$V = \delta/\delta_h = \delta/(1 - \eta_h) \tag{3}$$

The correct use of the scale effect formula for Kaplan turbines needs to know the values δ_M, δ_{ns}, and V for the model for its whole operating domain.

2.2. THE SCALABLE, δ_{Mo} AND NONSCALABLE, δ_{nso} HYDRAULIC LOSSES AT THE OPTIMUM WORKING REGIME

The solution of this problem, i.e. to establish relations of the form $\delta_{Mo} = f(\eta_{qo})$ and $\delta_{nso} = f(\eta_{qo})$, according to the method presented in what follows, it is possible when one has the hill diagram of various Kaplan turbine types, to be found in the computer program libraries of the great turbine manufacturing companies.

For example, in Fig. 1 it is represented the diagram of Kaplan turbine types developed by NOHAB Co. [14]. The closed curves are equal efficiency curves, $\eta_M = 0.91$, for various Kaplan turbines.

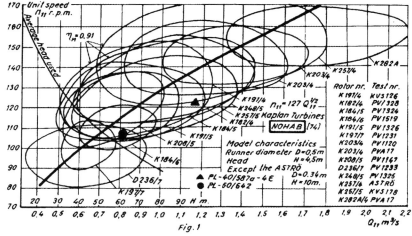

Fig. 1

The locus of the central points of the surfaces delimited by these curves represents the curve of optimum regimes, which has the form [15-17]

$$n_{1\,lok} = a_1 \cdot (Q_{1\,lok})^{p/q} \tag{4}$$

According to NOHAB this curve may expressed as [15,16]

$$n_{1\,lok} = 127.5 \cdot (Q_{1\,lok})^{1/2} \tag{4a}$$

Here n_{11o} and Q_{11o} are the unitary rotation speed and unitary flow rate, respectively, corresponding to the optimal working regime, were k is the rank of the turbine from the series considered. Taking into account the specific speed $n_q = n_{11}\sqrt{Q_{11}}$, relation (4a) becomes

$$n_{1\,lok} = 11.284 \cdot n_{qok}^{1/2} \tag{5}$$

When one of the Kaplan turbines developed by the company (in this case NOHAB) is considered as reference turbine, defined by n_{11oref} and Q_{11oref}, then it follows from (3) that

$$\frac{V_{ok}}{V_{oref}} = \frac{\delta_{Mok}/(1-\eta_{Mok})}{\delta_{Moref}/(1-\eta_{Moref})} \tag{6}$$

Without introducing significant errors, it may be admitted that $\eta_{hMok} \cong \eta_{hMoref}$. In this case (6) becomes

$$\frac{V_{ok}}{V_{oref}} = \frac{\delta_{Mok}}{\delta_{Moref}} \tag{7}$$

However, the current in Kaplan turbine models at the optimum operating regime, may be considered hydraulically smooth [7, 9, 12], therefore it may be written

$$\lambda_{ok} \equiv \delta_{Mok} = C/(Re_{Mok})^{1/n} \tag{8}$$

and

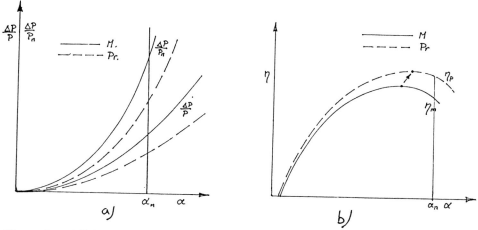

Figure 1. a) Schematic illustration of the loss relative to the nominal max. output P_A and the regarded output P. b) Schematic illustration of the shift in efficiency from model (full line) and prototype (dotted line).

Here $\xi = f(\text{Re})$ and $Q = f(\alpha)$ where α is the guide vane opening angle. The loss in power $\Delta P = \rho Q \Delta(gH) = \rho \xi Q^3$ and then the relative power loss will be:

$$\frac{\Delta P}{P} = \frac{\rho \xi Q^3}{\rho g Q H} = \text{const} \cdot Q^2 \tag{6}$$

The loss relative to the nominal max. output will be:

$$\frac{\Delta P}{P_n} = \frac{\rho \xi Q^3}{\rho g Q_n H_n} = \text{const} \cdot Q^3 \tag{7}$$

Because the relative loss is increasing with the square of the flow the BEP of the efficiency will shift to a larger guide vane opening for the prototype than for the model caused by the reduced friction factor $\xi = f(\text{Re})$. The relative loss and shift in efficiency is illustrated schematically in fig 1a and fig 1b respectively.

3.2. REDUCED DISK FRICTION LOSS IN PROTOTYPE

The disk friction loss will be reduced in the prototype compared with the model turbine if both are homologous in shape and hydraulically smooth. It should be emphasized that the requirement of homologous geometry of prototype and model not always is fulfilled for the surfaces of the covers

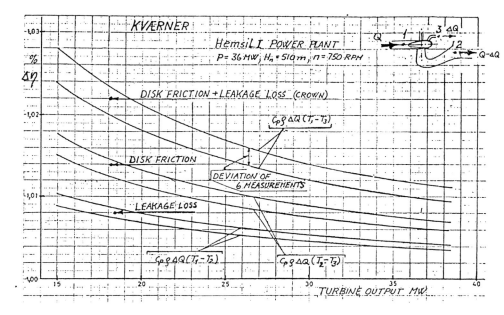

Figure 2.   Relative disk friction losses as function of turbine output for a high head Francis turbine. (Courtesy KVRNER.)

enveloping the runner and the outside surface of the runner. [Accept criteria for deviations are also not given with sufficient accuracy in the IEC code.]

The scalable disk friction loss has been measured during thermodynamic measurement in several high head turbines in Norway. The result shows that the powerloss is almost constant and not a pure function of the guide vane openings. In fig 2 is shown the result of 6 different measurements for a high head Francis turbine by means of the thermodynamic method from 1960 to 1963 during development of two different runners with homologous outside of crown and band.

When assuming the disk friction loss as a loss independent of the guide vane opening the relative loss will decrease with increasing output as shown in fig 2 and as illustrated schematically in fig 3a.

Because the disk friction loss decreases in the prototype due to increased Re, the BEP of the efficiency for the prototype will shift to a smaller guide vane opening than for the model . This is illustrated in fig 3b.

At the Norwegian University of Science and Technology research work

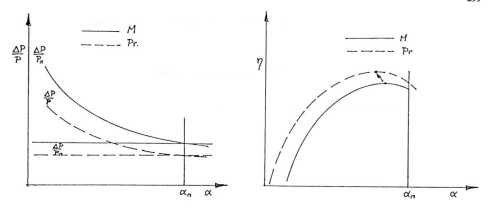

Figure 3. Relative loss of model and prototype (a), and increased efficiency and shift of BEP from model to prototype (b). Model full lines, prototype dotted lines.

is going on on disk friction losses. Of special interest is the influence of the geometry and leakage flow on the loss. Some data may be given during presentation of this paper.

4. The total losses in reaction turbines

In fig 4 is given an illustration of the different losses in a high head turbine. It can be found that the disk friction losses cover for approximately 1.5% of the total loss for a typical high head Francis turbine while 2.0% − 2.5% cover for the flow friction losses. Because the larger flow friction loss which gives a shift of BEP towards increased output for the prototype it should be expected a total shift towards higher output. However, another complex loss caused by the guide vane end clearance leakage has been measured to be as large as 1.5% on a prototype and approximately 0.5% for a model at BEP. See fig 5. The reason for the higher loss in the prototype may be found in the stronger guide vane end clearance leakage in the turbulent domain versus the laminar leakage in the model. An illustration of the end clearance losses in a model is given in fig 6 and for a prototype compared with models in fig 7.

The guide vane end clearance loss is increasing inverse proportional to the guide vane opening angle due to higher increasing leakage and stronger cross flow effect. For a prototype with homologous geometry to the model the relative clearance should be the same. Then the contribution from the

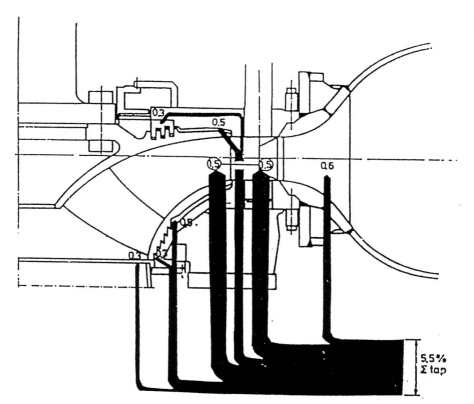

Figure 4. The different losses in a high head Francis turbine. Note the loss created by the guide vanes occures mainly in the runner and not 1.5% + 1.5% distributed as shown in the figure.

end clearance loss pushes the best efficiency point BEP towards higher guide vane openings for the prototype. In addition a "negarive scale effect" occures from the higher end clearance loss of the prototype.

Besides the described losses separation losses at the runner blade inlets, guide vanes and stay vanes occures. These losses gives normally not a contribution to the scale effect, but are losses which may be reduced to a minimum by an optimum design. An important parameter for an optimum design will be the blade lean angle θ, which is the blade angle normal to the stream line direction (Ref [2]). It should be emphasized on the fact that an analytical analysis must be made of the blade design prior to and in collaboration with modern CFD analysis. Without an analytical study of the influence of the geometry parameters prior to the CFD analysis for a direct solution it will be very difficult to obtain a sucesful design without an expensive "cut and try" period. However, CFD analyses are very useful

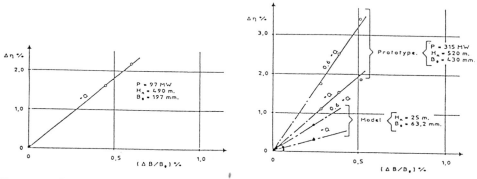

Figure 5. a) Efficiency loss versus guide vane clearance gap for prototype turbine as given by Mc.Hamish BOVING in IAHR Symp. Fort Collins 1978. b) Efficiency end clearance loss versus clearance gap for model and prototype at Kvilldal Power Station.

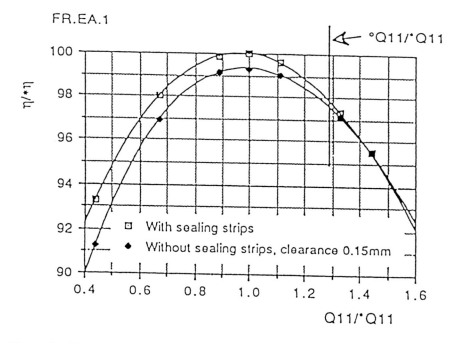

Figure 6. The measuring result of model turbine with sealing strips on the guide vane faces and compared with 0.15mm clearance gaps measured at 25m net head.

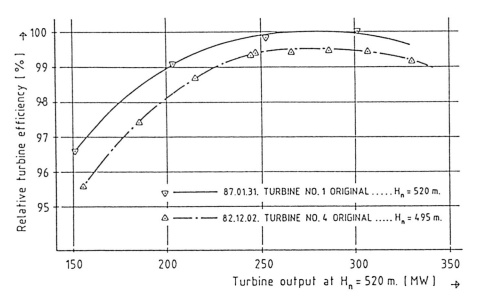

Figure 7. NVE Kvilldal Power Plant. Difference in efficiency of turbine No. 1 and turbine No. 4 with a difference in the guide vane clearance gap of 0.3*mm*.

in order to optimize a given blade structures within a turbine manufactorers geometry variation in the runner collection. Because a complete CFD analysis of the draft tube flow has so far not been available, there are still a need for laboratory work to be made on turbine models.

For low head turbines the lack of recovery of the runner outlet energy in the draft tube is the largest contribution to the losses.

5. Losses in impulse turbines

For impulse turbines i.e. Pelton turbines loss analysis and scale effect formulas have been discussed for years. However, up to now scale effect formula seems to be perfect even if the work by Dr. Grein and prof. Spurk [3] seems to be a promising approach.

The reason for failour to calculate the losses are the complex outlet flow. A typical example is outlet water in the bucket which may escape through the buckets entrance cut out when the last water is leaving the bucket. Secondary effect of such high velocity outlet water for a 6 jet turbine is fatal for the efficiency and the scale effect formula is not valid.

Following conclusion may be drawn. There is a possibility to make a useful loss analysis for a well designed low specific speed Pelton turbine, but a useful loss analysis to be used in a scale effect formula in general will fail.

6. Conclusion

For reaction turbines the understanding of the losses is improving by use of modern analyses and testing technology. However, computation of the draft tube flow is still not solved.

For impulse turbine the non stationary free surface flow is difficult to handle theoretically, and no CFD analysis has so far been succesfully presented.

However, high speed video technique and analytical flow analysis has improved the performance of Pelton turbines.

Scale effect formulas cannot be regarded to be valid for all types of Pelton turbines in general.

Aknowledge

The author is grateful for admission to measuring results from Norwegian Power Plant owners and Kvœrner.

References

1. Spurk, Joseph H. ", Dimensionsanalyse in der strohmungslehre", Springer Verlag, 1992, Germany. (German language.)
2. Brekke, H. ",A parameter study on cavitation performance of Francis turbines", ASCE Water Power Conf., 25-28 July 95, San Francisco.
3. Grein, H., Meier, J., and Klicov D. ", Efficiency scale effects in Pelton turbines", Proceedings IAHR Symposium, Montreal, 1986.

Symbols

u = circumferential speed
c = absolute velocity
H = Head in mWC (gH specific energy (J/kg))
Q = Flow (m^3/s)
n_s = Specific speed
g = gravity accelaration (m/s^2)
η = efficiency
Ω = Speed number
ω = angular velocity (rad/s)

SCALING-UP HEAD-DISCHARGE CHARACTERISTICS FROM MODEL TO PROTOTYPE

Michel COUSTON,Robert PHILIBERT
GEC-ALSTHOM NEYRPIC
82, Rue Léon Blum BP 75, 38041 Grenoble Cedex, FRANCE

We present here a method to calculate the scale effect on discharge and output that is correlated with the scale effect on efficiency . A theoretical analysis is proposed and general formulas are induced . Different comparisons betwen model and prototype measurements confirm the proposed analysis.

1-INTRODUCTION

Often the acceptance tests of an hydraulic turbine are based on model tests . So prototype characteristics have to be determined from model results . The scaling-up for head , discharge , output and efficiency is made according to IEC standards :output is corrected by efficiency , head and discharge are not corrected. But on-site measurements for different types of hydraulic turbines are not in good agreement with this evaluation [1,2,3].

So it seems necessary to foresee the prototype characteristics with a better accuracy from model tests .

In this paper we present a method that has been validated by comparisons between model results and on-site measurements .

304

E. Cabrera et al. (eds.), Hydraulic Machinery and Cavitation, 304–312.

3-2 SHIFT OF CHARACTERISTICS -SIMPLIFIED FORMULAS

3-2-1 TURBINE MODE

The formulas of §3-1-1,while not really complicated, are not so convenient to handle as α coefficient depends on both the specific speed and the operating point .However or FRANCIS type turbines and pump-turbines operating as turbine the coefficient $\dfrac{1}{2}\dfrac{Q_{11} - \alpha n_{11}}{Q_{11}}$ is quite close to 0,5 .

So in case of FRANCIS turbine the simplified formulas are :

$$\frac{\Delta Q}{Q} = \frac{\Delta Q_{11}}{Q_{11}} = 0,5\frac{\Delta \eta}{\eta}$$

$$\frac{\Delta P}{P} = \frac{\Delta P_{11}}{P_{11}} \approx 1,5\frac{\Delta \eta}{\eta}$$

However for low head turbine , such BULB units the α value becomes important . Thus the above simplification is not valid , for example at large blade angle (as can be seen on § 4) we get :

$$\frac{\Delta Q}{Q} = \frac{\Delta Q_{11}}{Q_{11}} = 0$$

$$\frac{\Delta P}{P} = \frac{\Delta P_{11}}{P_{11}} \approx \frac{\Delta \eta}{\eta}$$

3-2-2 PUMP MODE

As for FRANCIS turbines ,There is **an important case for pump-turbine in pump mode** where the operating point is close to :

$$H_{i0} = \frac{1}{2}\frac{U_2^{\,2}}{g}$$

and if the curve Hi(Q) has the value 0,85 U^2/g (statistical value)   for   Q=0 we obtain the simplified formulas:

$$\frac{\Delta Q}{Q} = 1,4\frac{\Delta \eta}{\eta}$$

and $$\frac{\Delta P}{P} \approx 0,4\frac{\Delta \eta}{\eta}$$

4-SOME PROTOTYPE RESULTS

Several cases of BULBS , FRANCIS , PUMP-TURBINES in pump and turbine mode are presented and demonstrate the validity of our analysis .

4-1 FRANCIS

To demonstrate the applicability of the method we present hereafter two comparisons for generating operation . The first case corresponds to the medium head ITAIPU turbine .Achieved results are conforting our assumptions .

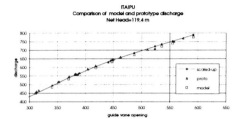
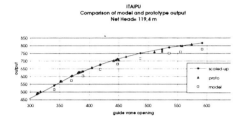

The second case corresponds to the high head MUJU pump-turbine in generating mode . While not as satisfactory as ITAIPU, is comparison is still good and full load operations are well reproduced .

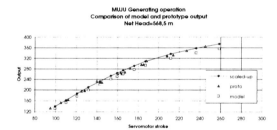

4-2 BULBS

For low head machine it is very difficult to have prototype results with enough accuracy .However results obtained on a model with different testing conditions give indications on the scale-up of discharge and output .On the following diagrams we can see two different results that were obtained on our test rig.

We can observe that the application of the formulas give a better evaluation than the usual way in which discharge is unchanged .

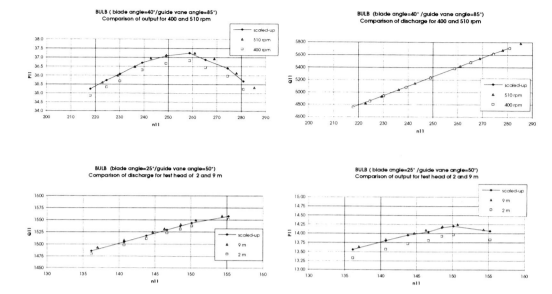

4-3 PUMP-TURBINE in PUMP MODE

High-head pump-turbines correspond to the straighforward application of the method presented in § 3-1-2 as such machines are almost non regulated .

Hereafter we present the achieved results for two recent single stage pump-turbines : GUANGZHOU in CHINA and MUJU in KOREA .

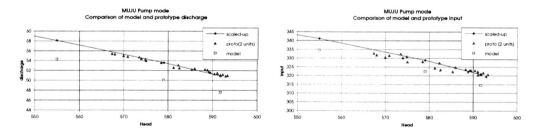

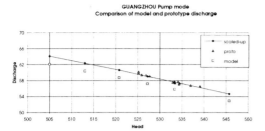

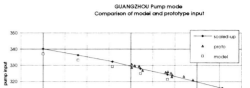

In both cases the shift of characteristic at a given head is very accurately predicted by the proposed simplified formula . However additional results for lower head pump-turbines have to be investigated to confirm the range of validity of the proposed approach .

5-CONCLUSIONS

The main conclusions of this paper are :

-Whatever the type of turbine the "efficiency scale-effect" induces a "shift of the head-discharge characteristic "and by a "shift of the head-output characteristic "

-The proposed formulas give an accurate evaluation of this shift for single regulated turbines as well as for double regulated turbines ; in pump mode as well as in turbine mode , for most of the cases available . Additional results have to be analysed in order to confirm the proposed approach.

6-REFERENCES

1-IDA,T. New formula for efficiency step-up of hydraulic turbines
          XVII AIRH Symposium (1994)   BEIJING   CHINA
2-SPURK,J.H. and GREIN,H. Performances prediction of hydraulic machines by dimensionnal considerations
          Water Power & Dam Construction (1993-11) pp 42-49
3-COUSTON,M. ,MACHADO FERNANDES FILHO A. and TEMEZ RUIZ DIAZ P.P. Hydraulic performances of ITAIPU Francis turbines .
          XVII AIRH Symposium (1994)   BEIJING   CHINA

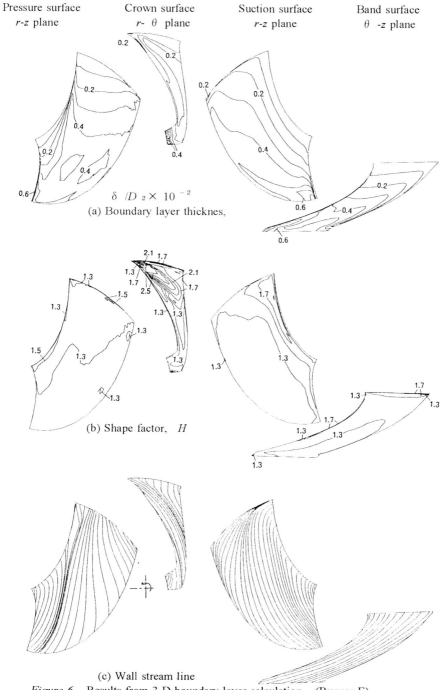

Pressure surface
r-z plane

Crown surface
r-θ plane

Suction surface
r-z plane

Band surface
θ-z plane

δ /$D_2 \times 10^{-2}$
(a) Boundary layer thicknes,

(b) Shape factor, H

(c) Wall stream line

Figure 6. Results from 3-D boundary layer calculation (Runner E)

5. Results and Discussions of Boundary Layer Calculation

To discuss the boundary layer characteristics, the contours of the equal boundary layer thickness δ, of the equal shape factors H, and the wall stream lines are illustrated in Fig.6 for the case of runner E. The boundary layer calculation were performed without any trouble, although reverse flow occurs near the crown wall of the pressure side in the main flow.

On the vane suction surface and on the band surface of the boundary layer grows from the leading edge, but on the vane pressure surface it grows from the entrance of the crown surface, since the main flow enters into the pressure surface through the crown surface as shown in Fig.6(a).

The Corioli's force works strongly in the main flow where the velocity is large, and balances with both the pressure gradient and the inertia force. Because the Corioli's force works weakly near the wall where flow is slow, the wall stream line turns to the opposite direction of Corioli's force, or turns to the direction in which the real pressure decreases if the pressure rise due to centrifugal force is removed as shown in Fig.4(c).

On the crown surface shown in Fig.6(b), the shape factor becomes very large near the corner between the inlet and the pressure surface where flow velocity is slow, and the boundary layer fluid is transferred along the wall stream line which deviated from the pressure side to the suction side between vanes. This flow behavior is mainly caused by the pressure gradient from the pressure side to the suction side rather than the Corioli's force.

In the suction surface of the vane, the boundary layer gradually grows toward the downstream near the leading edge, but the stream lines turns to the band side with an increase in the boundary layer thickness. This is because the pressure decreases from the crown side to the band side due to the meridian curvature of channel.

Using the present method, the hydraulic energy loss due to friction can be predicted by use of energy thickness of the boundary layer. The friction loss on the wall can be estimated as a energy defect at the channel exit. The energy defect is calculated from the energy thickness $\theta *$ at the exit surface for the case of three dimensional boundary layer.

$$\theta * = \frac{1}{U^3} \int_0^v V' \, (U^2 - V^2) \, dy \qquad (13)$$

Here, V' is the velocity component perpendicular to V, which is velocity in the boundary layer(Fig.2), on the exit section.

On the other hand, in the performance conversion, the ratio δ_{Er} of friction head to effective head is very convenient for standardization, as adopted in IEC Code or JSME Code. The ratio δ_{Er} is obtained by the following equation.

$$\delta_{Er} = \oint (U^3 \theta *) \, dl \, / \, gQH \qquad (14)$$

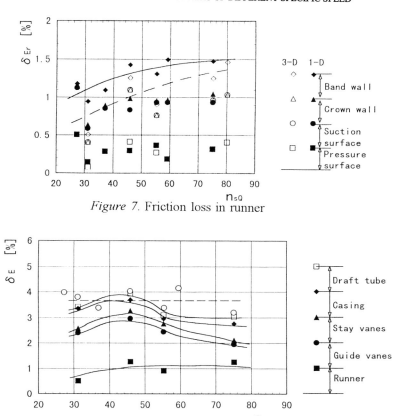

Figure 7. Friction loss in runner

Figure 8. Friction loss of turbine

Here, l is the distance along the contour of the exit wall. The curvilinear integration concerning l is performed along the exit wall edge of all channels of the runner.

The present calculation was applied to all turbine runners A-E and the calculated friction loss ratio δ_{Er} are plotted against specific speed n_{sQ} Fig. 7. The notations $\bigcirc\diamond\square\triangle$ are the results from the present calculation, and the notation $\bullet\blacklozenge\blacksquare\blacktriangle$ are those of the conventional two dimensional analysis, in which the flow is assumed to be along the grid line, and the energy equation and Rotta's equation are used.

The total friction loss in a runner becomes larger with an increase in the specific speed , but the scatter of the data is large because of the difference in the design philosophy and manufacturing tolerance of each turbine. The solid line indicates total friction loss obtained by two dimensional analysis.

Figure.7 shows that the total friction loss from two dimensional analysis becomes larger than that from the present analysis. Due to the secondary flow, the boundary layer thickness must be thinner in the thee-dimensional analysis than in the two dimensional analysis, because the fluid in the boundary layer is removed by the secondary flow in the 3Danalysis.

Figure.7 also reveals that the friction loss on the vane occupies large part of the total friction loss of a runner. On the other hand, on the crown surface, the main flow velocity is slow and the wall stream line leans. Therefore the boundary layer thickness in the exit section is thin and the friction loss becomes thus very small.

The friction loss ratio δ_E of the casing, stay vanes, guide vanes, and draft tube of turbines from A to D are shown in Fig.8. All the results except for loss of the runner are obtained by two dimensional calculation. Figure.8 reveals that the friction loss on guide vanes becomes smaller with an increase in specific speed. The notation $\bigcirc$ shows the sum of friction losses on all surfaces obtained by using two dimensional analysis. The dotted line is the averaged total scalable loss and is seen to be almost independent of the specific speed, and δ_E takes the value around 0.037.

6. Conclusions

The calculation method of three dimensional boundary layer developed by use of upstream FDM with local curvilinear coordinate system, is applied to five types of Francis turbines. The following conclusions are obtained.
(1) The present method can be solved without trouble, even if complicated flow including reverse flow is induced such as the flow in the pressure surface.
(2) The present method is able to estimate the influence of the secondary flow, which is perpendicular to the main flow. The friction loss obtained by this method becomes smaller than that obtained by two dimensional method.
(3) The averaged total friction loss ratio on all surfaces obtained by using two dimensional analysis is about 0.037 in Francis turbine, and it is almost independent of the specific speed.

Acknowledgements

Calculation was performed as a part of the activity of JSME WG "Performance Conversion Method for Hydraulic Turbines and Pumps". The main flow data for 3D Euler, 3D potential flow and 2D boundary layer data were offered by the committee members, Dr. Nagafuji and Mr. Suzuki in Toshiba Co.Ltd., Mr. Miura in Hitachi Co.Ltd. and Mr. Miyagawa in Mitsubishi Heavy Industry Co.Ltd.. The authors would like to express sincere gratitude to the WG members.

References
1. Arakawa, C., Samejima, M., et al., A 3D Euler solution of Francis runner using pseudo-compressibility, 3D-Computation of Incompressible Internal Flows (NNFM39), (1989),65-69
2. IEC draft code 4(CO)47-1987,Determination of the prototype performance from model tests of hydraulic machines with consideration of scale effects
3. JSME S008(1989), Performance conversion method for hydraulic turbines and pumps
4. Suzuki, R., An assessment of the loss distribution in Francis Turbines, Proc. of 18th IAHR Sympo., (1996)
5. Ida T., et al., Recent development of studies on scale effect in Japan, Proc. of 18th IAHR Sympo.,

SCALE EFFECT OF JET INTERFERENCE IN MULTINOZZLE PELTON TURBINES

YUJI NAKANISHI
TAKASHI KUBOTA
Kanagawa University
3-27-1, Kanagawa-ku, Yokohama 221, Japan

1. Introduction

The IEC model acceptance test code stipulates that there is no scale effect in Pelton turbines[1]. One of the primary reasons why the scale effect in Pelton turbines is ignored, so far, is that the complicated unsteady free water-sheet flow on Pelton buckets prevents to numerically analyze the loss mechanism. Another reason is that the positive scale effect in the closed conduit flow from the turbine inlet to the nozzle outlet tends to compensate for the negative scale effect in free water-sheet flow on the buckets.

In 1986, Grein and Spurk proposed a formula for efficiency conversion of Pelton turbines[2]. The formula directly converts the hydraulic efficiency considering the scale effects of Froude, Reynolds and Weber numbers. The origin of scale effect on the hydraulic efficiency of Pelton turbines, however, results primarily from the scale effect of energy coefficient ψ, and secondarily of the discharge coefficient ϕ and power coefficient Π. Therefore, the conversion of specific hydraulic energy, discharge and power shall theoretically precedes the conversion of the hydraulic efficiency. The Grein/Spurk formula can not convert the performance shifts neglecting the background of fluid dynamics, and the most detrimental scale effect of the jet interference.

The jet interference is strongly related to the spilt flow from the cutout. The free water-sheet flow on a bucket is apt to spill out of the cutout (CutFlow) under the operating condition of the higher speed factor with the larger needle stroke. This spilt flow may collide with the approaching jet out of the succeeding nozzle in the case of multi-nozzle units, namely, the jet interference. The collision of CutFlow with the approaching jet amplifies CutFlow with domino effect. Eventually, CutFlow is a trigger for the jet interference. Since the prototype CutFlow is larger than the model due to the scale effect of the jet, the jet interference shows very strong negative scale effect. Seriously, therefore, the jet interference can not be predicted by the model test.

The CutFlow increases with weakening of free jet in the course of nozzle outlet

E. Cabrera et al. (eds.), Hydraulic Machinery and Cavitation, 333–341.
© *1996 Kluwer Academic Publishers. Printed in the Netherlands.*

to the entrainment on the bucket. The weakened free jet leads to the decrease of jet velocity and the increase of jet diameter at the bucket landing. In the present study, the intensity of CutFlow from a bucket is tried to predict using the relative jet paths considering the weakened free jet velocity based on the scale effects of the conduit flow and of the free water-sheet flow on the buckets. The proposed CutFlow Intensity correlates well with the negative scale effect of the jet interference.

2. Closed Conduit Flow and Contraction Jet Diameter

2.1 POSITIVE SCALE EFFECT IN CLOSED CONDUIT FLOW

The mono-phase water flow is steady in the closed conduit of Pelton turbines from its inlet to the nozzle exit. All the hydraulic energy losses appear as a pressure loss in the closed conduit. While the kinetic energy deficiency $\delta_{C\text{-}ns}$ due to the branch loss and bend loss in the distributing pipe etc. is non-scalable, the wall friction deficiency δ_C is scalable depending on the Reynolds number R_e and relative roughness R_a/L as follows:

$$\delta_c = \Delta E_c / E = f(R_e, R_a / L) \tag{1}$$

where, ΔE_c and E denote the friction loss energy in the conduit and the specific hydraulic energy of the turbine. The prototype has smaller δ_c than the geometrically homologous model due to the difference in Reynolds number and relative roughness, in other words, the conduit flow has the 'positive' scale effect.

2.2. CONTRACTION DIAMETER OF FREE JET

The boundary layers developed in the closed conduit flow are confined close to nozzle and needle walls due to strong acceleration through the nozzle. The acceleration leads to the core flow structure at the exit of the nozzle as shown in Fig. 1 [3]. Since the most specific energy in the closed conduit flow is transformed to the kinetic energy, the core velocity becomes approximately $(2E)^{1/2}$. After that, the flow leaves the nozzle as a free jet. Since the jet becomes free from the solid wall, the low velocity region is rapidly accelerated up to the core velocity toward downstream. On the other

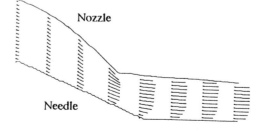

Figure 1. Velocity profiles in jet

hand, the boundary layer on the needle surface continues to develop in the jet, because the needle tip protrudes out of the nozzle, and finally the boundary layer is concentrated at the needle tip. Downstream of the needle tip, the low velocity region is also accelerated because of becoming free from the solid wall. Eventually, the uniform velocity distribution appears downstream of the needle tip at the circular cross-section of the jet. The uniform velocity is nearly equal to the throat core velocity, because the loss in the flow downstream the throat is negligibly small.

In the free jet, the increase of the jet velocity results in the decrease of the jet diameter from the mass continuity. This is the reason why the contraction of jet is generated at a little downstream of the needle tip. The boundary layer thickness at the nozzle tip is thinner in the prototype than in the model, because the thickness or the friction loss is dependent on the Reynolds number R_e, so that the prototype jet is not so contracted as the model jet under the same dimension less needle stroke $s_n(=S_n/D_T)$.

3. Scale Effect of Free Jet and Water-Sheet Flows

3.1. INCREASE OF FREE JET DIAMETER

Since the free jet travels through the air in the housing, the surface velocity of the jet is decelerated by the shear stress due to contact with the air flow. Since the free jet trys to retain its uniform velocity distribution across its cross section by dissipating its energy, the air resistance causes the jet velocity to weaken.

In addition, the water droplets splashed with the Pelton buckets exists around the free jet. Entraining the water droplets into the surface of the jet, also causes the kinetic energy of the jet to weaken. Since the amount of the water droplets rotating in the housing is proportional to the nozzle discharge, the kinetic energy loss in the jet depends on the needle stroke. After all, the jet velocity is gradually decreased toward downstream and the diameter is simultaneously increased to fulfill the mass continuity.

3.2. OPTIMUM JET DIAMETER AND INHERENT SCALE EFFECT OF JET

In general, a Pelton bucket has an optimum jet diameter d_{opt}, relating to its inner width B at optimum needle stroke $s_{n\text{-}opt}$. Although the prototype and its model have the same optimum diameter d_{opt}, the $s_{n\text{-}optP}$ for the prototype is smaller than $s_{n\text{-}optM}$ for the model because the jet diameter of the prototype is increased due to the scale effect in the closed conduit flow. Under the operating condition with smaller s_n than $s_{n\text{-}optP}$, the increased jet diameter of the prototype affects positively by approaching close to the optimum one. On the other hand, for large s_n, the increased jet diameter causes the spilt flow from the cutout CutFlow. Since the blade angle at the cutout is quite large than the angle at the brim, the amount of angular momentum $u_{cut}v_{u\text{-}cut}$ increases with increase of CutFlow. This leads to the decrease in specific energy of runner, or the appearance of the negative scale effect.

3.3. JET INTERFERENCE AND DOMINO EFFECT

Under the operating condition at the higher speed factor with the larger needle stroke, the free water-sheet flow in the bucket is apt to spill from the cutout. This CutFlow may collide with the approaching jet out of the succeeding nozzle in the case of a multinozzle unit more than 4 nozzles. The collision of the spilt flow with the approaching jet is called as the jet interference. The jet interference weakens the approaching jet, leads to the decrease in the velocity or the increase in the diameter. The weakened jet causes the delayed landing on the bucket, and that at the position near the cutout. This results in the more spilt flow from the cutout. Therefore, once the jet interference is occurred, the amount of the CutFlow is strikingly amplified with this chain reaction, namely, the domino effect.

Eventually, CutFlow is a trigger for the jet interference. Since the prototype CutFlow is larger than the model, the jet interference shows very strong negative scale effect. Seriously, therefore, the jet interference can not be predicted by the model test.

4. Modeling the Diameter of the Weakened Free Jet

In order to predict the intensity of CutFlow from a bucket using relative jet paths considering the scale effects of the conduit flow and of the free water-sheet flow on the buckets, the weakened jet diameter model is introduced as follows:

The free jet spouted from a nozzle, loses its kinetic energy or velocity gradually toward downstream. Then, the jet increases its diameter d_j ($\equiv D_j/D_T$, D_T denotes the nozzle tip diameter) to hold the continuity law. In this study, the increase in the jet diameter is represented by the following equation;

$$d_j = \left[1 + 0.001 \frac{\left(x_0 - x_j\right)}{d_0} + 0.001\left(\frac{d_0}{B/D_T}\right)^2 + 2.2 \times 10^3\left(\frac{V_{res}}{B^3}\right)^2\right] d_0 \qquad (2)$$

where, d_0 and x_0 denote the contraction diameter and its position respectively. The jet velocity c_j ($\equiv C_j/(2E)^{1/2}$) is calculated from the continuity equation using the jet diameter given by Eq. (2).

4.1. SCALE EFFECT ON CONTRACTION DIAMETER

As described in 2.2, the contraction diameter of the prototype d_{0P} is larger than the model due to the positive scale effect in the conduit flow. To take into account the effect of Reynolds number, the prototype contraction diameter is given by following equation;

$$d_{0P} = \left(\frac{R_{eP}}{R_{eM}} \right)^{\frac{1}{7}} d_{0M}$$

(3)

The weakened jet diameter for the prototype is calculated from Eq. (2) by substituting the obtained d_{0P} into the d_0.

4.2. EFFECT OF CONTACT WITH AIR ON FREE JET VELOCITY

Since the free jet travels through the air in the housing, the surface velocity of the jet is decelerated by shear stress due to contact with the air and the jet diameter is bloated toward downstream. Assuming the contribution of the travel to the increase in the jet diameter to be linearly proportional to the distance from the contraction position, the effect of resistance of the air on the jet velocity is considered as the second term in the right side of Eq. (2)

4.3. INCREASE IN JET DIAMETER DEPENDENT ON THE NEEDLE STROKE

The water droplets splashed with the Pelton buckets exists around the free jet. Entraining the water droplets into the surface of the jet, causes the jet diameter to bloat. Since the amount of the water droplets rotating in the housing is proportional to the nozzle discharge or the square of the jet diameter, the increase in the jet diameter dependent on the needle stroke is taken into account by the third term of Eq. (2).

4.4. MODELING OF THE DOMINO EFFECT CAUSED BY CutFlow

In a multinozzle turbine, once the spilt flow from the cutout CutFlow is occurred, CutFlow interferes with the succeeding jet and causes the jet diameter to bloat. The bloated jet diameter leads to the much more spilt flow from the cutout due to the domino effect. In order to consider the domino mechanism triggered by CutFlow, the contribution of CutFlow to the increase in the jet diameter is modeled by the fourth term of Eq.(2) so as to be proportional to the square of the residual jet volume V_{res} will be defined below.

4.5. DEDUCTION OF EMPIRICAL EQUATION
FOR MODELING THE DIAMETER OF THE WEAKENED JET

As described in (4.1) through (4.4), the increase in jet diameter is modeled by considering the effects of Reynolds number of the conduit flow, of contact with the air flow, of entrainment of the rotating water droplets and of the domino mechanism. Each coefficient of the terms in the Eq.(2) was determined so that the variation of the

CutFlow Intensity I_{cut} versus the speed factor and the needle stroke could represent the tendency of the deterioration of the model efficiency at the test head of 115m as shown in Fig. 2. To realize the remarkable jet inference, 17 buckets was selected. The specific speed of the model turbine B/D_{ref} was 0.35, where D_{ref} denotes the jet pitch diameter.

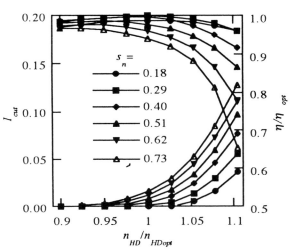

Figure 2. I_{cut} and model efficiency (H=115)

5. Relative Jet Path Considering the Weakened Jet Velocity

Suppose a splitter tip of a rotating bucket exits at an arbitrary position **M** in a jet as shown in Fig. 3. The position $\mathbf{I_r}$ of the jet particle of the impinging jet (ImpJet) which should catch up the splitter tip at the position **I**, can be calculated by integrating backward the jet velocity given from Eqs. (2) and (3) in the time $dt=MI/u$, where u denotes the peripheral velocity of the splitter tip. Similarly, The position $\mathbf{D_r}$ of the jet particle which passed the splitter tip at the position **D** just dt ago, can be also obtained. Since the

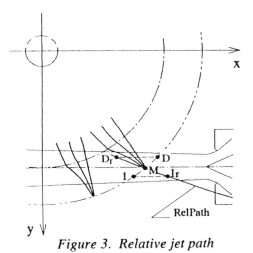

Figure 3. Relative jet path

passed jet is a little deviated due to contact with the bucket, we call this the deviated jet (DevJet). $\mathbf{MI_r}$ and $\mathbf{MD_r}$ both represent the parts of the relative jet path for the given bucket. While $\mathbf{MI_r}$ means the head of ImpJet which has not entered by this instant, **MDr** the head of ImpJet which has already entered by this instant. As for DevJet, while $\mathbf{MI_r}$ means the tail of the DevJet which will pass the bucket, $\mathbf{MD_r}$ the tail of DevJet which has already passed.

Let us consider the two adjacent buckets to investigate the relationship between the jet and the two buckets. If the following bucket of the two adjacent buckets is abbreviated to FolBuc, the preceding bucket to PreBuc. We can get FolPath for FolBuc and PrePath for PreBuc as shown in Fig. 4. Considering these two relative paths, the

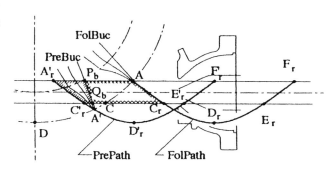

Figure 4. Fol- and Pre- relative paths

ImpJet for a bucket can be obtained as $A_r'C_r'C_rAA_r'$. The jet portion is designated as the responsible jet for a bucket V_{duty}. The portion $A_r'C_r'Q_bP_bA_r'$ of the responsible jet represents the entrained jet on the PreBuc at this instant. On the other hand, the remaining portion $P_bQ_bC_rAP_b$ represents the residual jet V_{res} that has not been entrained yet.

In order to predict the intensity of the jet interference, it is important to predict the amount of CutFlow quantitatively. However, the prediction of CutFlow is difficult and time-consuming. In this study, the residual jet related to the jet interference is to predict in lieu of the CutFlow. To predict the amount of the residual jet, the relative jet paths for PreBuc and FolBuc are drawn at a flash when the PreBuc first touches the second jet as shown in Fig. 5. Since the second jet starts to enter into the PreBuc after the flash, the residual portion $B_rQ_bC_rB_r$ of the first jet collides with the second jet in the PreBuc, and may be discharged as CutFlow. Therefore, A new term of CutFlow Intensity I_{cut} that is defined as the ratio of the residual jet volume V_{res} to the responsible jet volume V_{duty} is introduced to assess the jet interference.

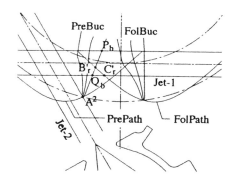

Figure 5. Residual jet volume

6. Comparisons of CutFlow Intensity with the Model Efficiency

The CutFlow Intensity I_{cut} was calculated using the relative jet paths considering the weakened jet diameter by Eqs. (2) and (3) under the operating conditions of the needle strokes s_n of 0.4 and 0.73, the heads of 40, 60, 90 and 115m, and 9 speed factors n_{HD}/n_{HDopt} from 0.9 through 1.1. The variations of obtained CutFlow Intensity I_{cut} versus n_{HD}/n_{HDopt} are shown in the lower parts of Fig. 6 for $s_n = 0.73$ and Fig. 7 for $s_n = 0.4$.

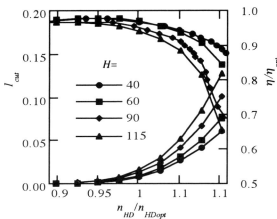

Figure 6. I_{cut} and model efficiency (s_n=0.73)

From the both figures, it is followed that the larger the n_{HD}/n_{HDopt} is, the larger Cutflow intensity I_{cut} becomes. The reason is that the large n_{HD}/n_{HDopt} caused the residual jet V_{res} to increase and the contribution of fourth term in the right side of Eq. (2) to the increase of the jet diameter was strongly intensified. In other words, the fast rotation of the runner made the jet catching up the bucket difficult. Since the higher heads causes the larger contraction diameter d_0 of Eq. (2), this tendency became outstanding at the larger heads. The CutFlow Intensity I_{cut} is larger at s_n=0.73 than at s_n=0.4 due to the effect of entrainment of

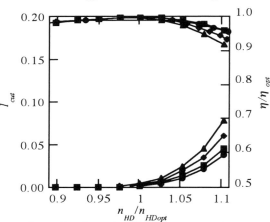

Figure 7. I_{cut} and model efficiency (s_n=0.40)

water droplets considered by the third term of Eq. (2).

In order to ascertain the relationship between I_{cut} and the intensity of the jet interference, the CutFlow Intensity I_{cut} was compared with the model efficiency deteriorated by the jet interference. The jet pitch diameter D_{ref} of the model runner is 0.325m, the bucket inner width B 0.114m and the number of the bucket 17. The uncertainty of the efficiency measurements is +-0.2% with the 95% confidence level.

In the upper parts of Figs. 6 and 7, the variations of measured efficiency in the model tests were shown, where n_{HDopt} and η_{opt} denote the speed factor and the efficiency at the optimum operating condition, respectively. The jet interference was not recognized only at the operating condition of H=40m and s_n=0.40. In that case, the efficiency curve is roughly symmetric to the vertical line through the best efficiency point.

The model efficiency is deteriorated at the operating condition of high speed factor with high test head due to the jet interference. The higher the test head is, the stronger the deterioration of the efficiency is. This reveal that the jet interference has the strong negative scale effect. This tendency is remarkable at the operating condition of s_n=0.73. The variation of the CutFlow Intensity I_{cut} well demonstrates the tendency of the model efficiency deteriorated due to the scale effect.

7. Conclusions

The scale effect in a Pelton turbine has not been investigated so far. Since the jet interference has very strong negative scale effect, the intensity of it in the prototype can not be predicted by the model tests. The jet interference is strongly related to the spilt flow from the cutout CutFlow, and so the intensity of CutFlow from a bucket was tried to predict using the relative jet paths considering the weakened free jet velocity based on the scale effects of the conduit flow and of the free water-sheet flow on the buckets. The proposed CutFlow Intensity correlated well with the negative scale effect of the jet interference in model tests.

References

1. IEC (1994), 4/111/CDV .
2. Grein,H. et al. (1986), Efficiency scale effects in Pelton turbines, IAHR-Montreal, 76.
3. Nonoshita T.,Matsumoto Y., Haneda Y. and Kubota T. (1993), Behavior of the Jet from a Pelton Turbine Nozzle, Proc. 4th Asian Int. Conf. Turbo Machinery, Suzhou, 499-503.

FURTHER DEVELOPMENT OF STEP-UP FORMULA CONSIDERING SURFACE ROUGHNESS

Alois Nichtawitz
Voest Alpine M.C.E.
Linz, Austria

Abstract: The paper is based on the proposal for a new step-up formula made by the author at the IAHR Symposium 1994 in Beijing. In its first part an upgrading of proposed formula to also consider influence of surface roughness is derived. In a second step it is demonstrated that the scalable loss S_0 which is originally a constant can be extended to a function of flow and specific hydraulic energy. Efficiency step-up at a constant flow coefficient and geometry automatically involves a change of specific hydraulic energy, so a formula to describe this effect is also given in the paper, which is concluded by a set-up for the power shift which can be derived from the above. Summarizing the paper offers a complete set of formulae describing a method for the transposition of characteristics from model to prototype.

1. Introduction

A large variety of step-up procedures has been developed in the past. With the code IEC 995 a procedure for efficiency step-up in reaction machines is effective since 1991. It seems that it is not accepted on a worldwide basis mainly because of its complicated structure.

IEC 995

$$\Delta\eta = \delta_{ref} \cdot \left[\left(\frac{R_{ref}}{Re_M} \right)^{0.16} - \left(\frac{R_{ref}}{Re_P} \right)^{0.16} \right]$$

$$\delta_{ref} = \frac{1 - \eta_M}{\left(\dfrac{Re_{ref}}{Re_M} \right)^{0.16} + \dfrac{1 - V_{ref}}{V_{ref}}}$$

During the IAHR Symposium 1994 in Beijing various existing step-up formulae and new approaches were discussed by the author. In the course of the paper a new step-up formula was presented and discussed:

NEW PROPOSAL

$$\Delta\eta = S_0 \cdot \left[\left(\frac{Re_0}{Re_M} \right)^{n} - \left(\frac{Re_0}{Re_P} \right)^{n} \right]$$

342

E. Cabrera et al. (eds.), Hydraulic Machinery and Cavitation, 342–351.
© 1996 *Kluwer Academic Publishers. Printed in the Netherlands.*

Based on the set-up as presented attempts were made by the author to further develop the step-up procedure in four directions.

- Implementation of different surface roughness at model and prototype.
- Discussion on the scalable loss S_0 and its upgrading to a function dependent on φ and ψ.
- Development of a step-up formula for the energy coefficient ψ.
- From the above a step-up formula for the power output / consumption can be derived.

2. Surface Roughness

As a starting point to extend the formula presented in Beijing the following equation by Colebrook and White was taken which is describing the friction coefficient also in the transition zone between smooth and fully rough pipe flow.

$$\frac{1}{\sqrt{\lambda}} = -2 \cdot \log\left(0.27 \cdot \frac{K_S}{D} + \frac{2.51}{Re \cdot \sqrt{\lambda}}\right)$$

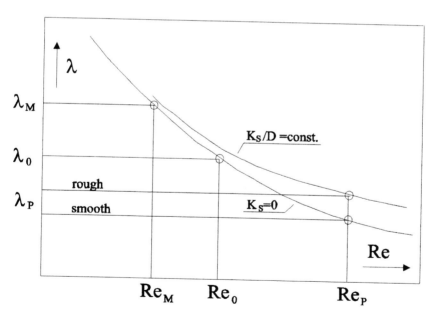

Fig. 1 Friction coefficient λ depending on Re and K_S/D

This is an implicit formulation for λ which can be converted by some tricky transformations and small simplifications to the following explicit shape as demonstrated in the following.

$$\frac{1}{2 \cdot \sqrt{\lambda_0}} - \frac{1}{2 \cdot \sqrt{\lambda}} = \log\left(0.18 \cdot \frac{K_S}{D} \cdot Re_0 \cdot \sqrt{\lambda_0} + \frac{Re_0}{Re} \cdot \frac{\sqrt{\lambda_0}}{\lambda} \right)$$

$$\frac{10^{\left(\frac{1}{2 \cdot \sqrt{\lambda_0}} - \frac{1}{2 \cdot \sqrt{\lambda}} \right)}}{\sqrt{\frac{\lambda_0}{\lambda}}} = 0.108 \cdot \frac{K_S}{D} \; Re_0 \cdot \sqrt{\lambda} + \frac{Re_0}{Re} \cong \left(\frac{\lambda - \lambda_\infty}{\lambda_0 - \lambda_\infty} \right)^n$$

a best fit is given by $\lambda_\infty = 0.00222$ and $n = 5.0$

$$\frac{\lambda}{\lambda_0} = 0.74 \cdot \left(0.108 \cdot \frac{K_S}{D} \cdot Re_0 \cdot \sqrt{\lambda} + \frac{Re_0}{Re} \right)^{0.2} + 0.26$$

$$\lambda = 0.00632 \cdot \left(C_R + \frac{Re_0}{Re} \right)^{0.2} + 0.00222 \quad \text{with} \quad C_R = 0.108 \cdot \frac{K_S}{D} \cdot Re_0 \cdot \sqrt{\lambda}$$

Friction losses in a hydraulic machine cannot be modeled just by frictional losses in a pipe but also friction at a flat plate and disc friction contribute to the total scalable losses due to friction. In a quite similar way a formulation by A. Fay describing friction coefficients for a flat plate can be simplified and brought to an explicit structure too.

$$\frac{1}{C_f^{0.4}} = -1.62 \log\left(\frac{1}{14.7} \cdot \frac{K_S}{L} + A \cdot \frac{0.241}{Re \cdot \sqrt{C_f}} \right)$$

$$\frac{C_f}{C_{f0}} \cong 0.81 \cdot \left(C_R + \frac{Re_0}{Re} \right)^{0.2} + 0.19$$

$$C_f \cong 0.0025 \cdot \left(C_R + \frac{Re_0}{Re} \right)^{0.2} + 0.0006 \quad \text{with} \quad C_R = 0.28 \cdot \frac{K_S}{L} \cdot Re_0 \cdot \sqrt{C_f}$$

To the knowledge of the author there is no similar formula for the disc friction but it is reasonable to assume that disc friction also can be described with sufficient accuracy in an explicit shape in quite a similar way.

$$\frac{C_m}{C_{m0}} = B \cdot \left(C_R + \frac{Re_0}{Re} \right)^{0.2} + (1 - B)$$

Taking the same explicit structure for all types of friction (pipe, plate, disc) and with $\Delta \eta \hat{=} (\lambda_M - \lambda_P)$ but also $\Delta \eta \hat{=} (C_{fM} - C_{fP})$ and $\Delta \eta \hat{=} (C_{mM} - C_{mP})$ there follows

$$\Delta \eta = S_0 \cdot \left[\left(C_{R,M} + \frac{Re_0}{Re_M} \right)^{0.2} - \left(C_{R,P} + \frac{Re_0}{Re_P} \right)^{0.2} \right]$$

In case of hydraulic smooth surfaces at model and prototype the terms C_{RM} and C_{RP} become zero and the formula is reduced to the one as presented in Beijing. $C_R=1$ corresponds to a roughness where no step-up will occur even at an infinite Reynolds number of prototype. It has to be mentioned that there is a difficult question how to determine an arithmetic mean roughness equivalent to a sand roughness K_S.

The quality of surface roughness of model machines normally is excellent so C_{RM} can be set zero. For the prototype machines it is proposed to introduce classes of roughness similar to the ones established for machining / grinding. It is advisable to follow a geometric sequence starting with $C_{RP} =1$, 0.5, 0.25, 0.125, 0.063, 0.032, 0.016 and finally zero.

Particularly at runner replacement projects the surface roughness of existing machine components may be far away from hydraulically smooth conditions which makes it absolutely necessary to also consider surface roughness.

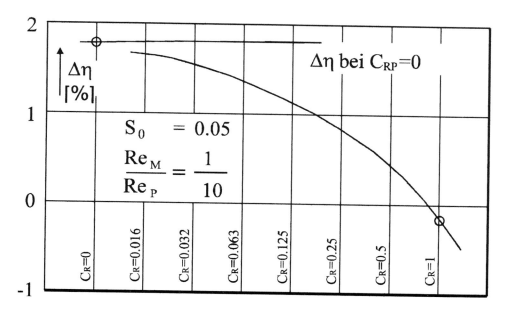

Fig. 2 Efficiency step-up for different surface roughness

Also this chart is a helpful tool for judgment if an improvement of roughness of remaining components should be made due to economic reasons. In many cases it will lead to the conclusion that it is economic to improve surface roughness of old machines. The increase in efficiency reached then could be even higher than by a newly designed runner however at lower costs.

The new formula also opens the possibility to calculate a step up for the frequent case that just the runner is replaced but nothing is done at surrounding components.

$$\Delta\eta = \left(S_0 - S_{0,RU}\right) \cdot \left[1 - \left(C_{R,P} + \frac{Re_M}{Re_P}\right)^{0.2}\right] + S_{0,RU}\left[1 - \left(\frac{Re_M}{Re_P}\right)^{0.2}\right]$$

Numerical example:

$S_0 = 0.05$ $\dfrac{Re_M}{Re_P} = \dfrac{1}{10}$ $C_{RP} = 0.5$ in general, just the runner with $C_{RP} = 0$

$S_{0,RU} = 0.02$

$\Delta\eta = 0.0029 + 0.0074 = 0.0103 \Rightarrow \Delta\eta = 1.03\ \%$

compared to

$\Delta\eta = 1.85\ \%$ at smooth surfaces at all components including runner
$\Delta\eta = 0.49\ \%$ at surfaces with class 0.5 at all components including runner

3. Scalable loss S_0

The scalable loss So is related to flow and specific hydraulic energy at peak efficiency. This value is a constant in the original approach depending on specific speed of machine which means that the step up derived therefrom is a constant adder independent of point of operation. Obviously it is very unlikely that this is the case in reality. In fact S_0 depends on coefficients of flow and specific hydraulic energy as well.

The losses due to friction in a hydraulic machine occur in all of the components and can be classified after the place where they are originated.

These main components
and losses are:

Losses due to main flow in the various components contribute to a loss in specific hydraulic energy. A different class of losses is given by disc friction and seal leakage resulting in a power loss.

- spiral case   - draft tube
- stay vanes
- wicket gates   - leakage
- runner   - disc friction

For sake of simplicity the procedure is outlined in the following by example of scalable losses in a spiral case.

$$\Delta\eta_{SP} = \left(\lambda_M - \lambda_P\right) \cdot \left(\frac{L}{D}\right)_{SP} \cdot \left(\frac{D}{D_{SP}}\right)^4 \cdot \frac{\varphi}{\psi_0}^2 \quad \left(\frac{L}{D}\right)_{SP} \text{ equivalent relative length of spiral case}$$

The friction coefficient λ has to be calculated according to the Reynolds number in the spiral case

$$Re_{SP} = \frac{c_{SP} \cdot D_{SP}}{\nu} \qquad\qquad Re = \frac{u \cdot D}{\upsilon}$$

$$\frac{\lambda}{\lambda_0} = 0.74 \cdot \left(0.108 \cdot \frac{K_S}{D_{SP}} \cdot Re_0 \cdot \sqrt{\lambda} + \frac{Re_0}{Re_{SP}}\right)^{0.2} + 0.26$$

$$\frac{\lambda}{\lambda_0} = 0.74 \cdot \left(\frac{D_{SP}}{D} \cdot \frac{1}{\varphi}\right)^{0,2} \cdot \left(\underbrace{0.108 \frac{K_S}{D_{SP}} \cdot \frac{D}{D_{SP}} \cdot Re_0 \cdot \sqrt{\lambda} \cdot \varphi}_{C_R} + \frac{Re_0}{Re}\right)^{0.2} + 0.26$$

The term $0.108 \cdot \frac{K_S}{D} \cdot \left(\frac{D}{D_{SP}}\right)^2 \cdot Re_0 \cdot \sqrt{\lambda} \cdot \varphi$ corresponds to a class of roughness C_R.

For the example of a spiral case the step-up can be written in the following way:

$$\Delta\eta_{SP} = S_{0,SP}\left(\varphi, \Psi\right) \cdot \left[\left(C_{RM} + \frac{Re_0}{Re_M}\right)^{0.2} - \left(C_{RP} + \frac{Re_0}{Re_P}\right)^{0.2}\right]$$

The term $S_{0,SP}$ in front of the brackets represent the scalable losses of a spiral case at the reference Reynolds number Re_0 and with hydraulically smooth surfaces. Assuming that various components show equivalent quality of roughness the term C_R is a constant. Equivalent means that with smaller hydraulic diameters of a component the surface roughness must be smaller which is in agreement with the practice. Therefore the roughness has to be chosen carefully respecting the individual Reynolds numbers and hydraulic diameters within the water passage. In this case the individual partial step-ups can be summed up to

$$\Delta\eta = S_0\left(\varphi, \Psi\right) \cdot \left[\left(C_{RM} + \frac{Re_0}{Re_M}\right)^{0.2} - \left(C_{RP} + \frac{Re_0}{Re_P}\right)^{0.2}\right]$$

4. Energy Coefficient

In the next step an attempt is made to develop a formula describing the step-up of energy coefficients which is strictly linked to the efficiency step-up at a given flow coefficient φ and a given geometry.

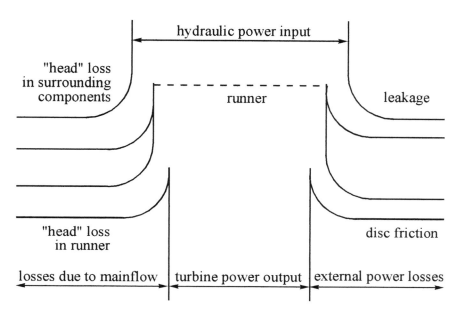

Fig. 3 Flow chart for losses in a turbine

$$g \cdot H = \Sigma\, g \cdot \Delta H_{HL} + g \cdot H_{RU} \qquad\qquad \Delta u \cdot c_u = \frac{M_{RU} \cdot \omega}{\rho \cdot Q_{RU}} = g \cdot H_{RU} \cdot \eta_{RU}$$

$$M = M_{RU} - M_{DF} = M_{RU} \cdot \eta_{DF} \qquad\qquad Q = Q_{RU} + Q_{LF} = \frac{Q_{RU}}{\eta_{LF}}$$

$$1 = \frac{\Sigma\, g \cdot \Delta H_{HL}}{g \cdot H} + \frac{\eta}{\eta_{RU} \cdot \eta_{DF} \cdot \eta_{LF}} \qquad\qquad \begin{array}{cc} \text{main flow} & \text{external} \\ \Rightarrow \eta = (\eta_{HL} \cdot \eta_{RU}) \cdot (\eta_{DF} \cdot \eta_{LF}) \end{array}$$

HL...   head losses in surrounding components
RU...   head loss in runner
DF...   disc friction
LF ...   leakage flow

Starting from Euler's equation and introducing φ and ψ one can derive from simple velocity triangles at runner inlet and outlet the following equations.

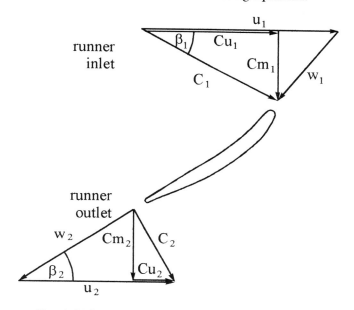

Fig. 4 Velocity triangles at runner inlet and outlet

$$u_1 \cdot c_{u1} - u_2 \cdot c_{u2} = g \cdot H_{RU} \cdot \eta_{RU} = g \cdot H \cdot \eta_{HL} \cdot \eta_{RU} = g \cdot H \cdot \frac{\eta}{\eta_{DF} \cdot \eta_{LF}}$$

$$tg\beta_1 = \frac{c_{m1}}{u_1 - c_{u1}} \qquad tg\beta_2 = \frac{c_{m2}}{u_2 - c_{u2}}$$

$$ctg\beta_2 \cdot \frac{c_{m2}}{u} \cdot \frac{u_2}{u} - ctg\beta_1 \cdot \frac{c_{m1}}{u} \cdot \frac{u_1}{u} + \frac{u_1^2 - u_2^2}{u^2} = \frac{g \cdot H}{u^2} \cdot \frac{\eta}{\eta_{DF} \cdot \eta_{LF}}$$

$$\varphi \cdot \left(a \cdot ctg\beta_2 - b \cdot ctg\beta_1\right) + c = \frac{\psi}{2} \cdot \frac{\eta}{\eta_{DF} \cdot \eta_{LF}}$$

$$\varphi \cdot f\left(C_R + \frac{Re_0}{Re}\right) + c = \frac{\psi}{2} \cdot \frac{\eta}{\eta_{DF} \cdot \eta_{LF}}$$

The constants a, b and c depend on the geometry of a hydraulic machine. On the other hand $f(\beta_1, \beta_2)$ is a function of flow angles in front of and behind a rotating runner and is assumed to depend slightly on Reynolds number Re and surface roughness C_R. This function changes slightly only when varying surface roughness and Reynolds number therefore a Taylor progression can be carried out.

$$\varphi \cdot \left[\frac{f(1)}{0!} + \frac{f'(1)}{1!} \cdot \left(C_R + \frac{Re_0}{Re} - 1 \right) + \frac{f''(1)}{2!} \cdot \left(C_R + \frac{Re_0}{Re} - 1 \right)^2 + .. \right] + c = \frac{\Psi}{2} \cdot \frac{\eta}{\eta_{DF} \cdot \eta_{LF}}$$

The Taylor progression can be stopped after the second term.

$$\varphi \cdot \left[\frac{f(1)}{0!} + \frac{f'(1)}{1!} \cdot \left(C_R + \frac{Re_0}{Re} - 1 \right) \right] + c = \frac{\Psi}{2} \cdot \frac{\eta}{\eta_{DF} \cdot \eta_{LF}}$$

Applying above equations to model and prototype conditions one gets

$$\frac{1 + h \cdot \left(C_{R,P} + \frac{Re_0}{Re_P} \right)}{1 + h \cdot \left(C_{R,M} + \frac{Re_0}{Re_M} \right)} = \frac{\Psi_P}{\Psi_M} \cdot \frac{\eta_P}{\eta_M} \cdot \frac{(\eta_{DF} \cdot \eta_{LF})_M}{(\eta_{DF} \cdot \eta_{LF})_P}$$ and consequently

$$\frac{\Psi_P}{\Psi_M} = \frac{\eta_M}{\eta_P} \cdot \frac{(\eta_{DF} \cdot \eta_{LF})_P}{(\eta_{DF} \cdot \eta_{LF})_M} \cdot \frac{1 + h \cdot \left(C_{RP} + \frac{Re_0}{Re_P} \right)}{1 + h \cdot \left(C_{RM} + \frac{Re_0}{Re_M} \right)}$$

in case of $C_{RM} = 0$, $Re_M = Re_0$ and $\eta_{LF,P} = \eta_{LF,M}$ there follows

$$\frac{\Psi_P}{\Psi_M} = \frac{\eta_M}{\eta_P} \cdot \frac{\eta_{DF,P}}{\eta_{DF,M}} \cdot \frac{1 + h_{TU} \cdot \left(C_{R,P} + \frac{Re_M}{Re_P} \right)}{1 + h_{TU}}$$ for generating mode

$$\frac{\Psi_P}{\Psi_M} = \underbrace{\frac{\eta_P}{\eta_M} \cdot \frac{\eta_{DF,M}}{\eta_{DF,P}}}_{\text{head loss}} \cdot \underbrace{\frac{1 + h_{PU} \cdot \left(C_{R,P} + \frac{Re_M}{Re_P} \right)}{1 + h_{PU}}}_{\text{change of runner/impeller head}}$$ for pumping mode

It is assumed that the coefficient h strongly depends on the specific hydraulic design e.g. number and geometry of wicket gates and of runner vanes/blades as well.

It is also suggested to analyze and compare corresponding points from model and field tests in order to determine the value of h. Excellent homology between model and prototype is necessary to get reliable results.

5. Power

Knowing the step-up of efficiency together with the shift of specific hydraulic energy between model and prototype conditions one can calculate now the power at a given value of phi. The power to be achieved operating a turbine at corresponding points can be written in the following way

$$P_P = \varphi \cdot \Psi_P \cdot \eta_P \cdot n_P^{3} \cdot D_P^{5} \qquad P_M = \varphi \cdot \Psi_M' \cdot \eta_M \cdot n_M^{3} \cdot D_M^{5}$$

Using formula for step-up of specific hydraulic energy coefficient one gets

$$P_{P,TU} = P_{M,TU} \cdot \left(\frac{n_P}{n_M}\right)^{3} \cdot \left(\frac{D_P}{D_M}\right)^{5} \cdot \frac{1 + h_{TU}\left(C_{RP} + \dfrac{Re_M}{Re_P}\right)}{1 + h_{TU}} \cdot \frac{(\eta_{DF} \cdot \eta_{LF})_P}{(\eta_{DF} \cdot \eta_{LF})_M}$$

For many types of hydraulic reaction machines differences in disc friction and in leakage losses can be neglected. In this case power step-up can be well estimated by testing homologous machines at a given φ-value just at two different Reynolds numbers whereby efficiencies do not appear in above equation (!)

6. Conclusion

The paper offers a complete set of formulae for the problem of transposition from model to prototype conditions. The procedures cover efficiency step-up considering the influence of surface roughness and demonstrate how the proposed formula can be extended in order to calculate efficiency step-up also in regions off the peak condition. It also outlines a new approach how to predict the shift in specific hydraulic energy between model and prototype machines together with ist effect to the power output respectively power consumption.

The presented paper is considered also to stimulate the work in WG 5 of IAHR. Based on the proposals worldwide experiences from various manufacturers should be collected in order to determine the parameters for an optimum fit with the experience.

The main goal of all step-up activities is to better predict the prototype performance, however, dealing with this subject certainly also will improve knowledge in loss distribution of hydraulic machines, which again will stimulate the future progress in technology of hydro machinery.

7. References

- IEC Publication 995: Determination of the prototype performance from model acceptance tests of hydraulic machines with consideration of scale effects, 1991
- Nichtawitz, A.: Discussion on step-up procedures in hydraulic machines, IAHR Symposium, Beijing 1994

NUMERICAL SIMULATION OF JET IN A PELTON TURBINE

TOMOYASU NONOSHITA
Dept. of Mechanical Engineering, Sophia University
7-1, Kioicho, Chiyoda-ku, Tokyo 102, Japan
YOICHIRO MATSUMOTO
Dept. of Mechanical Engineering, The University of Tokyo
7-3-1, Hongo, Bunkyo-ku, Tokyo 113, Japan
TAKASHI KUBOTA
Dept. of Mechanical Engineering, Kanagawa University
3-27-1, Rokkakubashi, Kanagawa-ku, Yokohama 221, Japan
HIDEO OHASHI
Dept. of Mechanical System Engineering, Kogakuin University
1-24-2, Nishi-shinjuku, Shinjuku-ku, Tokyo, 163-91, Japan

1. Abstract

In case of predicting the performance of prototype Pelton turbine from a model test, a scale effect has been ignored so far.   However, it is getting clear that there is an outstanding deviation from a classical similarity laws.   The direction of the deviation is completely inverse against the case of reaction turbines, namely, the performance of prototype is worse than the predicted one.   In order to analyze this negative scale effect, the behavior of jets should be considered more precisely. In this study, the unsteady jet issuing from a Pelton turbine nozzle was calculated numerically using HSMAC method under the assumption of axisymmetric flow. The calculation shows that the needle considerably affects the velocity distribution in the jet and the velocity distribution in the jet changes by the head.

352

E. Cabrera et al. (eds.), Hydraulic Machinery and Cavitation, 352–360.
© *1996 Kluwer Academic Publishers. Printed in the Netherlands.*

2. Nomenclature

D	: diameter of nozzle exit	ρ	: density
h	: free surface height from axis	σ	: surface tension
p	: pressure		
Q	: volumetric flow rate	subscript	
r, z	: radial and axial coordinates	a	: air
t	: time	d	: droplet
v	: velocity	s	: free surface
θ	: incline angle of nozzle exit	j	: jet
μ	: viscosity	nz	: nozzle
ν	: kinetic viscosity	r, z	: radial and axial component

3. Introduction

In Pelton turbines, the mechanism of motion has been simply explained by using the angular momentum theory for time averaged flow. However, the actual flow is very complicated due to the periodic change of relative position between a jet and a bucket(Bachmann et al., 1990) and the interaction among jets(Kubota, 1989). It is found that there is an outstanding deviation from the classical similarity law between the model test data and the prototype performance. Various reasons can be considered with respect to the scale effect which is caused by the complicated flow. Grein et al.(1986) made it clear that the specific flow rate, Reynolds number, Froude number and Weber number have to be considered for the scale effects in Pelton turbines by dimensional analysis. Kubota and Nakanishi(1994) classified the flow in Pelton turbines for the sake of studying the scale effect.

In this study, we focus on the jet from a Pelton turbine nozzle as a basic research in order to obtain a hint about the scale effect. The diameter of the jet is getting larger after the contraction because of the change of velocity distribution. Additionally, there are lots of water droplets in the casing which are generated by the reflection of jet at the buckets. Some of them are entrained into the jet, and it causes the momentum loss and the enlargement of jet diameter.

Behavior of jets has been investigated concerning with not only Pelton turbines, but also water jet cutting.   The length of jet which maintains its continuity
(so-called breakup length) was investigated by Phinney(1973).   He extended the
laminar jet stability theory, which is derived from Weber's theory, into the turbulent jet.   Hoyt et al.(1977) observed the complicated jet surface using high-speed
photography.   They showed that the instabilities occurs in high Reynolds number water jet.   These instabilities include the axisymmetric mode accompanying
the transition from laminar to turbulent flow at the nozzle exit, spray formation as
a culmination of the axisymmetric disturbances, and further down stream, helical
disturbances which result in the entire jet assuming a helical form.   In regard to
a Pelton turbine jet, Guilbaud et al.(1992) measured the head distribution inside
the jet under two kinds of conditions.   One is the jet from a secondary branch of
the wye-piece and another is the jet from a straight nozzle.   The former one has a
curvature effect imposed to the flow (deviated flow) and the latter is axisymmetric
flow (non-deviated flow).   The measurements disclosed that the jet from a deviated flow leads to a divergence located on the curvature inner side and it induces
a smaller head pressure than that measured on the curvature outer side.

4. Numerical Analysis Method

4.1. GOVERNING EQUATIONS

The jet issuing from a Pelton turbine
nozzle which is shown in Fig.1 is
numerically analyzed in this study.
The shape of the nozzle is as same
as that of the former report
(Nonoshita et al., 1994).   The diameter of the nozzle exit is 52mm
and that of the needle at the nozzle
exit is 20mm when the needle is in
its normal position. The conical angle of the needle tip is 50 degrees,
and the incline angle of the nozzle

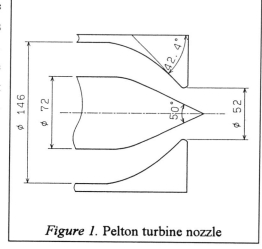

Figure 1. Pelton turbine nozzle

TABLE 1. Scale-up Formulae

Proposer	Formula	
JSME S008 (1989)	$$\frac{1 - \eta_{E\,P}}{1 - \eta_{E\,M}} = \frac{\eta_{E\,P}}{\eta_{E\,M}}\{(1 - V) + V\Lambda\}$$	(1)
	$$\Lambda = (D_P/D_M)^{0.18}\left\{(1 - \beta) + \beta\,(e_P/e_M)^{0.16}\right\}$$	(2)
IEC 995 (1991)	$$\Delta\eta_h = \delta_{ref}\left[\left(\frac{Re_{u\ ref}}{Re_{u\ M}}\right)^{0.16} - \left(\frac{Re_{u\ ref}}{Re_{u\ P}}\right)^{0.16}\right]$$	(3)
	$$\delta_{ref} = \frac{1 - \eta_{h\ opt\ M}}{\left(\frac{Re_{u\ ref}}{Re_{u\ opt\ M}}\right)^{0.16} - \frac{1 - V_{ref}}{V_{ref}}}$$	(4)
Spurk/Grein (1992)	$$\Delta\eta_h = (\eta_{h\ \infty} - \eta_{h\ M})\left[1 - \left(\frac{Re_M}{Re_P}\right)^{\alpha}\right]$$	(5)
Ida (1993)	$$\frac{\eta_{E\,P}}{\eta_{E\,M}} = \frac{1}{1 - \delta_E\left(1 - \Lambda'\right)}$$	(6)
	$$\Lambda = (D_P/D_M)^{-0.18}\left\{(1 - \beta) + \beta\left(e_P/e_M\right)^{0.18}\right\}$$	(7)
Nichtawitz (1994)	$$\Delta\eta_h = S_0\left(\frac{Re_0}{Re_M}\right)^n - \left(\frac{Re_0}{Re_P}\right)^n$$	(8)

Note; Some formulae include the conversion of operating condition as well as the efficiency step-up procedure for pumping operation. But only the efficiency step-up formulae for turbine are contrasted here.

η_E in the formulae by JSME and Ida is the specific energy efficiency, the ratio of the specific hydraulic energy available for the runner to the specific hydraulic energy of the machine (see equations (9) and (10) below).

cific speed $n_{sQ} = n\,Q^{0.5}/H^{0.75}$ is around 40 (min^{-1}, m^3/s, m). Runners A and B have the same number of blades but Runner A shows higher efficiency. Runners C and D have smaller number of blades, and shows lower efficiency than Runners A and B. Here, η_E denotes specific energy efficiency;

$$\eta_E = E_m / E = 1 - E_{L12} / E \qquad (9)$$

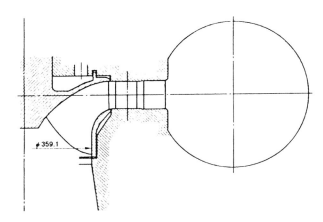

⌀ 359.1

FIGURE 1. Sectional View of the Specified Model Turbine

TABLE 2. Specifications of Runners

Runner	A	B	C	D
D (mm)	359.1	359.1	359.1	359.1
Re	6.3×10^6	6.3×10^6	6.3×10^6	6.3×10^6
n_{SQopt}	40.4	38.3	37.0	35.6
$n_{ED} / n_{ED\,(A)}$	1.00	1.00	1.00	1.00
$Q_{ED} / Q_{ED\,(A)}$	1.00	1.00	0.963	0.969
n_{ED} / n_{EDopt}	1.00	1.03	1.00	1.05
Q_{ED} / Q_{EDopt}	1.00	1.04	1.15	1.12
η_E (%)	95.0	93.6	92.0	91.3
η_{Eopt} (%)	95.0	93.7	92.4	91.9
Z_R	13	13	9	7

where E_m, E_{L12} and E indicate specific hydraulic energy of the runner, specific energy loss of the machine and specific hydraulic energy of the machine respectively. Therefore;

$$\eta_E = \eta_h \, / \, \eta_Q \, \eta_R \qquad (10)$$

where η_h, η_Q and η_R are hydraulic (internal) efficiency, discharge (volumetric) efficiency and power efficiency (related to disk friction) respectively.

The values n_{ED} and Q_{ED} listed in Table 2 without subscript are the speed factors and discharge factors of the operating conditions analyzed. Those with subscript (A) indicate the values for Runner A, and subscripts opt denotes the best efficiency point of each runner. Runner A is analyzed at the best efficiency point. Analyses for other runners are carried out at the same guide vane opening and the same speed factor as Runner A. The discrepancy between the analyzed point and the best efficiency point for each runner is not so large, and the maximum difference in efficiency is 0.6% in Runner D.

3. Numerical Analysis of Scalable Loss

3.1 INTERNAL FLOW ANALYSES IN CASCADES

The flow in stay vanes, guide vanes and runner blades are analyzed with 3D Euler code with pseudo-compressibility and implicit formulation of finite difference. This Euler code has the same numerical schemes as the Navier-Stokes code [6], but the effect of viscosity is neglected and the boundary condition on the wall surfaces is different. Figure 2 shows the flow field for Runner A obtained from the analysis.

FIGURE 2. Flow Velocity Vectors in Cascades

3.2 CALCULATION OF SCALABLE LOSS

3.2.1 *Scalable Loss in Runner*

Three Dimensional Boundary Layer Analysis. Three dimensional boundary layer is calculated on the wall surface in runner using the velocity distribution obtained by 3D Euler flow analysis. Coriolis force and centrifugal force are considered. The numerical scheme of calculation is described in [7]. Figure 3 illustrates the result of calculation for Runner A.

The scalable loss is obtained from the energy thickness of boundary layer at the outlet of blade surfaces and of crown and band surfaces.

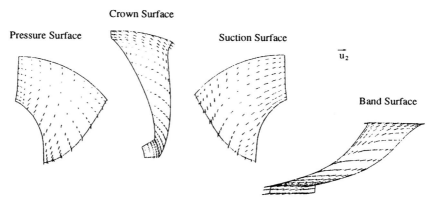

(a) Main Flow Velocity on the Walls

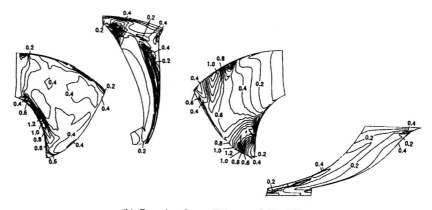

(b) Boundary Layer Thicness δ / D_2 (%)

FIGURE 3. Three Dimensional Boundary Layer on Runner Surface

One Dimensional Boundary Layer Calculation. One dimensional approximate boundary layer calculation is made along the grid lines using the velocity distribution obtained by 3D Euler flow analysis. Coriolis force and centrifugal force are not considered. The discrepancy in flow direction from the grid lines are also neglected.

The scalable loss is obtained from the energy thickness of boundary layer at the outlet of blade surfaces and of crown and band surfaces.

Integration of Shearing Stress. The scalable'loss is approximately estimated integrating the local shearing stress of a flat plate, which is obtained from the local velocity and the distance from the inlet. The velocity distribution on the wall surfaces obtained from 3D Euler flow analysis is used.

3.2.2 Scalable Loss in Stay Vanes and Guide Vanes

The scalable losses in stay vanes and guide vanes are obtained only by the integration of shearing stress mentioned above, because the flow distortion is not so large.

3.2.3 Scalable Loss in Spiral Casing and Draft Tube

The scalable losses in spiral casing and draft tube are calculated with Colebrook's equation by analogy of pipe friction.

4. Results and Discussions

Table 3 shows the results of scalable loss analyses. δ_E denotes the total scalable deficiency (relative scalable loss of specific hydraulic energy). δ_{ES}, δ_{ER} and δ_{ED} are the scalable deficiencies in the upstream parts of runner (spiral casing to guide vanes), in the runner and in the draft tube respectively. δ_{Ens} denotes the non-scalable deficiency which is obtained subtracting δ_E from the total specific hydraulic loss $E_{L12} = 100 - \eta_E$. The loss distribution coefficients are also tabulated. The subscripts S, R and D correspond to those of scalable deficiencies. Three values in array divided by slash marks correspond to the methods of scalable loss calculation.

Figure 4 illustrates the variation of the total scalable deficiency. Three kinds of bars correspond to the loss calculation methods. In spite of large difference in efficiency among four runners, the scalable deficiency is almost constant.

Figure 5 shows the variation of the non-scalable deficiency. δ_{Ens} steeply increases from Runner A to D, indicating that the discrepancy in efficiency is mainly caused by non-

TABLE 3. Scalable and Non-scalable Deficiencies

Runner	A	B	C	D
η_E	95.04	93.60	91.97	91.25
δ_{ES} (%)	1.75	1.75	1.65	1.64
δ_{ER} (%)	0.71 / 0.96 / 1.00	1.01 / 1.49 / 1.20	1.24 / 1.72 / 1.40	1.26 / 1.67 / 1.35
δ_{ED} (%)	0.04	0.04	0.04	0.04
δ_E (%)	2.50 / 2.76 / 2.79	2.80 / 3.28 / 2.99	2.91 / 3.39 / 3.07	2.95 / 3.35 / 3.04
δ_{Ens} (%)	2.46 / 2.20 / 2.17	3.60 / 3.12 / 3.41	5.12 / 4.64 / 4.96	5.80 / 5.40 / 5.71
V_S	0.35	0.27	0.20	0.19
V_R	0.14 / 0.19 / 0.20	0.16 / 0.23 / 0.19	0.16 / 0.21 / 0.17	0.14 / 0.19 / 0.16
V_D	0.01	0.01	0.01	0.01
V	0.50 / 0.56 / 0.56	0.44 / 0.51 / 0.47	0.36 / 0.42 / 0.38	0.34 / 0.38 / 0.35

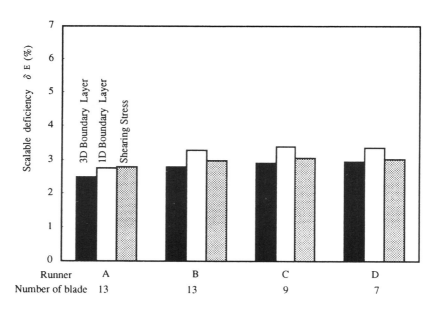

FIGURE 4. Scalable Deficiency

scalable loss. From Figure 6 illustrating the variation of the loss distribution coefficient, the same conclusion can be drawn.

Changing the point of view to the effect of the number of blades, the loss distribution coefficient decreases apparently with decreasing number of blades. If this result is extrapolated toward larger number of blades, it would be concluded that the increase in the number of blades would increase the value of V.

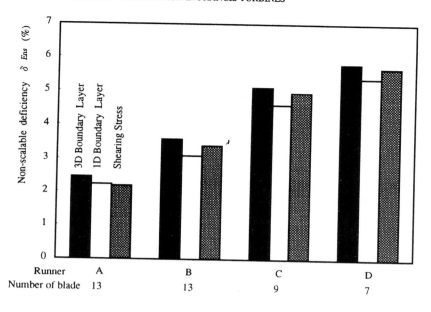

FIGURE 5. Non-scalable Deficiency

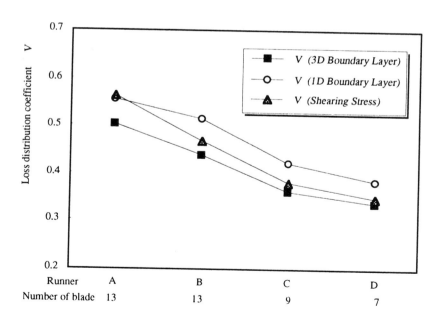

FIGURE 6. Loss Distribution Coefficient

Irrespective of the above mentioned tendency in loss distribution coefficient, however, the discussion about the effect of the number of blades might be disposed of when scalable loss is prescribed in scale-up formulae.

5. Conclusion

Results of scalable loss analyses for four Francis runners tested in the same model turbine are presented.
Analyzed results showed best regularity in scalable loss. Non-scalable loss steeply increased and loss distribution coefficient considerably decreased in proportion as efficiency dropped. Then lower peak efficiency seems to be mainly caused by non-scalable loss. This means that scalable loss is most suitable for prescription in scale-up procedures. Although the possibility could be seen on the increase in loss distribution coefficient with increasing number of blades, yet the problem would be solved by prescribing scalable loss instead of loss distribution coefficient.

6. Acknowledgement

This study was strongly motivated by the activities of the committee on preparing the revised JSME S008. The 3D boundary layer analyses are performed by Mr. Matsumoto, the graduate student in Yokohama National University.

7. References

1. JSME Standard S008, Performance Conversion Method for Hydraulic Turbines and Pumps (1989).
2. IEC Publication 995, Determination of the Prototype Performance from Model Acceptance Tests of Hydraulic Machines with Consideration of Scale Effects (1991).
3. Spurk, J.H. and Grein, H., Performance Predictions of Hydraulic Machines by Dimensional Considerations, Water Power & Dam Construction (1993-11), 42-49.
4. Ida, T., New Formula for Efficiency Step-up of Hydraulic Turbines, Proc. 17th IAHR Symposium (Beijing, 1994), 827-840.
5. Nichtawitz, A., Discussion on Step-up Procedures in Hydraulic Machines, Proc. 17th IAHR Symposium (Beijing, 1994), 841-852.
6. Qian, Y., Arakawa, C., Kubota, T. and Suzuki, R., Numerical Flow Simulation on Channel Vortex in Francis Runner, Proc. 17th IAHR Symposium (Beijing, 1994), 237-247.
7. Kurokawa, J., Kitahora, T., et al., Prediction of Scalable Loss in Francis Runners of Different Specific Speed, Proc. 18th IAHR Symposium (Valencia, 1996).

UNSTEADY FLOW CALCULATION IN A CENTRIFUGAL PUMP USING A FINITE ELEMENT METHOD

P.F. BERT, J.F. COMBES
EDF-DER, 6 quai Watier
78400 Chatou, FRANCE

J.L. KUENY
INPG-ENSHMG, St Martin D'Hères
38400 Grenoble, FRANCE

1. Abstract

In order to predict rotor-stator interactions in hydraulic turbomachinery, a new multidomain method was implemented in a finite element code developed in the Research Division of Electricité de France. This code deals with 2D or 3D laminar or turbulent flows in complex geometries. This method was used to study the unsteady flow in a simplified centrifugal pump model. The model consists in a 420 mm diameter unshrouded centrifugal impeller with seven untwisted constant thickness backswept blades and a radial vaned diffuser with twelve vanes and a six percent vaneless radial gap in which important unsteady interactions are expected. Unsteady numerical results are compared with the experimental results published by M. UBALDI et al. and also with a multistage calculation on the same geometry using a mixing plane to communicate circumferential average of flow properties.

2. Introduction

Centrifugal turbomachines with vaned diffuser or scroll have always a vaneless radial gap lower than the length scale of the blade row interactions. These interactions generate very important unsteady flows that extend both upstream and downstream of the gap. The two major sources of unsteadiness are potential interaction and blade/wake interaction (Dring et al. 1982). In centrifugal turbomachines, the effects of these sources of unsteadiness become comparable (Arndt et al. 1990). This remark is very important for numerical analysis of unsteady flows. Simplified approach used by Yu et al. (1995) taking into account bladed row effects as an unsteady boundary condition cannot be applied. A full modelling of the flow in the machine must be undertaken, that suppose a computational code able to solve rotor-stator problems with sliding meshes.

To study the unsteady flow in a centrifugal pump with vaned diffuser, experimented at the University of Genova (Italy), a multidomain method was implemented in a finite element industrial code. A full bidimensional discretization of the machine is used to calculate the flow and to analyse the basic fluid phenomena. Major objectives of this

E. Cabrera et al. (eds.), Hydraulic Machinery and Cavitation, 371–380.

paper are to complement the unsteady flow analysis performed by Ubaldi et al. (1994) and to show that the computational method developed is able to apprehend unsteady flow effects.

3. Simulation Test Case

The simplified model of centrifugal turbomachine studied by Ubaldi et al. (1994) has been utilised firstly to validate unsteady calculations and secondly to improve the understanding of basic fluid dynamic phenomena. The model consists of a 420 mm diameter unshrouded centrifugal impeller and a 664 mm diameter radial vaned diffuser. The geometry of the machine is shown in figure 1.a/ and the co-ordinates of the impeller blade and diffuser vane profiles are given in Ubaldi et al. (1994). The impeller has seven untwisted constant thickness backswept blades with rounded off leading and trailing edges. The numerical model cannot describe the manufacturing rounded off : the leading edge is pointed and the trailing edge is squared. The diffuser used has 12 vanes and a 6 percent vaneless radial gap. The tip clearance is set at a value of 0.4 mm, 1 percent of the blade span. Even if the geometry of the blades is bidimensional, the tip leakage effects and the fixed casing lead to tridimensional secondary flows that create difficulties for comparisons with bidimensional computations.

The measuring techniques used were hot wire anemometer and fast response pressure transducers. The hot wire probe was used to measure the unsteady tridimensional flow in the vaneless gap at a radial distance of 4 mm from the blade trailing edge and 8 mm from the vane leading edge. The unsteady static pressure was measured at the front cover facing the unshrouded impeller passages.

The model operates in an open circuit with air directly discharged into the atmosphere from the radial diffuser. The inlet air temperature was 298 K and the air density 1.2 kg/m^3. The results were obtained at the nominal operating conditions, at a constant rotational speed of 2000 rpm. The main geometric data of the model and the operating conditions are summarised in table 1.

Table 1 : Geometric data and operating conditions

Impeller			
inlet blade diameter	D_1 =	240	mm
outlet diameter	D_2 =	420	mm
blade span	b =	40	mm
number of blades	z_i =	7	
Diffuser			
inlet vane diameter	D_3 =	444	mm
outlet vane diameter	D_4 =	664	mm
vane span	b =	40	mm
number of vanes	z_d =	12	
Operating conditions			
rotational speed	n =	2000	rpm
flow rate coefficient	φ_e =	0.048	
total pressure rise coefficient	ψ_e =	0.65	
Reynolds number	Re =	6.5×10^5	

4. Numerical Approach

In order to solve industrial flow problems in complex geometries, a finite element code, N3S, has been developed at Electricité de France. It allows the computation of a wide variety of 2D or 3D unsteady incompressible flows, by solving the Reynolds-averaged Navier-Stokes equations together with a k-ε turbulence model.

The space discretization uses a standard Galerkin finite element method. In the present calculations we used isoP2 6-noded triangles verifying the LBB condition. The time discretization is based on a fractional step method ; at each time step, one has to solve successively :

- an advection step for the non-linear convection terms of the Navier-Stokes and k and ε equations. It is treated by a characteristics method.
- a diffusion step for the remaining part of the k and ε equations : the finite element discretization leads to linear systems solved by a preconditioned conjugate gradient algorithm.
- a generalised Stokes problem for the velocity and the pressure. It is solved by a projected gradient Uzawa or Chorin algorithm preconditioned in order to ensure a fast convergence.

The numerical method is described in details by Chabard et al (1991).

The most recent development of this code concerns the extension to multidomain problems. The method has to solve the N3S system equation on several domains with possibility of translation/rotation with or without periodic boundaries. The resolution of the problem in each mesh reference frame with sliding boundary was considered as the better solution. The fractional step method, used in N3S code, compels us to implement two different coupling methods. The first one assumes the coupling of the convection step and the second assumes the coupling of the diffusion and Stokes step.

The continuity of the non-linear convection terms is achieved through the integration of characteristics through subdomain boundaries. The second method is the most important step in solving the incompressible Reynolds-averaged Navier-Stokes and k-ε equations on multidomain meshes. A mortar element was used to assume the continuity of aerodynamics quantities on sliding meshes. It assumes also the local and global divergence of velocity to stretch to zero. The mortar element method is a non-conforming discretization based on the explicit construction of an optimal approximation space. This method preserves element-based locality, which distinguishes it from other more global techniques.

5. Steady computation

The 2D computational domain correspond to a meridian plane with a radial inlet. The total domain was treated by two parallel computations which exchange boundary informations. At the impeller outlet, an azimutal average of the pressure calculated at the previous step at the diffuser inlet was used. At the diffuser inlet, we use an azimutal average of the velocity and the turbulent quantities estimated at the previous step at the

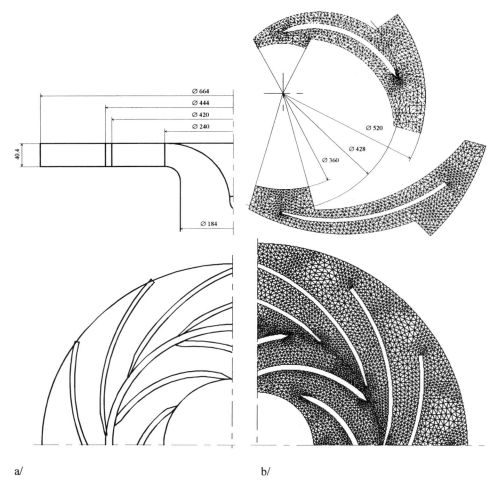

Figure 1 : a/ Experimental radial geometry pump ; b/ Navier-Stokes simulation meshes (each element is divided in 4 sub-elements), upper figure : steady calculation mesh, lower figure : unsteady calculation mesh.

impeller outlet. These dynamic type boundary conditions are naturally imposed for an incompressible computation. For a compressible computation these conditions are not suited, energetic conditions must be implemented. The average process and the exchange of information is consistent with the conservative equations and they integrate the change of frame.

The impeller finite element mesh consists in 1610 triangular elements and 3478 nodes, for the diffuser a 1398 triangular elements and 3066 nodes mesh is used (figure 1.b/). The impeller mesh was extended downstream about 46 mm. It was estimated that a constant pressure on this new outlet boundary does not modify the flow around the blade and the gradient constraint calculated on the real frontier was physically better. The same line of argument can be used for the diffuser with the velocity at the inlet where we have extended the mesh upstream about 34 mm.

The other boundary conditions are :

- at the inlet of the impeller, a uniform velocity was prescribed, according to the flowrate measured. Also k and ε were prescribed, corresponding to a turbulence level of 2%,

- at the outlet of the diffuser, a null constraint was prescribed. It correspond to a constant static pressure if the shear stress is negligible (atmospheric pressure),

- on the walls Reichardt's function was used,

- periodic conditions were also used to limit the calculation at one inter-blade channel.

6. Unsteady computation

The unsteady computation was achieved on the same meridian plane used in the steady case. The whole domain, impeller and diffuser, was treated on a global computation using a sliding mesh interface. In this global modelling the geometric periodicity does not exists, periodic boundary conditions can not be applied and the seven inter-blade channels of the impeller and the twelve inter-blade channels of the diffuser must be discretized. The total 2D finite element mesh is formed of 23000 triangular elements and 49800 nodes (figure 1.b/). The boundary conditions applied at the impeller inlet and at the diffuser outlet are the same as in the steady calculation.

7. Comparison results

We can make two kinds of comparisons between the steady, the unsteady calculations, and the experimental results : comparison of the temporal average values and comparisons of instantaneous values. Naturally, comparisons of instantaneous values have been only performed between unsteady computational results and experimental ones. Time averaged unsteady results are not equal to the steady calculation results because non-linear unsteady quantities are not evaluated by the second approach. As it is shown in figure 2, the differencies are not uniformly distributed in the azimutal direction.

In figure 2, we represent the time average relative velocity at the impeller outlet, normalized with the rotor tip speed U_2, in function of the normalized rotor circumferential co-ordinate. The pressure side peaks and the wake deficit are correctly represented around the blade trailing edge ; both calculations are able to estimate the maximum of the azimutal fluctuations generated by the blade passage. The large smooth maximum measured in the mid-passage, at mid-span ($z/b=0.5$), is underestimated by the calculations, but a detailed analysis of the experimental results

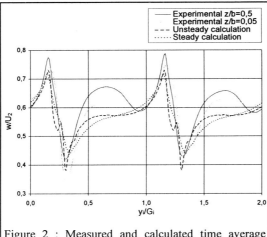

Figure 2 : Measured and calculated time average relative velocity at the impeller outlet.

show that this maximum disappears when z decreases. The relative velocity, near the hub (z/b=0.05), increases continuously from the wake suction side toward the pressure side peak as it was calculated. In the mid-passage, the velocity gradients seem to be better predicted by the unsteady calculation.

The instantaneous distributions, at the impeller outlet, of the radial velocity, measured (on the top) and calculated, are plotted with dark line in figure 3. The dotted line represent the temporal average of the velocity. The conservation of the flowrate prescribes in the calculation the global level of the radial velocity. This level is a little less than that measured because we consider a uniform distribution in the axial pump direction. In reality the effect of the hub and casing boundary layers is to

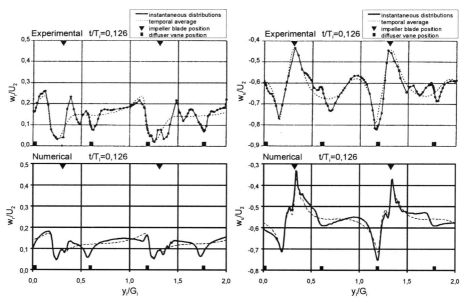

Figure 3 : Instantaneous distributions of the radial velocity at the impeller outlet.

Figure 4 : Instantaneous distributions of the relative tangential velocity at the impeller outlet.

increase the flowrate in the mid-passage. The amplitude of the fluctuations calculated is underestimated about thirty percent without any explication except a bad representation of the velocity on the walls by the Reichardt's wall function. The most important is that the peaks corresponding to the wake velocity deficit and the leading edge vaned diffuser interaction are well represented and positioned. Figure 4 shows the instantaneous distributions, plotted as in figure 3, of the tangential relative velocity. The weak mean level calculated is directly a consequence of the weak radial velocities. The amplitude of the fluctuations calculated is well estimated as the position of the peaks. The absence of a large smooth maximum in the mid-passage on the calculated results can be explained by the same line of arguments used in figure 2 for the temporal average of the relative velocity.

8. Flow Analysis

For incompressible fluid, the unsteady total pressure is representative of the local instantaneous fluid energy. Figure 5 show that impeller wakes are high energy region because the tip speed $U_2=\omega R_2$ is higher than the outlet absolute velocity.

The wakes decay very rapidly in the gap, but after their chopping by the vanes leading edge, they are convected in the vanes passage and we can clearly notice their presence at a radial distance of $1.4 R_2$. In the impeller, the total pressure is dominated by the pressure gradient except in the blade boundary layers where friction losses are important.

In figure 6.a/, we represent the temporal average relative velocity in the impeller (unsteady calculation). The maximum of velocity is located on the leading edge suction side, the minimum is just on the other side of the blade. The adverse pressure gradients decelerate the flow from the leading edge to the trailing edge on the suction side and

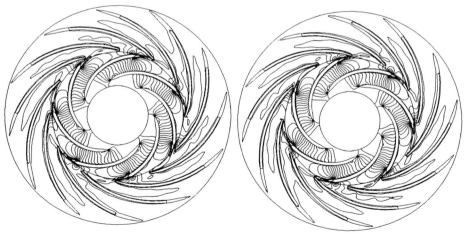

Figure 5 : Isovalues of instantaneous total pressure at $t/T_i=0$ and $t/T_i=0.5$.

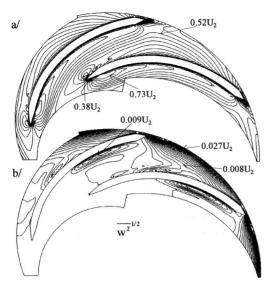

Figure 6 : a/ Isovalues of the temporal average relative velocity in the impeller ; b/ Isovalues of stator induced velocity fluctuations.

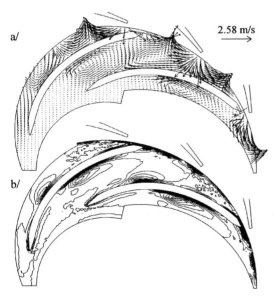

Figure 7 : a/ Unsteady secondary relative velocity vectors $w(t) - \overline{w}$; Unsteady vorticity contours.

between 10% and 50% of the chord on the pressure side. In these decelerate flow regions, the vanes passage induce a high level of fluctuations, highest than in the other regions (figure 6.b/). Apart from these two regions, the level of fluctuation decreases with the distance from the vanes, but the azimutal distribution is not uniform. On the trailing edge suction side gradients fluctuations are very concentrated, on the pressure side the fluctuations seem to decrease rather slowly.

The most important secondary instantaneous velocities are located in the regions of highest fluctuations (figure 7.a/). The secondary velocity is calculated as the difference between the instantaneous velocity and the time average one. We have plotted the velocity scale, it must be compared to the tip speed $U_2=43.98$ m/s and to the discharge velocity $v_{r2}=Q/(2\pi R_2 b)= 5.47$ m/s. The structure of the secondary flow induced by the stator is very complex ; the vortex, located near the suction surface is always present but its centre position change with time. The second vortex structure noticeable in figure 7.a/ disappears periodically with the vane passage. The leading edge acts as a source point and any vortex structure is associated to the potential interaction. We can remark that the vorticity is not the best parameter to analyse this mode of interaction (figure 7.b/).

In figure 8, the temporal average relative velocity and rotor induced rms velocity fluctuations are plotted

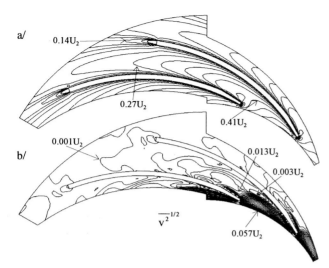

Figure 8 : a/ Isovalues of the temporal average absolute velocity in the vaned diffuser ; b/ Isovalues of stator induced velocity fluctuations.

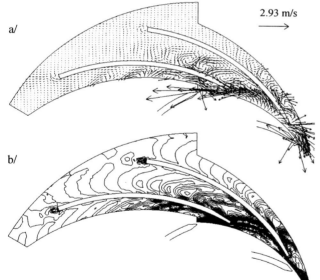

Figure 9 : a/ Unsteady secondary absolute velocity vectors $v(t) - \bar{v}$; b/ Unsteady vorticity contours.

in the vaned diffuser. The gradients of fluctuation are concentrated in the zone of maximum mixing process, before the wakes were chopped by the vanes. At the inlet, the fluctuations are maximum on the blades walls, in the inter-blade channel this observation seems to be inverted and the maximum is located on the centre passage. This centre passage maximum is associated with the secondary vortex velocity structure convected in the channel (figure 9.a/). The presence of wakes in the diffuser is clearly noticeable and it results in two counter rotating vortices on either side as it was indicated by several earlier investigators in axial turbomachines. Each wake is separated by large positive and negative vorticity regions (figure 9.b/), the isovalue zero representative of the wake structure is always present at the outlet.

On the contrary of in the axial machines (Yu et al. 1995), the velocity incidence angles tends to increase before the wake passage, it decreases when the wake impiges the leading edge and it increases again after its passage. An other difference with the kinematics of the wake in an axial machine (Hodson 1984) is that the wake is bowed only at the inlet in the maximum fluctuation region. In the diffuser channel, the wake becomes progressively perpendicular to the pressure side wall and we observed a concentration of the main structure vortex (line of iso-maximum velocity in the vortex).

9. Conclusion

To study the rotor-stator interactions in a centrifugal pump with vaned diffuser, a steady and an unsteady simulation were carried out using a finite element Navier-Stokes code. The results were compared with the experimental investigation of Ubaldi et al. (1994), good agreements were found and the main differencies can be explained by the 3D secondary flows generated by the unshrouded impeller. The stator induced flow in the impeller and the rotor induced flow in the diffuser are studied : rms fluctuations, secondary instantaneous velocity and unsteady vorticity are plotted. In the impeller, it was shown that high fluctuation regions exist in the centre of the channel induced by the periodic passage of the leading edge vane which acts as a source point. In the diffuser, it was observed that the wake mechanisms are different in an axial and in a radial machine.

10. References

Anagnoustou, G., Maday, Y., Patera, A.T., 1991, "A Sliding Mesh Method for Partial Differential Equations in Nonstationary Geometries : Application to the Incompressible Navier-Stokes Equations", Publ. of Numeric Lab., Université Pierre et Marie Curie

Arndt, N., Acosta, A.J., Brennen, C.E., Caughey, T.K., 1990, "Experimental Investigation of Rotor-Stator Interaction in a Centrifugal Pump with Several Diffusers", A.S.M.E. Transactions, Vol. 112, pp 98-108

Chabard, J.P., Metivet, B., Pot ,G., Thomas, B., 1991, "An efficient finite element method for the computation of 3D turbulent incompressible flows", Finite Element in Fluids, vol.8

Chen, Y.S., 1988, "3-D Stator-Rotor Interactions of the SSME", A.I.A.A. Paper 88-3095

Dring, R.P., Joslyn, H.D., Hardin, L.W., Wagner, J.H., "Turbine Rotor-Stator Interaction", Journal of Engineering for Power, Vol. 104, pp 729-742

Gallus, H.E., Grollius, H., Lambertz, J., 1982, "The Influence of Blade Number Ratio and Blade Row Spacing on Axial-Flow Compressor Stator Blade Dynamic Load and Stage Sound Pressure Level", Journal of En gineering for Power, Vol. 104, pp 633-641

Giles, M.B., 1988, "Stator/Rotor Interaction in a Transonic Turbine", A.I.A.A. Paper 88-3093

Hodson, H.P., 1984, "Measurements of Wake-Generated Unsteadiness in the Rotor Passages of Axial-Flow Turbines", A.S.M.E. Paper 84-GT-189

Korakianitis, T., 1991, "On the Propagation of.Viscous Wake and Potential Flow in Axial-Turbine Cascades", A.S.M.E. Paper 91-GT-373

Krain, H., 1981, "A Study on Centrifugal Impeller and Diffuser Flow", Journal of Engineering for Power, Vol. 103, pp 688-697

Rai, M.M., 1987, "Unsteady Three Dimensional Navier-Stokes Simulations of Turbine Rotor-Stator Interaction", A.I.A.A. Paper 87-2058

Ubadi, M., Zunino, P., Barigozzi, G., Cattanei, A., 1994, "An Experimental Investigation of Stator Induced Unsteadiness on Centrifugal Impeller Outflow", A.S.M.E. Paper 94-GT-105

Yu W.S., Lakshminarayana, 1995, "Numerical Simulation of the Effects of Rotor-Stator Spacing and Wake/Blade Count Ratio on Turbomachinery Unsteady Flows", Journal of Fluid Engineering, Vol. 117, pp 639-646

One can notice that over a period 15°. corresponding to a canal between the guide vanes, the pressure evolution is important and represents the guide vanes wake.

4.2. TOTAL PRESSURE

The pressure probe has shown, according to figure 3, a vertical distribution of the bidimensional and non - uniform velocity. Indeed the component as per the height (z) is negligible as against the two others - from 1% to 3% of the total velocities. The static pressure profiles are uniforms as per the height but a variation is noticed according to the azimuth position and the operating point - see figure 3.

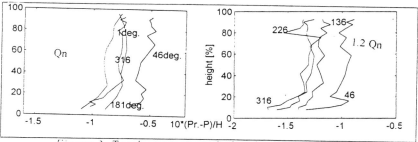

Figure 3. Total pressure evolution for Qn and 1.2 Qn

4.3. MEAN VELOCITY

Two velocity components, for each measurement point, were measured and the corresponding average and root-mean-square were calculated. The two components: radial and circumferential are considered according to the figure 1.

The means have been performed on 7000 acquisition points. One can notice the non - uniformity vertical distribution of the velocity field which was noticed previously by total pressure probe. Following an interval of 15° one can notice a wake. zone (the velocity decrease), as well as maximum velocities in the proximity of the wall zones - see figure 4.

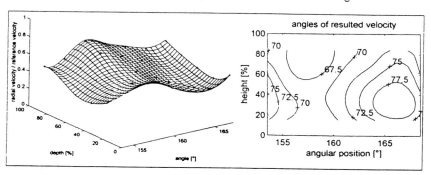

Figure 4. Velocity and angles of resulted velocity
distribution in a canal between guide vanes

Moreover we find a periodic distribution as per the azimuth position, but with the different flow rates in guide vanes canal, for a same functioning point. The non - uniformity of the velocity field is emphasised by its removal from the nominal point. The distribution of the resulted velocity angle is the same with that found by total pressure probes.

The same vertical distribution can be noticed at the exit of stay vanes - see figure 5.

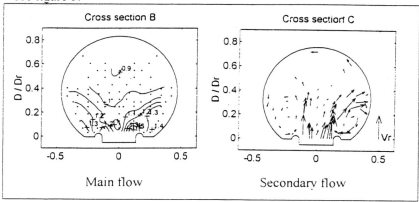

Figure 5. Example of the velocity distribution in the spiral casing

4.4. PERIODIC UNSTEADY VELOCITY SYNCHRONISED WITH RUNNER ROTATION

In the periodic turbulent flows, the statistical averages depend, besides the space variables, on the angle of phase Φ. The idea of this approach is therefore to separate the determining part, linked to the runner wake - the periodic fluctuation of the velocity from the totality of the velocity fluctuations. This technique, called synchronisation means, allows to represent the velocities according to the runner phase. Therefore we shall have:

$$v(i) = \bar{v} + v^{\sim} + v'$$

where: $v(i)$ is the instantaneous velocity of a particle; $\bar{v}$, the mean velocity ($\Sigma v(i)/N$); $v^{\sim}$, synchrony unsteady velocity; v', turbulent unsteady velocity with $< v'2 > = 0$. To calculate $v^{\sim}$ we take into account a time interval (window), small enough against the periodic flow variations - see figure 6.

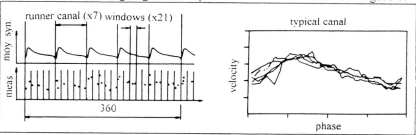

Figure 6. Synchrony unsteady velocity calculation

For a turbomachine, the flow being periodic, this window becomes an interval of angle corresponding to an angular variation of the runner position (therefore a cannel becomes a runner canal). Consequence of several tests, 21 windows were chosen per canal, which correspond to a runner rotation angle of 2.44°. On the other hand the points number of each window has to be sufficient to have a correct average. Thus, the tests have been performed and the best compromise between the points acquired number and the acquisition time has been ~300 points by window. The good flow symmetry between runner canals has allowed to use a representative canal, called typical canal - see figure 6.

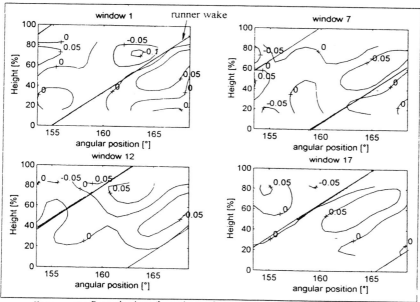

Figure 7 v˜ evolution function of runner phase

4.5. TURBULENCE INTENSITY

The calculation of the turbulence intensity was performed by carrying out the isotropy hypothesis according to the 3 directions. Since the size order of the root-mean-square for the two measured is the same, the root-mean-square for the 3rd component (which was not measured by LDV) was considered the arithmetic average of the other two. This calculation can be carried out by and without overlooking v˜ due to the runner.

The root-mean-square of the mean velocity can by calculate as:

$$\sigma^2 = [\Sigma(v(i) - \bar{v})^2]/(N-1), \text{ overlooking } \tilde{v}$$

or

$$\sigma^2 = [\Sigma(v(i) - \bar{v} - v^\sim)^2]/(N-1)$$

Therefore the turbulence intensity was considered:

$$k = 3(\sigma_r^2 + \sigma_t^2)/4$$

Where: σ is root-mean-square; N - number of measurements for each point; σ_r - root-mean-square of radial velocity; σ_t - root-mean-square of circumferential velocity; k - turbulence intensity. There is between the two calculations a difference of up to 10% for the same space distribution.

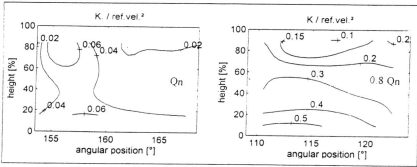

Figure 8. Turbulence intensity for Qn and 0.8 Qn

Mention should be made of the fact that any supply difference between the runner canal adds a false information to the calculation of the turbulence intensity.

5. Discussion

The distribution of the velocity and pressure field allow to analyse the flow at this level. Firstly the velocity field is marked by the guide vanes wake, which gives it a periodical evolution following the angular position. It finds the same characteristic in the static pressure evolution. To this a flow rate evolution between guide vanes canal is added which can be a pressure gradient effect induced at this level by the spiral casing geometry. This can be noticed in the static pressure profiles evolution already found by Sideris and al [9] which also show the influence on the spiral casing form on the velocity at the runner outlet profile, imposed by the pressure distribution in the spiral casing. The effect of the guide vanes wakes is also confirmed by the distribution of the angle of the resulting velocity.

The distribution of the velocity profile following the height, with the on - velocity up and down is in accordance with the distribution the runner outlet velocity profile. This non - uniformity increases with the removal from the nominal point.

This velocity distribution, which is transmitted to the entry of the volute, has a very large importance for the secondary flow formation in the volute. Thus in the tongue zone the velocity excess of the upper part of the section crushes the inferior component and creates a vortex which will be transported by the flow. Moreover the secondary flow in all sections of the spiral casing has as origin this velocity distribution. For details see Ciocan and al [4].

The results of the numerical simulation which uses a uniform velocity profile as against this distribution type can give a very different distribution of flow rate as per the azimuth position. In figure 9 the mean velocity evolution following the section for a calculation performed under the two conditions (with the good boundary layers this calculation gives the very good results - see Soares and al [10]) is shown.

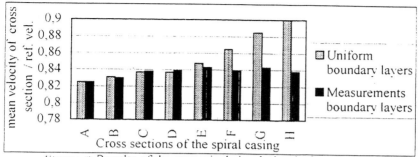

Figure 9 Results of the numerical simulation in function of boundary conditions

From the point of view of the synchrony component of the runner, one notices the propagation of the runner wake (the inclination of the synchrony wake is the same as that the angle of the runner trailing edge) up to this level (1.33 outlet runner diameter) - see figure 7. This is not the case of the configuration whiteout stay and guide vanes where the wake is crushed by the main flow in the spiral casing. This wake is due to the difference of the velocity between the suction side and the pressure side of the blade at the runner outlet. The turbulence level is raised enough and has the same intensity for the two velocity components. The turbulence increases strongly with the removal from the nominal point (which is in agreement with Casey and al [3]) - see figure 8. The obtained error by overlooking the determining part of the velocity synchrony component in the turbulence calculation is of 10%.

6. Conclusion

Despite the difficulty of the access, the complexity of the pump regime and the large complexity of adjustment for the measurement equipment to be implemented (very raised costs and implementation duration of the several months), the detailed data bank was achieved on the flow in the fixed parts of the turbine-pump tests pattern. That has allowed a precision analysis of the steady and unsteady flow, and the turbulence intensity of the flow between stay and guide vanes. That allows also to explain the secondary flow sources in the spiral casing.

Even the comparisons are difficult to carry out and they risk to be inconclusive taking into account the determining influence of geometrical characteristics of each machine (runner and spiral casing geometry), it is considered that there are representative results for global comparisons of qualitative order or of physical explanations. The 3D Navier-Stokes unsteady calculations with this data bank as validation conditions, will allow a detailed loss analysis.

8. Acknowledgements

We acknowledge with gratitude EDF (Electricité de France), NEYRPIC (GEC ALSTHOM) and Région Rhône-Alpes for their collaboration and financial support of this project.

9. References

1. Amarante Mesquita, A. L., (1993) Experimental Techniques for the Flow Characterisation in Hydraulic Turbomachines Models, *Proceedings of the 2nd Meeting of the Latin-American Division of the International Association of the Hydraulic Research - IAHR*, Vol. 1, pp. 193-209, Ilha Solteira, Brazil.
2. Amarante Mesquita, A. L., Philibert, R., Very, A., and Kueny, J. L., (1994) Laser Velocimetry Measurements in a Pump Turbine Spiral Casing, *Proceeding of the 19th Symposium of the International Association of the Hydraulic Research - IAHR*, pp. 395-404, São Paulo, Brazil.
3. Casey, M.V., Eisele, K., Zhang, Z., Gulich, J., Schachenmann, A., (1995) Flow analysis in a pump diffuser: LDA and PTV measurements of the unsteady flow, *Laser Anemometry, FED-Vol. 229, ASME, pp. 89-99*
4. Ciocan, G.D., Amarante Mesquita, A. L., and Kueny, J. L., (1996) Three-Dimensional Flow Patterns in a Pump Volute at Nominal and Off-Design Flow Conditions by LDA Measurements, *Proceedings of the 6th International Symposium on Transport Phenomena and Dynamics in Rotating Machinery - ISROMAC - 6*, Vol. 2, pp. 229-237, Honolulu, USA.
5. Deojeda, W., Flack, R.D., Miner, SM., (1996) Laser Velocimetry Measurements in a Double Volute Centrifugal Pump, *Proceedings of the 6th International Symposium on Transport Phenomena and Dynamics in Rotating Machinery - ISROMAC - 6*, Vol. 2, pp. 131, Honolulu, USA.
6. Elholm, T., Ayder, E., and Van der Braembussche, R., (1992) Experimental Study of the Swirling Flow in the Internal Volute of a Centrifugal Pump, *Journal of Turbomachinery*, Vol. 114, pp. 366-372.
7. Mofat, R. J., (1985) Using Uncertainty Analysis in Planning of an Experiment, *Journal of Fluid Engineering*, Vol. 107, pp. 173-178.
8. Shuliang, C., Qian, H., and Ruchang, I., (1986) Experimental Study on Flow Characteristics and Hydraulic Losses in a Pump Turbine Volute in Pump Operation, *Proceeding of the 13th Symposium of the International Association of the Hydraulic Research - IAHR*, paper N°. 46, Montréal, Canada
9. Sideris, M. T., and Van der Braembussche, R., (1987) Influence of a Circumferential Exit Pressure Distortion on the Flow in a Impeller and Diffuser, *Journal of Turbomachinery*, Vol. 109, pp. 48-54.
10. Soraes Gomes, F., Amarante Mesquita, A. L., Ciocan, G., and Kueny, J. L., (1994) Numerical and Experimental Analysis of the Flow in a Pump-Turbine Spiral Casing in Pump Operation, *Proceedings of the 17th Symposium of the International Association of the Hydraulic Research - IAHR*, pp. 249-258, Beijing, China

SELF-SUSTAINED OSCILLATION OF GAS-LIQUID FLOW IN A CENTRIFUGAL PUMP WITH SEMI-OPEN IMPELLER

J. KUROKAWA, J. MATSUI, H. TAKADA, and T. HIRAYAMA
Yokohama National University
156, Tokiwadai, Hodogaya-ku, Yokohama, 240 Japan

1. Introduction

When a centrifugal pump is operated at very low discharge with low suction head, the air dissolving in water merges to become bubble and stays inside an impeller channel. The staying bubbles accumulate to form a large bubble zone and decrease pumping head. This phenomena is called "air locking", and often arises when air-rich water is used in a pumping system, such as a water supply system of a building.

The air locking phenomena was also raised when air is supplied forcedly into a suction pipe instead of using air-rich water and low suction head. However, a sever self-sustained oscillation was encountered under the same operating conditions, when a semi-open type impeller was used. In this case the pump had stable head-capacity performance and was operated under non-cavitation condition.

There are many types of self-sustained oscillations caused by gas-liquid flow in pumping system, such as low frequency oscillation accompanying cavitation in [1], [2], the oscillation due to time lag between air inlet and impeller inlet as in [3], and so on. These phenomena is accompanied with severe oscillation of a suction pipe. However, the present oscillation is raised only in a semi-open impeller with no suction pipe oscillation.

In order to elucidate the mechanism of the present oscillation, the instantaneous measurements of blade-to-blade pressure were performed and the parameters influencing the oscillation were determined together with the theoretical considerations.

2. Nomenclature

p : static pressure
ρ : density of water
r, r_2 : radius and impeller radius
U : impeller tip speed
Q, q : volume flowrates of water and air, respectively (under suction pressure)
x : distance along vane center line from leading edge
ϕ : discharge coefficient, defined as $Q / (A_2U)$ (A_2: impeller outlet area)
ψ : head coefficient, defined as $H / (U^2/2g)$ (H: pump head)
ψ_s : coefficient of static pressure, defined as $(p - p_{s1}) / (\rho U^2/2)$
τ : power coefficient, defined as $P_w / (\rho A_2 U^3/2)$ (P_w: shaft power)
subscripts s1, s2 : 900mm and 16mm upstream of impeller inlet, respectively
d : discharge (400mm downstream of impeller outlet)

E. Cabrera et al. (eds.), *Hydraulic Machinery and Cavitation*, 391–400.
© 1996 Kluwer Academic Publishers. Printed in the Netherlands.

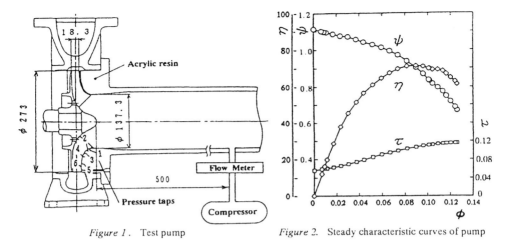

Figure 1. Test pump   Figure 2. Steady characteristic curves of pump

3. Experimental Apparatus and Dimensions

The pump tested is shown in Fig. 1. The impeller is a semi-open type, of which specific speed is 230[m, m^3/min, rpm] , blade outlet angle β_2=25°, radius r_2=137mm, blade number 5, and tip clearance 1mm. The suction cover and the discharge pipe are made of acrylic resin so that the bubble behavior can be observed. On the suction cover six pressure holes, No. 1~6 shown in Fig 1, are drilled and semi-conductor type pressure transducers (natural frequency of 10 kHz) are mounted for instantaneous measurements.

The test apparatus consists of a suction pipe of 2m length, a pump, a discharge pipe of 10m length and a reservoir. The compressed air is supplied at 500mm upstream of the impeller inlet, and is separated in the reservoir. A swirl stop is mounted at 600mm upstream of the impeller inlet. The pump rotational speed was varied from 500 rpm to 1800 rpm, corresponding Reynolds number ranging from 9.8×10^5 to 3.5×10^6.

4. Experimental Results and Discussions

4. 1 PUMP PERFORMANCE

The pump tested has stable characteristic curves as shown in Fig. 2 with a negative gradient of head-capacity curve over the whole operating range.

When the pump is operated at very low discharge and air is supplied into a suction pipe, the pumping head gradually decreases and begins to oscillate, as shown in Fig. 3. Not only pump head ψ but also discharge coefficient ϕ, discharge pressure

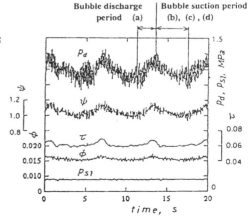

Figure 3. Oscillation of pumping head and others (Q /q = 0.014)

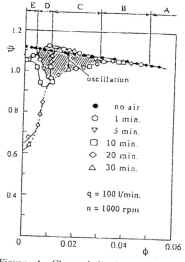

Figure 4. Change in head curve with time

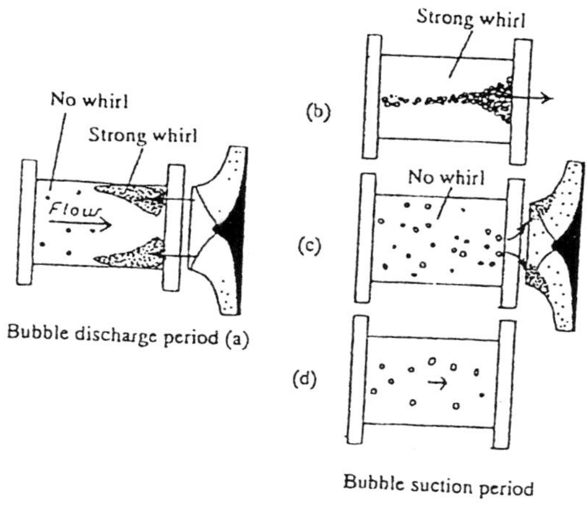

Figure 5. Sketch of bubble behavior in a suction pipe while oscillating (Pattern C)

p_d and shaft power coefficient τ oscillate with the period of 6.7s and in the same phase. However, the pressure p_{s1} at 900mm upstream of the impeller inlet does not oscillate.

Time dependence of head curve is shown in Fig. 4. Both the maximum and the minimum values are plotted when oscillation occurs. After 30 minutes from the start of air supply , the oscillation is still maintained only in the discharge range C shown in Fig. 4, and in the other range the head curve comes to be the dotted line showing a steady operation. From Fig. 4 it is recognized that there exists following five patterns ;

A: Pump operates steadily without decrease in pumping head, though a bubble zone exists in an impeller channel and many small bubbles discharge at the pump outlet.

B; Flow pattern is almost same as A, but the blade channel is full of small bubbles, the pumping head slightly decreases and the head-capacity curve becomes horizontal.

C; Sever oscillation occurs with periodical bubble discharge from the impeller inlet to the upstream suction pipe. The oscillation becomes larger with a decrease in pump discharge. Pressure fluctuation of this pattern is shown in Fig. 3 and the sketch of the flow behavior is illustrated in Fig. 5.

The bubbles accumulating in the impeller are suddenly discharged to the upstream along the casing wall with strong whirl, and liquid single flow enters into the impeller in the pipe center (Fig. 5(a)). Many small bubbles spreading in the pipe then gather to the pipe center due to a strong whirl, grow to large bubbles, form a large group of bubbles (Fig. 5(b)) and then come into the impeller. As the inlet whirl then gradually disappears, the residual bubbles spread to a whole pipe and come into the impeller together with those from the upstream (Fig. 5(c), (d)). The flow pattern again returns to Fig. 5(a).

Throughout one cycle of oscillation, small bubbles flows out steadily at the pump outlet, though the discharge pressure p_d fluctuates largely as shown in Fig. 3.

D; The impeller inlet flow is separated to liquid flow and gas flow. Soon after the start of air supply the oscillation occurs between two curves shown in Fig. 4, but after

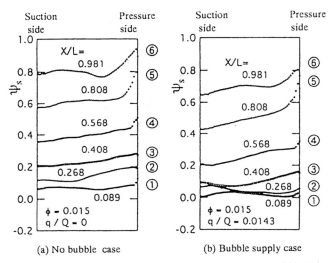

(a) No bubble case   (b) Bubble supply case

Figure 6. Blade-to-blade pressure distribution (ϕ =0.015, 1000rpm)

5 minutes it ceases and the head curve comes to the dotted line. The amplitude of oscillation decreases with a decrease in pump discharge.

E; Bubbles accumulate in the impeller inlet region without oscillation, and spread to occupy the upper part of suction pipe, which decreases the pumping head gradually.

4. 2 INSTANTANEOUS BLADE-TO-BLADE PRESSURE DISTRIBUTIONS

In the case of sever oscillation (pattern C), the shaft power also oscillates, which implies that the blade loading fluctuates. The instantaneous pressure of one impeller channel is then measured and was averaged over 50 rotations at each measuring point. The averaged values of 50-rotations didn't show significant difference from those of 200-rotations. The blade-to-blade pressure distributions thus obtained are shown in Fig. 6, in which the liquid single flow case in Fig. 6(a) is compared with the gas-liquid flow case in Fig. 6(b).

Comparison of Figs. 6(a) and 6(b) reveals that air supply decreases the pressure in the whole impeller channel, when the oscillation occurs. A remarkable difference caused by air supply is seen in the measuring points ①, ②and③. This shows that bubbles accumulate around the impeller inlet region and decreases the blade loading. However, the pressure at ⑤ is seen to recover largely, which shows that the blade loading near the outlet becomes larger than that in the no air case.

It is then concluded that the head drop due to air supply is caused by the loss of blade loading around the impeller inlet due to the accumulation of bubbles.

4. 2. 1 *Bubble discharge period and bubble suction period*

The above-described results are obtained by the 50-rotations-averaged values, corresponding to 3 cycles of oscillation, but the instantaneous blade-to-blade pressure distribution must be different from the averaged one especially in the bubble accumulating period. The data averaging period is thus divided to the bubble discharge period and the bubble

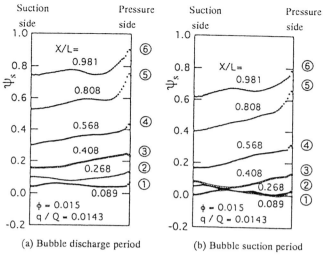

(a) Bubble discharge period   (b) Bubble suction period

Figure 7. Comparison of instantaneous blade-to-blade pressure distribution (ϕ =0.015, 1000rpm)

suction period by manually switching the data acquisition with observing the phenomena, and the results are compared in Figs. 7(a) and (b).

Comparison of Figs. 7 with Fig. 6 reveals that the instantaneous blade-to-blade pressure profile at bubble discharge period (Fig. 7(a)) is almost same as that of no air case (Fig. 6(a)) and that at bubble suction period (Fig. 7(b)) is almost same as that of the one-cycle averaged (Fig. 6(b)). That is to say, the bubble suction period occupies large part of one cycle of oscillation, and after that bubbles are rapidly discharged to the upstream. In the latter case the impeller is filled only with liquid and attains the maximum pumping head which is equal to that of no air case.

4. 2. 2 *Influence of Air Flowrate on Blade-to-Blade Pressure Distribution*

When the air flowrate ratio is increased, the instantaneous pressure distribution between blades changes as shown in Fig. 8, of which lower figures show the net pressure rise along a blade obtained by subtracting the pressure rise due to centrifugal force.

The upper figures reveal that the pressure decrease is considerable especially around the leading edge, and that the pressures at ①, ② and ③ decreases in order with an increase in air flowrate ratio, while the pressure at ⑥ changes little. This shows that the bubble accumulation zone gradually spreads to the downstream from the leading edge.

In the lower figures in Fig. 8, the pressure difference between pressure side and suction side expresses the blade loading and decreases rapidly in the upstream half while increases in the downstream half with an increase in air flowrate ratio.

Comparison of Figs. 8(a) with 8(b)~(d) also reveals that the air supply decreases the pressure especially in the blade pressure side. Visual observation revealed that bubbles accumulate much more in the pressure side. These two results reveals that the bubbles accumulating in the pressure side near the leading edge decreases the blade loading and makes the main flow deviate to the blade suction side. The deviation of main flow should decrease the pressure recovery in the suction side and the pressure at the downstream

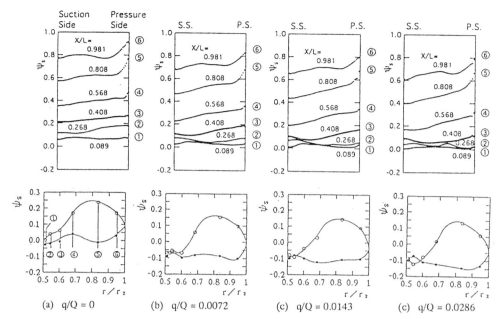

Figure 8. Change in instantaneous blade-to-blade pressure distribution by the change of air flowrate ratio
(φ =0.015, 1000rpm)

of bubble accumulation zone should recover rapidly due to sudden expansion of main flow, as is shown in Fig. 7. Thus the bubbles accumulate mainly in the blade pressure side at very low discharge, though they accumulate in the suction side at BEP as in [4].

Though sever oscillation is induced at impeller inlet region, the pressure p_{s1} at 900 mm upstream of the impeller is kept constant as shown in Fig. 3. This suggests that the flow oscillation is limited to the pump inlet region and influence little to the upstream suction pipe. To make this point clearer, the variation of pressure p_{s2} at 50mm upstream of the impeller inlet is shown in Fig. 9. The pressure p_{s2} is seen to oscillate in the same phase with the maximum value equal to that of no-air case, the minimum value equal to the upstream pressure p_{s1}. This pressure fluctuation is due to a strong periodical whirl of the recirculation flow which exists near the suction pipe wall of pump inlet.

4. 3 CHARACTERISTICS OF PRESENT OSCILLATION

The period and the amplitude of present oscillation depend largely on the air flowrate and the pump rotational speed. In order to reveal the characteristics of the present oscillation, the variations of amplitude and frequency are shown in Figs. 10 (a) and (b), when the pump rotational speed and the flowrate ratio q/Q of air to liquid are varied.

When the air ratio is increased under the same rotational speed, the amplitude ratio $\Delta\psi/\psi$ of pumping head becomes larger and the frequency becomes smaller. In this case the visual observation revealed that the bubble accumulation zone along a blade becomes larger and both the scale and the strength of oscillation become larger with an increase in air flowrate ratio. When the pump rotational speed is increased, both the amplitude and

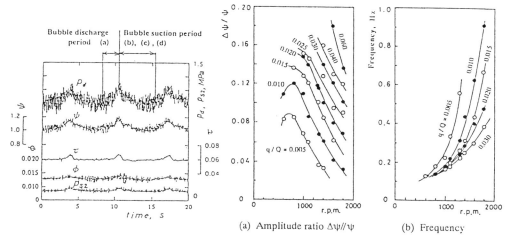

Figure 9.   Fluctuation of p_{s2}

(a) Amplitude ratio $\Delta\psi//\psi$ (b) Frequency

Figure 10.   Influence of rotational speed and air flowrate ratio

the frequency become smaller resulting in smaller-scaled oscillation.

The present oscillation is one of the unstable phenomenon accompanying gas-liquid flow in a pumping system. The characteristics of the present oscillation is that it occurs in a very low discharge range(10~30% of the designed) in which large recirculation flow should be induced if there were no oscillation, and that it occurs only by supplying more than only 0.5 % of air to liquid. The length of recirculation flow zone in a suction pipe is then considered to influence the phenomena, but the removal of the swirl stop at 600mm upstream in the suction pipe revealed no significant change in the phenomena.

Another remarkable characteristics of the present oscillation is that it occurs only in a semi-open type impeller and doesn't occur in a closed type impeller.   The tip clearance was thus considered to influence the phenomena, but the zero clearance test by attaching a rubber plate at the impeller tip revealed no significant change in the phenomena.

5. Theoretical Considerations on the Mechanism of Present Oscillation

In order to elucidate the mechanism and the cause of the present oscillation, the stability analysis based on Gleitzer [5] is applied to the present pumping system.

In the pumping system shown in Fig. 11(a) under non-cavitation condition, it is assumed that the fluctuations of pressure p, liquid flowrate q and gas flowrate q_G (these notations are different from the preceding chapters) be small compared with the mean pressure P and the mean flowrates Q, Q_G of liquid and gas, respectively. Under these assumptions a perturbation method can be introduced to linearize the equation. It is also assumed that the air flowrate at pump outlet is constant and subscripts 1 and 2 are given to the pump suction side and the discharge side, respectively.

Considering that the bubble flow rate is small compared with the liquid flowrate, and that the valve resistance is dominating in the discharge pipe, the linear equations of motion as for the fluctuation terms in the suction pipe, the discharge pipe and the pump are expressed as follows in the non-dimensional form;

$$p_1 = -L \dot{q}_1 \tag{1}$$
$$p_2 = Rv q_2 \tag{2}$$
$$p_2 - p_1 = R_P q_1 - \dot{q}_1 + R_G q_G \tag{3}$$

Here, the pressure p is normalized by $\rho U^2/2$, the flowrate by AU and the time t by $2L_p/U$, where U, A and L_p are the impeller tip speed, the sectional area of pipe and the equivalent length of pump, respectively. The notation $\dot{q}_1$ denotes the differentiation of q_1 as for time t. L is the length of the inlet recirculation zone (or the length of suction pipe if there is no reverse flow) normalized by L_p. The parameters R_p, Rv and R_G are the gradients of the characteristic curves of the pump system shown in Figs. 11 (b) and (c), and are defined as

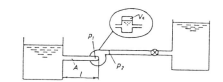

(a) Pump system

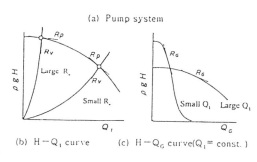

(b) $H - Q_1$ curve   (c) $H - Q_G$ curve(Q_1= const.)

Figure 11. Model and characteristic curves of pumping system and system resistance

$$R_P \equiv d(2gH/U^2) / d(Q_1/AU), \quad Rv \equiv d(2gHv/U^2) / d(Q_1/AU), \quad R_G \equiv d(2gH/U^2) / d(Q_G/AU)$$

where H and Hv are the pumping head and the system resistance(valve), respectively.

Putting the air volume accumulating in an impeller as V_c and its fluctuation as δV_c normalized by AL_p, the following equations are obtained in the non-dimensional form;

$$\dot{V}_c = q_2 - q_1 + q_G \tag{4}$$
$$\delta V_c = C_p p_1 + M_b q_1 + M_G q_G \tag{5}$$

where $C_p \equiv \partial V_c / \partial p_1$, $M_b \equiv \partial V_c / \partial Q_1$, $M_G \equiv \partial V_c / \partial Q_G$

The motion of a bubble is to be determined from the balance of forces acting on a bubble. Here for the simplicity, the fluctuation of the bubble volume at an impeller inlet is assumed as follows in the non-dimensional form;

$$q_G = \alpha \, p_1 + \beta q_1 \tag{6}$$

where $\alpha \equiv \partial Q_G / \partial P_1$, $\beta \equiv \partial Q_G / \partial Q_1$

Equations (1) to (6) are all linear with constant coefficients. Hence the solutions are of the form e^{st}. For stability the real part of s must be negative. Substituting expressions of the form e^{st} for the perturbations, the growth rate must satisfy the following equation, if there are to be non-tribial solutions to the resulting system;

$$s^2 + as + b = 0 \tag{7}$$

where
$$a = -\left\{ \alpha + (M_b + \beta M_G) / L + (1 + 1/L + \alpha R_G)/Rv \right\} / (C_p + \alpha M_G)$$
$$b = -\left\{ 1 - \beta - (R_p + \beta R_G) / Rv \right\} / L(C_p + \alpha M_G)$$

Generally speaking, the negative valve of b gives static instability, while the negative value of a gives dynamic instability if b is positive. In the present pumping system the parameters in the above-equations are considered to be as follows;

$$C_p < 0, \quad M_b < 0, \quad M_G > 0, \quad R_p < 0, \quad R_V > 0, \quad R_G < 0, \quad \alpha < 0, \quad \beta \sim 0$$

Thus the denominators of a and b are negative and b takes a positive value, but the quantity a can be positive or negative depending on the numerator of a. After all for the dynamic instability must satisfy the following equation.

$$\alpha + M_b/L + \left(1 + 1/L + \alpha\, R_G\right)/R_V < 0 \tag{8}$$

The first two terms are negative, and the last term in bracket is positive. The self-sustained oscillation could thus be caused depending on these parameters. If the liquid flowrate is large, the system resistance R_V becomes small, resulting that the system is stable. With an increase in system resistance, the flowrate decreases and the last term becomes very small, which satisfies Eq. (8), resulting in an unstable system. When the flowrate ratio of air to liquid is large, the value αR_G becomes large, resulting in dynamic stability.

From the above-described theoretical considerations, it is concluded that the present self-sustained oscillation could be caused by the unbalance between the system resistance rate R_V and the loss rate of blade loading R_G under the influenced of the behavior M_b of the bubble accumulating zone.

One of the other remarkable characteristics is that the oscillation occurs only in a semi-open type impeller. The careful observation revealed that the bubbles in an impeller channel become very small-sized, and accumulate mainly near the blade pressure side. This might be caused by the existence of recirculating flow zone induced by flow separation of blade suction side, as shown by the three-dimensional structurein in Fig.13(a). When the bubble zone further grows, the recirculation flow zone might be suppressed and disappear,because the recirculation flow of a semi-open type impeller is very unstable due to the lack of the outer shroud wall.

The visualization of flow behavior using tufts attached on casing wall revealed that the recirculation flow disappears instantaneously and at the same time the bubble accumulating zone rapidly grows to the upstream, and that the casing wall just in front of the impeller inlet is covered with air layer rotating much more slowly than the impeller.

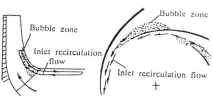

(a)   Recirculation flow region and bubble zone

When inlet recirculation flow disappears, both the meridional and the tangential velocities rapidly decreases at an impeller inlet, which results in large angle of attack at blade inlet. A large-sized recirculation flow is thus induced again, and the bubbles are suddenly discharged to the upstream with the recirculation flow. During the period from bubble discharge to bubble gathering to the pipe center, the liquid flowrate increases at impeller inlet, and the pumping head increases due to the recovery of

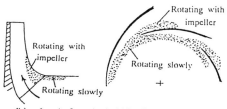

(b)   Just before the bubble dicharging

Figure 12. Sketch of bubble behavior and recirculation flow at pump entrance

blade loading, which results in positive slope of head-capacity curve. On the contrary in a closed type impeller the recirculation flow might form a steadily staying bubble zone which coexists stably with a recirculation flow.

From the above-described considerations the disappearance of the recirculation flow due to the growth of bubble zone might be responsible for the present oscillation, and such difference in the bubble behavior between a semi-open type impeller and a closed type could be expressed by the difference in α and M_b in Eq. (8).

When a pump is operated near the shut-off point, the impeller inlet recirculation flow develops to be so strong and large-sized that the growth of bubble zone cannot cause the recirculation flow to disappear. The pumping system thus does not come into a dynamic instability but rapidly come into a static instability, resulting in a large head-drop.

6. Conclusions

A low frequency oscillation caused by a semi-open impeller at low discharge operation in gas-liquid flow is studied. The characteristics of the oscillation are determined experimentally and its mechanism is studied theoretically by use of a linear stability analysis.

The measurements revealed that the head drop is caused by the accumulation of bubbles along a blade pressure side around the leading edge. Theoretical considerations suggest that the bubble accumulation in blade-to-blade channels could be responsible for the present oscillation. The theoretical model provides one possible explanation as for the excitation of the oscillation, that it is caused by dynamic unbalance between the head drop rate due to bubble accumulation and the system resistance rate under the influence of flow interference between the bubble accumulating zone and the recirculation flow.

The reason why the present phenomena occurs only in a semi-open type impeller might be that the recirculation flow at impeller inlet is so unstable that the growth of bubble accumulating zone near the blade leading edge suppresses the recirculation flow and makes it disappear, which rapidly causes a large recirculation flow.

Acknowledgements

The present authors would like to express their sincere gratitude to Dr. K. Yamamoto, Dr. M. Aoki and Mr. I. Ichiki of Ebara Research Co. Ltd. for their valuable suggestions and financial support in the present study.

References

(1) Yamamoto, K., Instability in a Cavitating Centrifugal Pump, *JSME International Jr.*, Ser. 2, Vol. 34, No. 1(1991), pp. 9-17.

(2) Brennen, C. and Acosta, A. J., The Dynamic Transfer Function for a Cavitating Inducer, *Trans. ASME*, Jr. Fluid Dynamics, Vol. 98, No. 2(1976), pp. 182-191.

(3) Kaneko, M. and Ohashi, H., , Self-Excited Oscillations of a Centrifugal Pump System under Air/Water Two Phase Flow Condition , *Proc. IAHR Symp.*(Amsterdam), Vol. 2 (1982), Paper No. 36.

(4) Ichiki I., Airlocking Phenomena at Partial Load Operation in Centrifugal Pump, *Ebara Research Report* (in Japanese), Vol.151(1991-4), p. 24.

(5) Gleitzer, E. M., The Stability of Pumping Systems, *Trans. ASME*, Jr. Fluid Engineering, Vol. 103, No. 2(1981), pp. 193-242.

MEASUREMENTS IN THE DYNAMIC PRESSURE FIELD OF THE VOLUTE OF A CENTRIFUGAL PUMP

J.L. PARRONDO, J. FERNÁNDEZ, C. SANTOLARIA & J. GONZÁLEZ
Área de Mecánica de Fluidos, Universidad de Oviedo
ETSIIG. Campus de Viesques. 33271 Gijón. Spain.

Abstract

This paper presents an experimental investigation into the dynamic pressure field existing in the volute of an industrial centrifugal pump in order to characterize the interaction phenomena between impeller and volute. For that purpose, pressure signals were obtained simultaneously at different points of the volute casing by means of two miniature fast-response pressure transducers. Particular attention was paid to the pressure fluctuations at the passing blade frequency, regarding both amplitude and phase delay relative to a reference point. The analysis of the dependence of the pressure fluctuations on both flow-rate and position along the volute clearly indicates the leading role played by the tongue in the impeller-volute interaction and the increase of the amplitude of the dynamic forces in off-design conditions.

1. Introduction

Centrifugal pumps with volute casing are usually subjected to static thrust in both the axial and radial directions. Radial thrust is particularly great in off-design conditions, because, for both low and high flow-rates, the presence of the volute provokes a non-uniform pressure distribution around the outlet of the impeller (Stepanoff 1957, Csanady 1962). Additionally, centrifugal pumps can also present non-steady radial forces, which are usually associated with the frequency of rotation or with the passing-blade frequency (and with their harmonics). Excitation at the frequency of rotation may be provoked by impeller whirling, when the impeller has an orbital motion coupled to the rotation, due to shaft misalignment for instance. An impeller with small manufacturing imperfections will also present some excitation at that frequency. Excitation at the passing-blade frequency (frequency of rotation multiplied by the number of blades of the impeller) is a consequence of the finite thickness of the blades, which causes flow disturbances in the volute associated with the passage of each blade. In the case of centrifugal pumps with vaned diffusers significant excitation may also exist at the vane passage frequency (frequency of rotation multiplied by the number of vanes) and at the blade-vane passage frequency (passing-blade frequency multiplied by the number of vanes).

E. Cabrera et al. (eds.), Hydraulic Machinery and Cavitation, 401–410.
© *1996 Kluwer Academic Publishers. Printed in the Netherlands.*

The effects of those excitation mechanisms may be substantially modified by rotor-stator interaction phenomena. Previous work in this field was mostly focused on axial machinery, either compressors (Gallus et al. 1980) or turbines (Dring et al. 1982). Regarding centrifugal pumps, Adkins & Brennen (1988) investigated both theoretically and experimentally the hydrodynamic forces arising from impeller whirling in a pump with volute casing, and found that those forces could be destabilizing under many operating conditions. Arndt et al. (1990) investigated the interaction of a centrifugal impeller with several vaned diffusers of different geometry by measuring pressure fluctuations in both impeller and diffuser. More recently Chu et al. (1993), also Katz (1996), used particle image velocimetry to measure the velocity distribution in the near-tongue region of the volute of a centrifugal pump, together with pressure and noise measurements. They concluded that dominating phenomena are associated with the interaction of the blades with the tongue, and that even a small increase in the space between impeller and tongue (up to 20% of the impeller radius) causes significant reductions in noise levels.

This paper presents an experimental study of the dynamic interaction between the volute and the impeller of an industrial centrifugal pump, with particular attention to the excitation at the passing-blade frequency. Pressure measurements were taken simultaneously at different positions of the volute casing by means of miniature fast-response pressure transducers. The amplitude and phase delay of the pressure fluctuations obtained for different positions and flow-rates clearly show that blade-tongue interaction is a primary contributor to the fluctuating pressure field in the volute.

2. Experimental Equipment

The pump tested in this investigation was a Worthington EWP-65-200, a centrifugal pump with axial suction and volute casing, equipped with an impeller of 190 mm in outside diameter (Figure 1). Other dimensions of the impeller were: diameter at inlet hub, 67 mm; diameter at inlet tip, 88 mm; discharge width, 16.8 mm; number of blades, 7; and angle of blades at impeller output, 29°. The diameters of the suction and impulse pipes were 107.1 mm and 64.2 mm respectively. Figure 2 shows the performance curves of this pump when driven at 1500 r.p.m. (Fernández & Santolaria 1993), with a best efficiency point at about 48 m^3/h. Manufacturer's data at 2900 r.p.m. indicated a maximum efficiency of 76% for a rate of 93 m^3/h at 43 m head. According to similarity laws, both best efficiency points nearly agree.

The pump was connected in closed-loop to a reservoir with a capacity of 10 m^3. The set-up was designed according to BS-5316/2 specifications. Flow-rate was regulated by means of a butterfly valve located at the reservoir inlet, together with a pinch valve at the outlet of the pump for fine adjustment. The flow-rate was measured with a previously calibrated orifice plate, connected to an inclined piezometric mercury manometer. Measurement uncertainty was estimated to be less than 4% for flow-rate values greater than 20 m^3/h. The pump was driven by a DC-motor governed by a regulation device that maintained a rotational speed of 1500 ± 1.0 r.p.m.

3 mm diameter pressure taps were located every 30° along the volute, at 7 mm

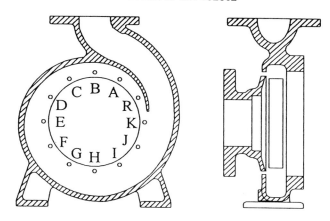

Figure 1. Centrifugal pump drawing and location of the pressure taps.

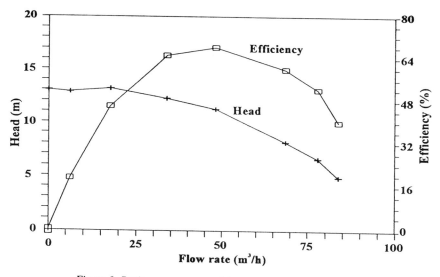

Figure 2. Performance curves of the pump tested at 1500 r.p.m.

from the outlet of the impeller (Figure 1). Static pressure at those positions could be obtained with a Kistler 4043A10 piezo-resistive pressure transducer and a K-4601 current amplifier, which provided absolute pressure values with an uncertainty of less than 0.5% (according to manufacturer's data). Fluctuating pressure could be also measured in the pressure taps along the volute by means of two Kistler 601A miniature fast-response piezo-electric pressure transducers, each of which was connected to a K-5007 charge amplifier (combined uncertainty of less than 1.5% according to manufacturer's data). Pressure signals were processed with an HP-3562A two-channel

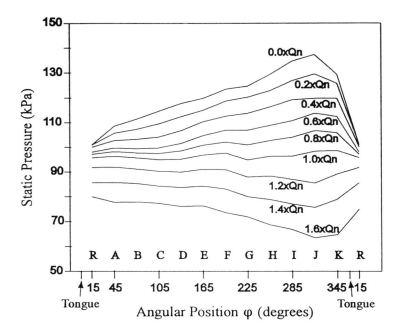

Figure 3. Static pressure distribution around the impeller outlet, for different flow-rates (Qn = 48 m³/h).

dynamic signal analyzer, which permitted either time or spectral averaging of the signals and the obtaining of various spectral functions, including power and cross-power spectra, and coherence.

3. Static Pressure Measurements

A preliminary series of tests was conducted to measure the static pressure along the volute, as a function of the flow-rate. For such purpose each of the pressure taps was connected to a collector chamber by means of a flexible hose with a switching valve. Proper operation of these valves permitted the pressure at the taps to be transmitted to the collector, where the piezometric pressure transducer was installed (Fernández & Santolaria 1993). The flow-rate was progressively increased from zero to the maximum value achievable in the set-up, and for each flow-rate the switching valves of the different pressure taps were sequentially operated. Measurements were taken some time after each valve switching to allow for regularization of the pressure in the collector.

Figure 3 shows the circumferential distribution of the static pressure so obtained for different values of the flow-rate; angular position φ is zero at the edge of the tongue and increases when rotating anti-clockwise. It may be observed that the static pressure along the volute is quite uniform for flow-rates around the best efficiency point. This is a

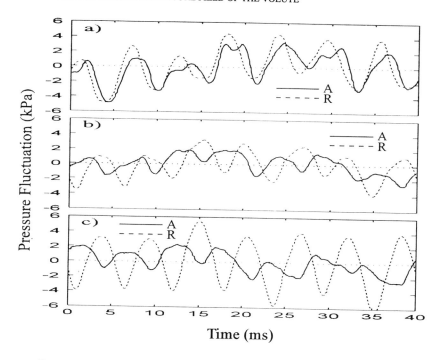

Figure 4. Pressure signals at positions 'A' and 'R' during one complete revolution.
Flow-rate: a) 17.1 m³/h; b) 48.0 m³/h; c) 67.2 m³/h.

foreseeable result, since an efficient design of the volute is associated with a minimum radial thrust.

In off-design conditions, however, pressure is non-uniform and achieves a maximum for low flow-rates or a minimum for high flow-rates at about the position $\varphi = 315°$. This result agrees well with the trends indicated in classical texts (Stepanoff, 1957). Such non-uniform pressure distribution is due to the direction of the absolute velocity at the outlet of the impeller, variable with the flow-rate, and the subsequent incidence of the flow on the volute. This interaction between impeller and volute may be said to be stationary, as opposed to the dynamic interaction referred to below.

4. Dynamic Pressure Measurements

For the fluctuating pressure measurements, both piezoelectric pressure transducers were connected to two of the pressure taps. A small adaptor was used so that the sensitive membrane of the transducers was 10 mm from the internal surface of the volute, thus ensuring that any disturbing resonance phenomena inside that cavity would only occur well above the range of frequencies of interest in this study. Simultaneous

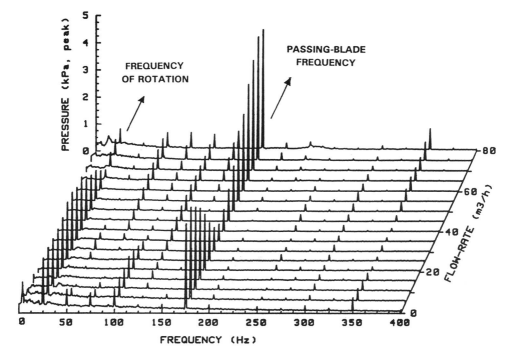

Figure 5. Pressure power spectrum at position 'R' as a function of flow-rate.

capture of the signals from both transducers permitted not only the amplitude but also the phase delay between the corresponding fluctuations to be obtained.

Figure 4 shows the evolution of the pressure at positions 'A' and 'R' (see Figure 1) during one complete revolution of the impeller for three different values of the flow-rate. These curves are the average of 100 trials, with each sample being triggered by the signal from an optical tachometer oriented to the pump shaft. A total of 7 hills and valleys can be appreciated in these curves, corresponding to the 7 blades of the pump as they pass in front of the pressure transducers. Such modulation is due to the finite thickness of the blades and the corresponding non-uniformity of the flow around the outlet of the impeller. However, it is clear from Figure 4 that: 1) the shape and the amplitude of the pressure fluctuations depend on the position of the transducers for a given flow-rate; 2) both shape and amplitude depend on the flow-rate for a given transducer position; and, 3) the time (or phase) delay between the pressure fluctuations at two positions depends on the flow-rate.

All this behaviour is a consequence of the dynamic interaction between impeller and volute, which is the principal objective of this study. A systematic series of experiments were conducted for which one of the transducers was located permanently at the tap with $\varphi = 15°$, the one referred to as 'R' in Figure 1, whereas the other transducer was progressively moved from position 'A' to position 'K' (Figure 1). For each pressure tap position and for stepped increments of the flow-rate, after some time for flow settling, the analyzer performed spectral averages of 100 trials of the in-coming signals (for 5 minutes) and obtained their power spectrum, cross-power spectrum and coherence.

Whereas the power spectrum provides a direct measure of the amplitude of the fluctuations at each frequency, the phase of the cross-power spectrum provides a measure of the phase delay between the fluctuations of two signals. The coherence function permits an evaluation of the degree of correlation between them at each frequency. Figure 5 shows the evolution of the power spectrum of the pressure at the reference position ('R'), with respect to the flow-rate. As expected (see Section 1), the predominant spikes correspond to the frequency of rotation of the impeller (25 Hz) and in particular to the passing-blade frequency (175 Hz), together with their respective harmonics. As stated above, the present study was focused on the excitation at the passing-blade frequency, as provoked by the finite thickness of the blades.

Since the power spectrum of the pressure at position 'R' was obtained several times for a given flow-rate (one series of measurements for each of the other pressure taps), examination of the spike amplitude at a given frequency permitted the checking of the maintenance of flow conditions between one test to another. The variations so observed at the passing-blade frequency were always less than 10% and usually less than 3%. At least two series of tests were conducted for each pressure tap position, on different days, and good repeatability was found with respect to both amplitude and phase delay at the passing-blade frequency. Phase delay variations were always less than 4%, referring to 360°. Coherence at the passing-blade frequency was always greater than 0.9 and usually greater than 0.99.

5. Experimental Results at the Passing-Blade Frequency

Figure 6 shows the amplitude of pressure fluctuations at the passing-blade frequency as a function of the flow-rate for different circumferential positions along the volute. The behaviour observed is very different from that of the stationary pressure distribution (Figure 3). In general the pressure fluctuations are greater in off-design conditions, i.e. for both low and high flow-rates. The maximum values, which correspond to the region of the tongue (position 'R') for very high flow-rates, are about 6% of the corresponding stationary pressure. Some of the curves exhibit a clear minimum, at a flow-rate dependent on the tap position ('R', 'B', 'D', 'F'...), and some other curves, however, are much flatter ('A', 'C', 'E'...). The continuous crossing of the curves of Figure 6 suggests the existence of node and anti-node positions along the volute, variable with the flow-rate, which is a typical effect of wave resonance phenomena.

Figure 7 shows the phase delay ψ between the pressure fluctuations at each pressure tap and those at position 'R', after subtraction of the nominal phase associated with the time lag that one blade needs for moving from one position to the other. This nominal phase, which is not dependent on the flow-rate, may be shown to be equal to $210 \cdot i$ degrees, for a position separated $30 \cdot i$ degrees from tap 'R'. Interestingly, most of the curves of Figure 7 exhibit a very similar behaviour: increasing the flow-rate results in

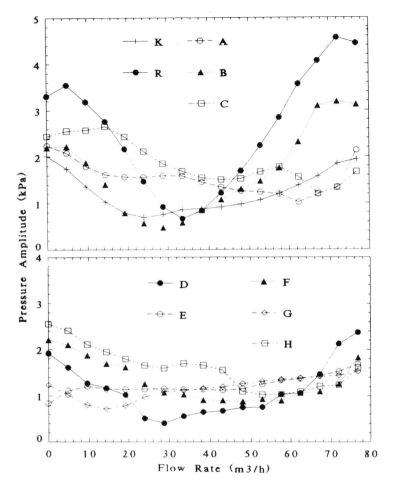

Figure 6. Amplitude of pressure fluctuation (passing-blade frequency)
as a function of flow-rate for several locations along the volute.

a reduction of the phase delay ψ, from about 180°-240° for the lower range of flow-rates to values surrounding 0° for flow-rates between 85% of the best efficiency point and the maximum available flow-rate. This means that the pressure fluctuation peaks are usually not in phase with the passage of the blades. Position 'B' is the only one to follow a different trend, which might be attributed to the existence of a node in its vicinity for small flow-rates.

The large relative phase delay at small flow-rates actually corresponds to low absolute phase delays, i.e., the pressure fluctuations around the impeller are synchronized with the passage of the blades in front of the tongue. This result indicates the great influence of the tongue-blade interaction on the dynamic pressure field existing

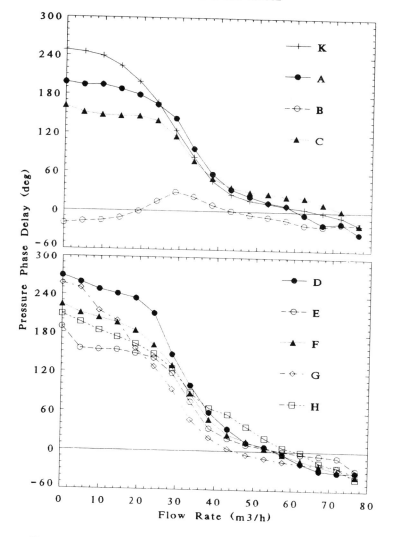

Figure 7. Pressure phase delay (at the passing-blade frequency) relative to position 'R' as a function of flow-rate for several locations along the volute.

within the volute for the lower range of the flow-rate, which agrees with the results of previous studies (Chu et al. 1993, Katz 1996). For increasing values of the flow-rate such influence is progressively reduced and partially counterbalanced by the pressure difference between both sides of the blades. The best efficiency point is also the point for which the dynamic excitation is minimum: minimum pressure fluctuations along the volute and phase delay $\psi \approx 0°$.

6. Conclusions

A systematic series of tests were conducted to measure the dynamic pressure field in a number of positions along the volute of a centrifugal pump. The analysis focused on the pressure amplitude and phase delay at the passing-blade frequency, depending on both angular position and flow-rate. The results obtained show that the distribution of pressure fluctuations greatly depend on the flow-rate. In general the amplitude of the fluctuations is greater in off-design conditions, in particular in the near-tongue region, which suggests the monitoring of the pressure in that zone as an effective means of observing the operation of the pump, for predictive maintenance applications. Minimum dynamic excitation approximately corresponds to the best efficiency point, for which the pressure fluctuations are approximately in phase with the passing of the blades. For low flow-rates, the interaction between impeller and tongue is dominant in the generation of the dynamic pressure field, as indicated by the analysis of the phase delay existing between the pressure fluctuations at different positions.

Acknowledgements

The authors gratefully acknowledge the financial support of the Dirección General de la Investigación Científica y Tecnológica (Spain) under Project TAP-94-0653 entitled "Optimización de técnicas de diagnosis y control de turbomáquinas".

References

Adkins, D.R. & Brennen, C.E. (1988) Analyses of hydrodynamic radial forces on centrifugal pump impellers, *ASME Journal of Fluids Engineering* **110**, 20-28.

Arndt, N., Acosta, A.J., Brennen, C.E. & Caughey, T.K. (1990) Experimental investigation of rotor-stator interaction in a centrifugal pump with several vaned diffusers, *ASME Journal of Turbomachinery* **112**, 98-108.

Chu, S., Dong, R. & Katz, J. (1993) The effect of blade-tongue interactions on the flow structure, pressure fluctuations and noise within a centrifugal pump, in G. Caignaert (ed.), *Pump Noise and Vibrations*, CETIM, Paris, pp.13-34.

Csanady, G.T. (1962) Radial forces in a pump caused by volute casing, *ASME Journal of Engineering for Power* **84**, 337-340.

Dring, R.T., Joslyn, H.D., Hardin, L.W. & Wagner, J.H. (1982) Turbine rotor-stator interaction, *ASME Journal of Engineering for Power* **104**, 729-742.

Fernández, J. & Santolaria, C. (1993) Esfuerzos de origen fluidodinámico en una bomba centrífuga con rodete recortado, *Actas del I Congreso Iberoamericano de Ingeniería Mecánica* **2**, AEIM, Madrid, pp.187-193.

Gallus, H.E., Lambertz, J. & Wallman, T. (1980) Blade row interaction in an axial flow subsonic compressor stage, *ASME Journal of Engineering for Power* **102**, 169-177.

Katz, J. (1996) On the relationship between volute geometry, flow structure, pressure fluctuations and noise in a centrifugal pump, *Flow in Radial Turbomachines*, Lecture Series 1996-01, Von Karman Institute for Fluid Dynamics.

Stepanoff, A.J. (1957) *Centrifugal and Axial Flow Pumps*, Wiley, New York.

FUNCTIONAL MODELLING OF PUMP VOLUTE GEOMETRY

P.R. THACKRAY, R.D. JAMES
Hydraulic Design Engineer, Kvaerner Boving Ltd, Doncaster, UK.,
Senior Lecturer, The University of Hull, UK

Abstract

The task of hydraulic design is usually undertaken using traditional techniques, coupled with many simplifying assumptions. This can lead to long design lead-times and under-utilisation of performance potential. New computational approaches are now required. In this resolve, some considerations of linking intelligent computer technology with fundamental aspects of hydraulic design, in particular pump volute geometry, are discussed. The paper firstly reviews volute flowfield measurements and the effects of geometry, and goes on to discuss some recent experiences in developing an experimental system to provide a functional modelling facility.

1. Background

The desire to understand the real flow characteristics in spiral casings of pumps and turbines stems from an intense need to produce higher hydraulic efficiencies. Early theoretical treatments of pump outflows by Kucharski and others focused on the impeller and assumed perfect fluid flow; the effects of blade curvature were often simplified, and volute effects were ignored completely [1]. Photographic flow studies by Fischer and Thoma [2] in 1932 using dye injection into an open impeller pump with glass sides showed recirculation at low flows. Binder and Knapp [1,3] determined that radial velocities and pressure distributions were non-uniform even at design flow. In 1956 Bowerman and Acosta [4] demonstrated that an impeller operates differently with different volute spirals. Efficiency variations of around 5% were reported for three log-spiral volutes of vane angles varying by 9°. The shape of the volute was also found to be prominent in determining the overall efficiency of the pump. In 1960, Iversen *et al* [5] presented a theoretical pressure distribution analysis which included an estimate for mixing losses. Csanady [6] obtained similar results by conformal mapping of the shed flow cascades.

In the 1960's, Worster [7,8] provided basic empirical guidelines for impeller-volute interaction and geometry. Potential flow theory was used to superimpose a source and free vortex to model the volute flow. The shape of the volute section was found to strongly affect pump performance, and the magnitude and position of the best efficiency flow point. A partly-trapezoidal section provided a smoother pressure distribution around the volute and marginally higher efficiency than a comparable

411

E. Cabrera et al. (eds.), Hydraulic Machinery and Cavitation, 411–418.

rectangular section. This was also confirmed in tests by Pekrun [9] and others. The shape and angle of the tongue was found by Hira and Vasandani [10] to affect the magnitude and position of the best efficiency flow point. Kurokawa [11] showed that a 2-d ideal frictionless flow analysis could still be used to accurately predict volute flow characteristics over a wide range of off-duty conditions. Imaichi [12] used a potential flow analysis accounting for the effect of vortex shedding to estimate unsteady forces; these were found to agree well with empirical formula. Streamline plots derived by streak photography by Brownell and Flack [13] showed stagnation points and zones of flow separation in the volute; streamlines were also found to deform in proportion to the magnitude of off-design flowrate.

The advent of laser technology meant that non-intrusive measurements of velocity fields became possible. Normalised non-rotating flow profiles taken by Thomas et al [14] in 1986 using laser velocimetry showed velocities at the impeller periphery up to 30% higher than the average volute core velocity. The nature of the volute flow was found to depend strongly on impeller outlet angle. Miner et al [15] showed that the momentum flux in the volute is always asymmetric, even at design flow. Volute tangential velocity profiles were found to be in close agreement with a free vortex flow, and the jet-wake phenomenon from the impeller was seen to rapidly dissipate within the volute. The turbulent flow structure of the volute was recently measured by Dong et al [16] using laser velocimetry. Phase-averaged vorticity distributions showed a flow structure strongly dependent upon geometry and flowrate.

Today many of the hydrogeometric features can be determined and optimised at an early stage of the design process. However, most hydraulic pump design is essentially empirically-based and therefore sub-optimal. Volute geometry is of fundamental importance in minimising losses and achieving an optimal outflow field. In addition to impeller design improvements, problems remain in modelling an optimum volute profile. Because of the large number of variables and design steps involved, a decision making process is obviously required. The following discussion relates to some recent experiences [17] in this field of turbomachinery design, focusing on the use of artificial intelligence (AI) to capture and model required hydraulic design functionality.

2. Modelling Design Intent

Nearly all existing computer-aided design systems suffer from an inability to work intelligently with the designer. On their own, these systems are not knowledgeable, nor gain knowledge, and may accept erroneous inputs without checking against some reference base for conformance and accuracy. A system which provides some inference capability is known as an intelligent CAD system (ICAD). Ideally such a system should be able to understand the intentions of the designer, detect errors, suggest alternatives, etc, by referring to a design knowledge-base. Currently there is very limited experience of geometrical reasoning in engineering design [18,19], and few examples are known of such working applications in the turbomachinery field. In the complex design cycle of a new pump or turbine, the conceptual design stage is currently in greatest need of rationalisation.

Decision making in a design process involves a sequence of conformance or alteration steps, performed within a framework of constraints or requirements. Along with a suit of alteration rules, a specific method for knowledge representation is required, ideally as an intelligent CAD system which is capable of interacting with the designer. "Intelligence" in this context describes the ability to solve problems by applying inference rules from a knowledge-base. This knowledge-base should consist of a matrix of rules and facts represented in a specific language. Amongst the number of different types of inference known, only deductive inference has been studied in sufficient detail for use as a general processing mechanism [20,21].

Design rules can be formulated in a knowledge-base by use of a procedural or declarative computer programming language. However the problem now lies in codifying the knowledge and testing and verifying the output. Additional problems exist with control of the inference process; this will affect the solving process as the characteristics of the designer's thought process will be replicated in the way their decision-making is undertaken. This thought process has been simulated on computers in simple terms using declarative sequences, although the inference mechanism must be regulated. This can be undertaken using a metamodel approach [18], ie, developing functions which govern rules about rules and interdependencies.

3. Functions and Symbols

In order to perform intelligently, ICAD systems must be able to integrate individual design operations into a design process. Such a system can achieve this by processing the input and transferring information internally from one operation to another. Each operation, or function, is defined independently and modelled where possible in terms of other functions. Different forms of information such as procedures, specifications, databases and other knowledge must be used together, coupled with a good user interface. The use of a declarative (or quasi-declarative) language greatly facilitates the modelling of entities and their functionalities in this manner [22]. Simple inferences can be modelled, for example:

if a and b represent parameters such

$$a = f(b) \text{ and } b = f(c), \text{ then } a = f(c) \text{ may be inferred} \qquad (1)$$

This concept can be extended to include basic hydraulics rules, for example:

if ($Ns<20$) then (use symmetric volute rules + Stepanoff progression rules) or,
if ((mixing losses>5%) then ((reduce sidewall angles) and (increase blend radius))) else (modify impeller outlet width) $\qquad (2)$

Such instructions may also be processed iteratively, for example:

if (flow area tolerance < 0.5%) and (low flow separation) then (accept profile)
otherwise (repeat by adjusting geometry per sub-rules) $\qquad (3)$

or recursively :

$$f\,(\text{rules }(\text{area}_{0\rightarrow 360^\circ})) = f((\text{rule area}_0)+(\text{rules b,c,d....})) \tag{4}$$

The following section relates to some recent experiences of an experimental system which was developed and implemented on a trial computer-aided design system [17]. Because of its interactive qualities, and its ability to process function-argument instructions [23], a LISP-type environment was specifically chosen to construct the knowledge base. Programmes were written in AutoLISP language, operating interactively with AutoCAD™ system.

4. Modelling Functions using AutoLISP

Hydrogeometric information can be represented symbolically using lists. A list consists of groups of elements treated as one expression and stored implicitly in memory; this facilitates the modelling of rules and relationships between parameters. AutoLISP algorithms consist of blocks of lists, arranged as constituent functions. The programmer-designer can also fashion algorithms to manage the inference process, for example cyclically or by conditional branching. Such functions may process any number of arguments. These consisted of variables, commands, constants, user inputs, rules and other functions. During computation, the evaluator reads a sequence of instructions, evaluates it and returns a result which performs a particular task.

5. Experimental System

5.1 INPUT PROCESSING

To deal with relationships between two or three dimensional geometric objects, the knowledge of analytical geometry which defines points, lines, curves, planes, intersections, transformations, blends and fillets was defined as lists. Mathematical knowledge is required implicitly to define blending forms and other trigonometric relationships of the hydraulic profiling functions. This information was supplied by the hydraulic designer. In its basic form, some 10^3 lines of AutoLISP were required to describe a simple impeller-volute design. The process was initiated by firstly evaluating user-inputs to establish fundamental hydraulic geometry and expected performance parameters for the impeller. A predicted throat area A_{th} subsequently formed the basis for required volute geometry.

5.2 HYDRAULIC FUNCTIONS

Individual hydraulic parameters were represented using functions. For example, the hydraulic function called *slipf* was used to define impeller slip factor μ :

```
(defun slipf ()
(setq μ (- 1(/(* pi (sin β2))z))))
```
                                                                                                           (5)

AutoLISP's (setq) function was used to define geometric constructions, either locally or globally. Along any volute periphery S, control functions were used to define points $S = f(p_1, p_2, p_3 \ldots)$ in terms of system x, y, θ coordinates. Points were joined using lines, arcs, curves etc, into closed complex areas A_s, and exported graphically (or as datafiles). A typical peripheral function used was:

(defun periphery (S))= (area p_1, p_2, p_3 ......p_i), where (setq p_1 rule a), (setq p_2 rule b).... (setq p_n rule n) (6)

Self-predictive-corrective algorithms generated complex volute profiles, iterating successively on area A_v until a suitable profile S_v was formed, meeting active hydrogeometric criteria. Two types of rule were entered into the knowledge base: (a) hydraulic assignments, parameters of head, flow, impeller characteristics etc, and (b) geometric parameters such as desired ratios between impeller outlet/volute inlet dimensions, sidewall angles, aspect ratios, fillets, swept centres, tangency allowances etc. Some geometric variables were only partly defined, providing flexibility during iteration. Friction losses, turbulence losses, radial forces and other parameters were estimated concurrently.

Theoretically, any number of sections S_n may be defined along the volute path, but values of n<10 proved adequate for small pump designs. To achieve an accuracy of outflow area A_v < 0.1%, (acceptable for most low Ns pumps), typically only four or five iterations per section S_v were required, each with a computing time of some 15 seconds operating on a 486-PC. For higher Ns pumps, smaller convergence tolerances, therefore higher flowarea accuracy, could be set at a cost of longer solving times. The remaining volute path was developed recursively:

$$\text{(defun 3-dvolute } (V_{Ath \to 0})) \tag{7}$$

During recursive operation, rules were only accessed once per cycle, thus requisitioning memory and computation only as needed. This allowed the program to tackle problems of varying size. Devised section areas were calculated internally by the foster CAD system and verified separately.

Because the modelling process describes the hydraulic profile in terms of inputs and modifying operations, this readily facilitates the handling of changes. Thus any of the inputs may be altered, and the process repeated parametrically. An example of a typical profile output is shown in Figure 1, and a typical AutoLISP listing is shown in Figure 2.

5.3 PARADOXES

During development, problems occurred with non-convergence and instability caused either by self-intersection of curves, oscillation or overplotting. For complex area shapes, non-convergence was offset by use of double iteration, this also reduced some convergence times. Average-damping was also used to force convergence. A compromise was apparent between definition of adequate number of control points along bounding curves, yet providing limited flexibility for modification purposes.

6. Summary

The method demonstrated a successful application of an artificial intelligence process during the hydraulic design phase of a pump. It was found that the codification procedure determines how effectively design knowledge can be modelled. Functionality has been derived from the knowledge-processing system by development of an additional processing mechanism on top of the conventional computing mechanism. In order to integrate into a mainstream turbomachinery design process, an ICAD system must:

- provide means of easily updating/entering new knowledge,
- be able to work interactively with the design engineer,
- be linked to analytical facility (FEA, CFD systems),
- have means of testing and verifying outputs,
- provide the knowledge-base with ability to adapt generatively.

Such methods can help design engineers during conceptual phases, thereby helping to shorten design lead times. Further work is recommended in the development of a tandem AI-CFD system; this would help integrate fundamental fluid parameters deeper into the root design process, and reduce the reliance of empiricism.

References

1. Binder, R.C., Knapp, R.T.: Experimental Determination of the Flow Characteristics in the Volutes of Centrifugal Pumps, *Trans. ASME*, HYD-58-4, 1936.
2. Fischer, K, Thoma, D.: Investigation of the Flow Conditions in a Centrifugal Pump, *Trans. ASME*, Vol 54., HYD-54-8, 1932.
3. Knapp, R.T.: Centrifugal Pump Performance as Affected by Design Features, *Trans. ASME*, April 1941.
4. Bowerman, R.D., Acosta, A.J.: Effect of the Volute on Performance of a Centrifugal Pump Impeller, *Trans. ASME*, July 1957.
5. Iversen, H.W., Rolling, R.E., Carlson, J.J.: Volute Pressure Distribution, Radial Force on the Impeller and Volute Mixing Losses of a Radial Flow Centrifugal Pump, *Trans. ASME, Jnl Eng for Power*, April 1960.
6. Csanady, G.T.: Radial Forces on a Pump Impeller Caused by a Volute Casing. *Trans. ASME, Jnl.Eng. for Power*, October 1962.
7. Worster, R.C.: The Interaction of Impeller and Volute in Determining the Performance of a Centrifugal Pump, *BHRA paper RR679*, November 1960.
8. Worster, R.C., The Flow in Volutes and its Effect on Centrifugal Pump Performance, *Proc. IMechE*, Vol. 177, 1963.
9. Pekrun, M., Flörkemeier, K.H.: Experimental Investigation on Volutes of Centrifugal Pumps with Radial or Tangential Diffusers, *Pumpentagung Karlsruhe '78*, September 1978, Germany.
10. Hira, D.S., Vasandani.,V.P.: Influence of the Volute Tongue Length and Angle on the Pump Performance, *Jnl. of the Institute of Engineers India*, Part M.E. No.56.
11. Imaichi, K., Tsujimoto, Y, Yoshida, Y.: A Two-dimensional Analysis of the Interaction Effects of a Radial Impeller in Volute Casing, Proc. *IAHR Symposium*, Tokyo 1980.
12. Kurokawa, J.: Theoretical Determination of the Flow Characteristics in Volutes, *Proc. IAHR Symposium*, Tokyo 1980.
13. Brownell, R.B., Flack, R.D.: Flow Characteristics in the Volute and Tongue Region of a Centrifugal Pump, ASME paper 84-GT-82, 1984.
14. Thomas, R.N., Kostrzewsky, G.J., Flack, R.D.: Velocity measurements in a Pump Volute with a Non-Rotating Impeller, *Intl. Jnl. Heat & Fluid Flow*, Vol.7 No.1, 1986.

15. Miner, S.M., Beaudoin, R.J., Flack, R.D.: Laser Velocimeter Measurements in a Centrifugal Flow Pump, *Trans. ASME, Jnl. of Turbomachinery*, Vol.111, 1989.
16. Dong, R., Chu, S., Katz, J.: Quantitative Visualisation of the Flow within the Volute of a Centrifugal Pump, FED-Vol 107, *General Topics in Fluids Engineering*, ASME 1991.
17. Thackray, P.R.: Computer-Aided Analytical Design of Centrifugal Pumps, *Ph.D. Thesis*, Hull University, UK, 1994.
18. Tomiyama, T., Yoshikawa, H.: Requirements and Principles for Intelligent CAD Systems, *Knowledge Engineering in CAD*, Elsevier 1985.
19. Simmons, M.K.: Artificial Intelligence for Engineering Design, *Computer-Aided Engineering Journal*, April 1984.
20. Jakiela, M.J., Papalambros, P.Y.: Design and Implementation of a Prototype Intelligent CAD System, *Trans. ASME Jnl. Mech. Trans. and Automation in Design*, June 1989.
21. Ohsuga, S.: Toward Intelligent CAD Systems, *Computer-Aided Design*, Vol.21, No.5, June 1989.
22. Gevarter, W.B.: The Languages and Computers and Computers of Artificial Intelligence, *Computers in Mechanical Engineering*, Nov. 1983.
23. Winston, P.H., Horn, B. K. P.: LISP, Addison-Wesley Publishing Co., 1981.

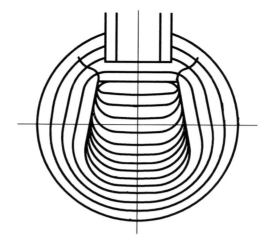

Figure 1.   Typical profile output (2-d superimposed).

```
(defun blendfunct1()
(if(< rar 0.57) (setq rfu 0.95)(setq rfu 1.00)) (if(< rar 0.57) (setq rfu bl1)(setq rfu bl2)))
(defun volute1()
(while(or(>=a1(* flowarea 1.001))(<= a1 (* flowarea 0.999))))
(if(< impb2b3 1.5) (setq rule10a rule10b)(setq rule10a rule10c))
(command "erase" t1a "")
(setq thrht (* thrht (sqrt(/ flowarea a1))))(setq thr1 (+ thrht nt))(setq deltax (/ (+ x1 x2) 2))
(setq orig2 (polar orig 1 (dtr 0) deltax))(setq origtop (polar orig1 (dtr 90) thrht))
(setq cltop (polar origtop (dtr 0) deltax))(setq cenc4l (polar orig 1 (dtr 0) x4))(.....)
(setq v1 (*(dist cendis c1)(sin(-(angle cendis c1)(dtr 90)))))
(setq rd1 (*(+(*(-(/ disd 2) rd1) rar) rd1)rfu))(setq x1 (+ x1 (* v1 rar)))
(setq v2 (*(dist cendis c2)(cos(angle cendis c2))))
(setq rd2 (*(+(*(-(/ disd 2) rd2) rar) rd2)rfu))(setq x2 (- x2 (* v2 rar)))(.....)
(command "pline" t4b t1a "")(setq t41 (entlast))
(command "pline" t3a t4a "")(setq t34 (entlast))(.....))
(command "pedit" k21 "y" "j" k22 t32 k33 t34 k44 t43 t41 k45 t46 k47 t48 k11 "" "x")
(command "area" "e" k11)(setq a1 (getvar "area"))
)
(if (< flowcon flowtol)((mixloss)(frictloss)(blend)(tabulate))(accrls))
)
```

Figure 2.   Typical AutoLISP computer listing.

ANALYSIS OF FLOW MEASUREMENTS IN THE IMPELLER AND VANED DIFFUSER OF A CENTRIFUGAL PUMP OPERATING AT PART LOAD.

Michel TOUSSAINT
CNAM, Laboratoire de la chaire de Turbomachines, Paris.

François HUREAU
CETIM, Laboratoire d'Hydraulique Industrielle, Nantes.

Abstract

In a recent european contract, four impellers with different specific speeds have been equiped with transparent front shrouds in order to obtain optical access to the flow between the blades. The presented results are the last analysis of internal flow measurements in one of these impellers (Ns32) fitted with a vaned diffuser. They are especially related with the interaction between impeller and diffuser, e.g. the phenomenon of flow separation, wakes, and backflow inside the pump. Unsteady velocity measurements were carried out in the hydraulic laboratory of CETIM with the help of laser doppler velocimetry technique. The data fields have been treated by the CNAM-LEMFI in order to give instantaneous view of the flow between the blades. From these pictures, a three dimensionnal movie has been built, that allows to explain some details in the behaviour of centrifugal impeller between 0.2Qn and Qn.

1. Introduction

If a centrifugal pump operates under its nominal flow rate (Qn), there comes a moment when the "incoming flow" fails to fill simultaneously all the channels of the impeller thereby generating counterflows at the impeller inlet and outlet. The experimental results given in this paper show for example how the flow "swings" from the shroud side to the hub one when the flow rate is reduced, and where are located the backflow areas inside the channels of the impeller. The comparison of results between vaneless and vaned diffuses gives some explanation on the nature of separated flows in this part of a rotodynamic pump. These detailed flow measurements complete other experimental results achieved on centrifugal pumps in Europe [2,3,4,11,12,13,14,18]. They should be extensively used to calibrate the computational fluid dynamics tools actually used in the design methodology of pump (see for exemple [1,5,6,7,10]).

E. Cabrera et al. (eds.), Hydraulic Machinery and Cavitation, 419–427.
© *1996 Kluwer Academic Publishers. Printed in the Netherlands.*

2. Experimental setup.

2.1. TEST EQUIPMENT

The test engine is a centrifugal pump which main characteristics at nominal point are: speed 1450 rpm, flowrate 136 l/s, static head 30 m, mechanical power 62 kW. The flow at the outlet of the impeller is drived in a main collector linked to the pipe with 4 branch'lines. This configuration give an axisymetric behaviour of the flow within the pump, in the case of vaneless diffuser. The geometrical data of the centrifugal impeller (6 blades) and the vaned diffuser (8 blades) are given on figures 1 and 2, with the locations of velocity measurements. The test rig, that has been built to fulfill the different tasks for the investigations, is more completely described in reference [17].

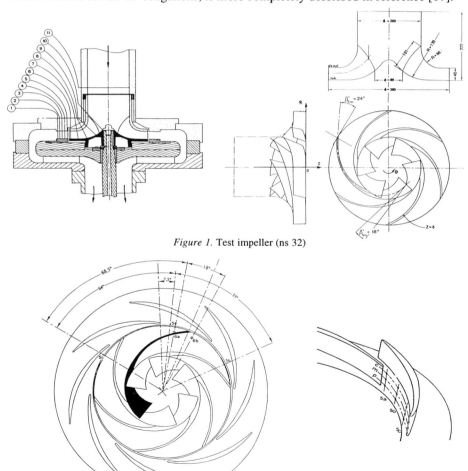

Figure 1. Test impeller (ns 32)

Figure 2. Vaned diffuser

2.2 INSTRUMENTATION.

In addition to conventional apparatus such as a test bed (rotational speed, torque, delivery rate, etc.), the equipment includes a slip ring collector allowing the measurement of pressure provided by built-in sensors. The scale model is also equiped to measure pressure fluctuations upstream and downstream from the impeller. The upper part of the model and the impeller flange are made of perspex to enable observation of any pre-rotation, recirculation or cavitation phenomena. Eleven glass ports, laid out as shown in Figure 1, provide access to impeller velocity measurements using the laser velocity measurement technique. Very fine particles (1 to 10 μm) of titanium dioxide act as tracers within the flow. The system used is a Doppler laser velocimeter with two interference fringe networks, including a COHERENT Argon laser source (6W output for all lines) and a TSI optoelectronic unit. Each reading of the two velocity components of these particles simultaneously leads to the input of the impeller's angle position using an optical encoder.

3. Analysis of experimental results.

3.1. RECIRCULATION IN THE IMPELLER.

The recirculation phenomena affecting the impeller and the diffuser in a plain diffuser type pump are summarised in the two drawings in Figure 3. It is noticeable that at delivery rates of between Qn and 0.61 Qn, slight signs of recirculation occur simultaneously in proximity to the impeller inlet and inside the diffuser. When the rate decreases, recirculation continues to spread, occurring in both the inlet and outlet of the impeller and much of the plain diffuser at very low delivery rates. Recirculation affecting the inlet originates at some distance in the impeller pipe. It takes up approximately 50% of the channel height and remains localised along the inner surface of the blade. This pocket of recirculation seems to "wear itself out" as it travels against the flow to reach the impeller outlet, which is logical since part of the recirculating fluid must be carried along as a result of the viscosity and set back on the right course by the "normal" delivery flow. These various observations confirm that recirculation at the impeller inlet is completely independent of the recirculation observed at the impeller outlet.

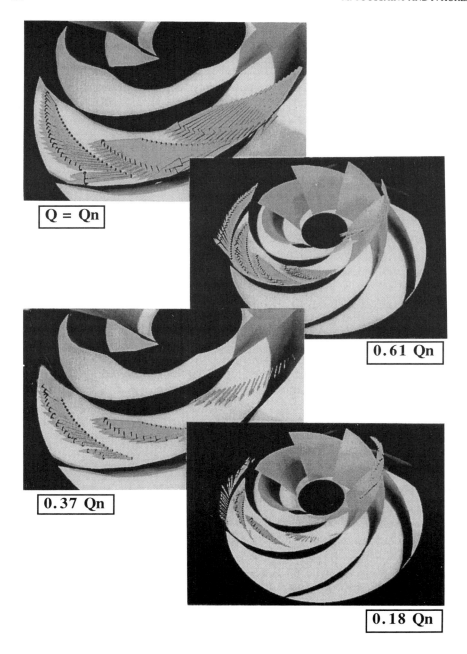

Figure 3: Flow measurements in the impeller.

3.2. INFLUENCE OF THE VANED DIFFUSER.

Although it is true that the relative flow in a centrifugal impeller may be "almost steady" [15] [16], this is not the case in an industrial operation where the impeller is always followed by a diffuser, return pipe or volute. The interaction between the fixed parts (diffuser vanes or volute tongue) and moving parts usually leads to a turbulent flow that is unsteady in both a Galilean referential system and in an impeller-linked relative reference system. Research carried out into this type of flow [8] [9] makes analysis of the results difficult (it is always somewhat specious), since the subsequent processing of pinpoint measurements of velocity and pressure values in the channels between the vanes of a turbomachine impeller always requires the use of extensive computer and graphics facilities. Using 3D images we shall, in this article, present an analysis of the flow in the area of interaction between impeller and diffuser, in a centrifugal pump operating with a vaned diffuser.

A comparison of the various velocity profiles recorded in the impeller in configurations with either plain or vaned diffuser shows that the vaned diffuser has an almost negligible influence on the flow in the inter-vane channels, except near the impeller outlet. Figure 4, which gives a comparison of the profiles recorded on axis 7 in the two operating configurations, confirms this initial analysis. As regards recirculation at the inlet and in the impeller, the presence of a vaned diffuser makes no notable difference compared to the system operating with a plain diffuser.

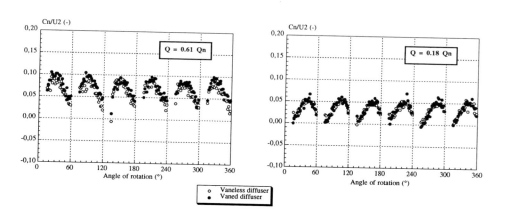

Figure 4: Comparison of velocity measurements with and without vaned diffuser.
Axis 7 - R=0.75 R2 - middle - Cn: normal to axis of measurement

Near the diffuser inlet the influence is more important, especially at part load operation. The figure 6 gives at three flow rates the radial component of velocity on Axis 6 (R=0.92 R2) with and without vaned diffuser. Taking into account the disymetry of flow around the impeller, two locations of measurements on the same axis are given (6a and 6b; see on figure 2)

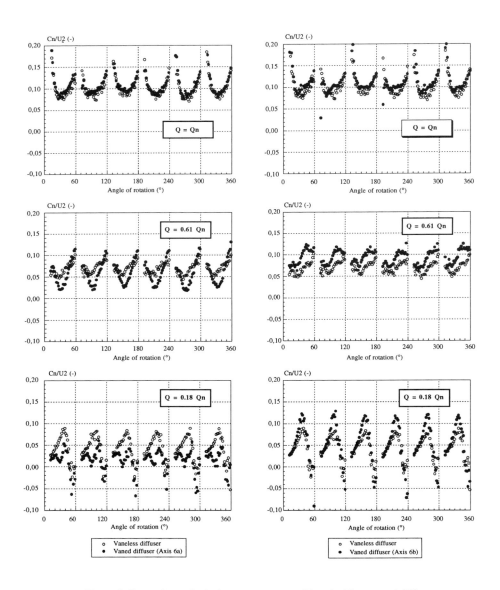

Figure 5: Comparison of velocity measurements with and without vaned diffuser.
Axis 6 - R=0.92 R2 - middle (b/b2=0.5) - left: axis 6a; right: Axis 6b

The three-dimensional and unsteady phenomena which occur when the impeller vanes pass in front of the diffuser vanes may be given in vector form. This operation was carried out with support of the LEMFI Laboratory in Orsay using IRIS Explorer software on a Silicon Graphics workstation. An animated film was built using these 3D images.

The film explains the following phenomena in particular:

- at the nominal rate and slightly below (up to 0.9 Qn), the flow is particularly "stable" in the area around the diffuser inlet. The velocity deviations are small over a period of time and distribution is almost uniform between the diffuser vanes.
- at 0.82 Qn, separation occurs in the corner between the bottom of the vane and the concave side of the diffuser vane. This separation is resorbed as the impeller vane approaches and creates a flow running into this area.
- at 0.61 Qn, separation occurs on the shroud side (top of the diffuser vane). This separation is located between the leading edge of the vane and the upper flange and it acts as a result of the significant wake generated at the impeller outlet at this flow rate.
- at 0.37 Qn, the separation again occurs at the bottom of the diffuser vane, owing to recirculation inside the impeller which completely alters the axial profile of the velocity field.
- at 0.18 Qn, the separation phenomenon at the leading edge of the diffuser vane is further accentuated. It occupies the full height of the blading when an impeller vane passes in front of the diffuser vane.

This analysis of the flow between the rotor and stator highlights the two main characteristics of the unsteadiness inside a pump (see ref [1]):

- potential interaction: this corresponds to the encounter between 2 distinct flows. Even in the case of an inviscid flow, a spiral flow passing in front of a flow between fixed channels leads to periodical fluctuations in local velocity values.

- wakes interaction: because of the viscosity of the water, separation arises in the impeller and generates a backwash in the outlet along the concave side of the vanes. The density of the wake varies depending on the pump operating condition. The wake reaches the diffuser vanes and creates further periodic local disturbances.

In this type of machine it may be assumed that the two types of interaction are of similar size. Both phenomena are superposed giving rise to considerable fluctuations in velocity, mainly near the leading edge of the diffuser vanes. These local hydrodynamic disturbances spread further along the flow and cause loud noises, mainly due to acoustic propagation.

4. Conclusions

An analysis of the test results obtained for a centrifugal impeller with or without vaned diffuser has enabled us to emphasise the following points.

* The phenomenon of the recirculation beginning inside the impeller leads to complete instability of the flow between the vanes and causes a significant increase in the overall level of velocity fluctuation. The three-dimensional aspect becomes predominant and everything occurs as if the relative flow in the impeller formed an eddy as a result of the effect of the various curvatures of the sides. Separation appears on the concave side of the vane near the front flange where the radius of curvature is the smallest. At the impeller outlet, the recirculation phenomena in the impeller is superimposed on the separation phenomena affecting the vaneless diffuser, and this complicate analysis at this particular spot. Reverse flows in the impeller are only visible at a very low flow rate (near the trailing edge they appear on the outer surface of the vanes) whereas at or near the nominal rate recirculation can already be observed in the diffuser.

* By comparison, the measurements taken with a vaned diffuser show that the recuperating part at the impeller outlet has very little effect on the flow inside. The area near the leading edge of a diffuser vane is the region where the flow is most disturbed when the rate decreases, in both the impeller and the diffuser. In particular, when the rate reaches the critical recirculation value in the impeller, it is observed that the profiles suddenly fail to match the delivery flow.

This study marks an end to the analysis of periodic flow in the ns32 impeller. All the results obtained will serve as a database for the definition of computation codes. Work on the subject is currently being carried out by the SHF's Turbomachine Working Group - Internal flows and related performances. Furthermore, the turbulent aspect of the velocity field may be dealt with in the course of further research work.

References

1. BERT P.F., COMBES J.F., KUENY J.L. (1996) *Unsteady flow calculation in a centrifugal pump using a finite element method*, XVIII IAHR Symposium on Hydraulic Machinery and cavitation, Valencia, Sep. 16-19.
2. BOIS G., RIEUTORD E. (1990) *Etude de l'écoulement en sortie de roue de pompe centrifuge depuis le débit nominal jusqu'au débit critique de recirculation*, Compte rendu MRT n° 88 H 0636, Octobre 1990.
3. CAIGNAERT G., BARRAND J.P. (1989) *Débits critiques de recirculation: synthèse de l'ensemble des résultats expérimentaux sur les roues SHF essayées en air (ENSAM Lille) et en eau (INSA Lyon, EPF Lausanne, Hydroart Milan)* 20èmes journées de l'Hydraulique, SHF Lyon, 4-6 Avril.
4. CASEY M.V., EISELE K., ZHANG Z., GULICH J., SCHACHENMANN A. (1995) *Flow analysis in a pump diffuser, part 1: LDA and PTV measurements of the unsteady flow*, ASME Paper, FED Summer Meeting, Symposium on Laser Anemometry.

5. CASEY M.V., EISELE K., MUGGLI F.A.., GULICH J., SCHACHENMANN A. (1995) *Flow analysis in a pump diffuser, part 2: Validation of a CFD Code for steady flow*, ASME, Numerical simulations in turbomachinery, Vol.227 pp.135-143

6. COMBES J.F., GRIMBERT I., RIEUTORD E. (1991) *An analysis of the viscous flow in a centrifugal pump with a finite element code*, Fluid Machinery Forum, 1st ASME-JSME Fluids Engineering Conference, Portland, Oregon, June 23-27.

7. CROBA D. (1993) *Modélisation de l'écoulement instationnaire dans les pompes centrifuges. Interaction roue et volute*, Thèse Doctorat, INP Grenoble, Juillet 1993.

8. HUREAU F, KERMAREC J., FOUCHER D. (1993) *Etude de l'écoulement instationnaire dans une pompe centrifuge fonctionnant à débit partiel*, 1st International Symposium on Pump Noise and Vibrations, Clamart, France, July 7-9, pp.

9. HUREAU F., KERMAREC J., STOFFEL B., WEISS K. (1993)*Study of internal recirculation in centrifugal impellers*, ASME, 2nd Pumping Machinery Symposium, Washington, D.C, June 20-24, FED Vol. 154, pp. 151-157.

10. LONGATTE F. (1996) *Analyse phénomènologique du couplage rotor-stator dans les pompes centrifuges. Etude sous Fluent d'une pompe 2D avec diffuseur aubé*, LEGI-CREMHyG, Mai 1996.

11. MOREL Ph. (1993) *Ecoulements décollés dans une roue de pompe centrifuge- Conception et réalisation d'un banc d'essai. Analyse des pressions pariétales*, Thèse Doctorat, Université de Lille, Decembre 1993.

12. PAONE N., RIETHMULLER M.L., VAN DEN BRAEMBUSSCHE R.A. (1989) *Experimental investigation of the flow in the vaneless diffuser of a centrifugal pump by particle image displacement velocimetry*, Experiments in Fluids 7, Springer Verlag 1989, pp 371-378

13. STOFFEL B., WEISS K. (1994) *Experimental investigations on part load phenomena in centrifugal pumps*, World Pumps, October 1994, pp 46-50

14. STOFFEL B., LUDWIG G., WEISS K. (1992) *Experimental investigations on the strucure of part-load recirculations in centrifugal pump impellers and the role of different influences*, XVIth IAHR Symposium, Sao Paulo, Brazil, September 14-18.

15. TOUSSAINT M. (1993) *Contribution à l'étude des recirculations dans les pompes rotodynamiques*. Thèse de doctorat, Université P et M Curie Paris VI, Paris, Décembre 1993.

16. TOUSSAINT M. (1994) *Etude des recirculations dans les pompes centrifuges roue Ns 32 diffuseur lisse*, Rapport interne CNAM/CETIM, Paris, Septembre 1994.

17. TOUSSAINT M., HUREAU F. (1993) *Etude expérimentale de l'écoulement instationnaire dans une pompe centrifuge fonctionnant à débit partiel*, IAHR, 6th International Meeting, The behavior of Hydraulic Machinery under steady oscillatory conditions, Lausanne, Sept. 1993.

18. UBALDI M., ZUNINO P., BARIGOZZI G., CATTANEI A. (1994) *An experimental investigation of stator induced unsteadiness of centrifugal impeller outflow*, ASME, La Haye, Paper 94-GT-105.

INFLUENCE OF THE BLADE ROUGHNESS ON THE HYDRAULIC PERFORMANCE OF A MIXED-FLOW PUMP. A VISCOUS ANALYSIS

S. UNDREINER, E. DUEYMES
EDF-DER, 6 quai Watier
78400 Chatou, FRANCE

1. Abstract

The aim of this work is to quantify the blade roughness influence on the hydraulic performances of a mixed-flow pump. In this way the flow in a hydrodynamic tunnel equipped with a blade profile was analysed by measurements performed by the CREMHyG and numerical simulations with the code N3S (developed by 'Electricité de France'). The roughness parameter z_0 used in the simplified rough wall law was calibrated from the measured roughness for three Reynolds numbers and three roughnesses. The application to the pump flow was obtained for the scale model conditions (0.41). For the prototype pump conditions, a calibration was suggested but has to be refined by some other simulations.

2. Nomenclature

Q_n	nominal flow rate	x_d	separation point abscissa
k	measured roughness	z_0	roughness parameter
u_*	shear velocity	U_t	tangential velocity on the wall
ν	kinetic viscosity	ρ	density
L	lenght scale (chord for the pump)	U_c	neck speed

$$U^+ = \frac{U_t}{u_*} \qquad y^+ = \frac{yu_*}{\nu} \qquad z_0^+ = \frac{z_0 u_*}{\nu}$$

3. Introduction

A shift of the hydraulic performances of the pump was detected during some specific tests that EDF performed on a test-ring. After a certain number of running hours, the pump head increased of about 5% (figure 1) of the nominal value predicted by the designer. Several hypothesis were analysed to explain this phenomenon. The last one retained was based on a possible modification of the blade roughness. Some theoretical investigations have been performed by computational modelling. The aim was to get a better understanding of the flow structure in the impeller and to know how it can be modified by the roughness. First computations have been performed with an inviscid flow code [1]. At nominal flow rate the results have pointed out an important load on the trailing edge of the pressure side. Indeed, a low velocity level had appeared which

E. Cabrera et al. (eds.), Hydraulic Machinery and Cavitation, 428–437.
© 1996 *Kluwer Academic Publishers. Printed in the Netherlands.*

could provoke a separation. The computations were completed by a Navier-Stokes approach with the code N3S [2]. Several simulations have been performed: at nominal flowrate, at below and above the nominal flowrate (respectively 0.9 and 1.2 Qn) for a smooth surface. The viscous simulations have given better predictions than the inviscid flow computations, particularly in case of non-nominal operations (0.9 and 1.2 Qn). The comparisons with the measurements were satisfactory. The viscous analysis was completed by a qualitative modelling of the roughness influence on the boundary layers. The results have confirmed the correlation between an important head evolution and a blade roughness modification. In an extreme situation, an evolution on the suction side from a rough to a smooth surface provokes an important increase of the head (about 4%). However it was not yet possible to quantify the relationship between the roughness and the head of the pump: the roughness parameter in the wall law did not represent the right roughness.

The target of this work was to evaluate the roughness corresponding to the parameter used in N3S. In this way an experiment had been performed by CREMHyG [3]. First the perspectives were to analyse the iinfluence of the roughness on the separation point position, second to calibrate the roughness parameter. The results from several N3S simulations were compared to the measurements.

This work is decomposed into six parts:
⇒ in the first part we will describe the CREMHyG experiments and the most important results,
⇒ in the second part, we will briefly describe the N3S code,
⇒ the third part concerns the laws of the wall commonly used in industrial codes for smooth or rough walls,
⇒ the fourth part is the N3S modelling of the water tunnel flow (CREMHyG) and the comparison with the measurements,
⇒ in the fifth part we will suggest a transposition of the hydrodynamic tunnel calibration to the pump flow before concluding in a sixth part.

4. The CREMHyG experiments

The hydrodynamic tunnel was equipped with a blade profile in order to partly reproduce the velocity profile on the pressure side of the pump blade, near the hub (it was obtained from the Euler simulations [1]). The geometry is presented figure (2): the 0.691m length corresponds to the pump chord where the velocity profile is reproduced, the roughness can be adjusted only in the first 0.324m. Three velocities at the neck and three roughnesses are studied: U_c=3, 8 and 24m/s and k=0, 3, 7μm. The separation point was determined either by the pressure gradient cancellation or by means of velocity measurements. From these experiments it appears that: the separation point goes upstream when decreasing the Reynolds number or increasing the roughness (table 1). The roughness influence on the separation point position is more important for the high velocities (U_c=24m/s)than for the low.

k (μm) Uc (m/s)	0	3	7
3 m/s	150 ($\pm$ 2)	148 ($\pm$ 2)	148 ($\pm$ 2)
8 m/s	151 ($\pm$ 2)	148 ($\pm$ 2)	148 ($\pm$ 2)
24 m/s	172 ($\pm$ 2)	166 ($\pm$ 2)	162 ($\pm$ 2)

Table 1 Evolution of the separation point abscissa (in mm) with velocity and roughness (CREMHyG [3])

5. The numerical method

The viscous computations are performed with the industrial code N3S (release 3.1) developed by 'Electricité de France'. A finite-element method solves the unsteady Navier-Stokes equations for an incompressible flow. The time discretization is based on a fractional step method which solves successively an advection step, a diffusion step and a generalized Stokes problem. The boundary conditions are based on a non-homogenous Dirichlet condition in the inlet (for the velocity and the turbulent quantities) and a null constraint in the outlet.

6. The wall law

The turbulence model used for these studies is the standard k-ε model associated to a wall law. For a smooth wall N3S solves the law of Reichardt [4] valid for values of y+ ranging from 5 to 200. The rough wall law was established for a pipe flow [5]:

$$U^+ = \frac{1}{\kappa}\ln(y^+) + B - \Delta B(z_0^+)$$ (1)

where B=5.5

ΔB is a constant which depends on the roughness and the shear-stress; for a sand roughness wall:

$$\Delta B = \frac{1}{\kappa}\ln(1 + 0.3z_0^+).$$ (2)

This law can be simplified for a fully rough flow ($z_0^+ > 60$):

$$U^+ = \frac{1}{\kappa}\ln(\frac{y}{z_0}) + 8.5$$ (3)

In the previous study on the pump flow with the N3S code the rough wall law was the simplified one with different constants ($\kappa = 0.42$ instead of 0.4):

$$U^+ = \frac{1}{\kappa}\ln(\frac{y}{z_0}) + 8.23$$ (4)

In our case, since z_0^+ can be inferior to 60, the z_0 parameter can no more represent the physical roughness. A calibration is then necessary.

7. The hydrodynamic tunnel simulation

The flow in the hydrodynamic tunnel of the CREMHyG [3] was simulated for the three neck velocities (3, 8 and 24 m/s). The roughness parameter of the simplified law of the wall (equation 4) was calibrated to the measured roughness (0, 3 and 7μm) essentially at 24m/s velocity.

A study of the finite-element mesh dependance showed that it was possible to use the same mesh for the three velocities. Indeed we did not modify the refinement for the rough simulations. The mesh includes 25492 P2 nodes.

First we studied the flow on a smooth surface to check the ability of N3S to predict the separation point evolution with the Reynolds number. The results are given in the table (2).

U_c (m/s)		3	8	24
x_d (mm) N3S		173	185	195
x_d (mm) Experiments		150 (±2)	151(±2)	172 (±2)
error		15.3 %	22,5%	13.4%
Evolution of the separation point abscissa (mm) compared to the reference case 3m/s	N3S	-	+12	+22
	Exp	-	+1 (<5 and >-3)	+22 (<26 and >18)

table 2 Comparison of computations/measurements for a smooth wall

The N3S predictions are globally correct for the separation point position and for the relative evolution with the Reynolds number. The comparison computations/measurements concerning the velocity profiles confirms this tendancy (figure 3 for the 24m/s velocity). The deviation can be essentially explained by the turbulence modelling (k-ε model and wall law). This approach can not take into account the presence of the adverse pressure gradient and then overestimates the separation point abscissa. Indeed the wall law is no more valid in a recirculating region.

The simulations for the rough walls cases have been performed with the simplified rough wall law. The modelling with the general law (equations 1 and 2) did not give any satisfactory results. Computations of a rough pipe with this law have shown that it was not adapted to the recirculating flow.

The main difficulty was to define a calibration criterion of the z_0 parameter with the experimental roughnesses. The only possibility was to consider the following hypothesis: the measured relative separation abscissa point evolution from the smooth to the rough surfaces had to be reproduced by N3S. Therefore, the abscissa that N3S has to predict is determined and z_0 is computed by successive simulations. With this method we obtained the following calibration table 3.

measured roughness k (m)	0	$3\ 10^{-6}$	$7\ 10^{-6}$
z_0 (m) Uc=24m/s	$5\ 10^{-6}$	$2\ 10^{-5}$	$5\ 10^{-5}$
z_0 (m) Uc=3m/s	$2\ 10^{-4}$	$3\ 10^{-4}$	$3\ 10^{-4}$

Table 3 Calibration of the z_0 parameter at 3 and 24 m/s

For the smooth surface (k=0μm) the z_0 parameter was calibrated with the N3S smooth computations.

The evolution of the z_0 parameter is more important for the higher Reynolds number: z_0 is multiplied by 10 between 0 and 7 μm at the 24m/s velocity and by 2/3 at the 3m/s velocity. This behaviour corresponds to the experimental measurements, table 1.

The suggested calibration shows that the z_0 parameter is a function of the physical roughness and the neck speed. This means in adimensional terms that z_0/L depends on the Reynolds number and k/L where L is a length scale.

In table 3, it appears that:
- the z_0 parameter evolution with the roughness is stronger at the high Reynolds number values,
- the z_0 parameter evolution with the Reynolds number is stronger at low roughness (for the smooth surface).
- the z_0 parameter tends toward a limit value at each Reynolds number.

8. The transposition to the pump flow

The aim is to quantify the roughness parameter z_0 used for the study in the pump flow [1]. The problem is now how to apply the previous calibrations in the water tunnel for the pump flow?

In table 5, the Reynolds numbers are given for each case: the hydrodynamic tunnel, the prototype pump, and the scale model conditions; for the pump flow the Reynolds number is based on the blade chord and the mean velocity at the leading-edge.

		Reynolds number
Water tunnel	3m/s	$2\ 10^6$
	8m/s	$5.3\ 10^6$
	24m/s	$1.6\ 10^7$
pump: scale 1 (hot water)		$1.5\ 10^8$
pump: scale 0.41 (cold water)		$3\ 10^6$

Table 5 Reynolds number for each configuration

8.1. The scale model

In table 5 it appears that the Reynolds number of the flow in the scale model is comparable to the 3m/s velocity in the water tunnel. It is then possible to apply the water tunnel calibrations to the scale model conditions : the ratio z_0/L (L is the length scale) is equal. The results are given in table 6.

k (µm)	3
z_0 (m)	10^{-4}

Table 6 Transposition to the scale model flow

However, scale model flow simulations have shown that:

♦ the roughness effects are opposite to those obtained for the prototype pump conditions (the head increases with the roughness),

♦ the previous calibration needs a finer analysis of the water tunnel calibration applicability to the scale model flow.

8.2. The scale 1

With results concerning only two Reynolds numbers it is difficult to extrapolate the z_0 values for the pump flow corresponding to 3 and 7 µm: the pump Reynolds number is 10 times higher than the Reynolds number in the hydrodynamic tunnel. In figure (4) we tried to represent from our calibration the z_0 evolutions with the roughness k for two Reynolds numbers. The values z_{01} and z_{02} in figure (4) were used for the pump simulations [2]. If we extrapolate the difference between the velocities 3 and 24 m/s to the difference between 24m/s and the pump flow conditions, we can suggest that the $z_0=10^{-5}$ (m) corresponds to a maximum of 7µm. However, the Reynolds number effects have to be taken into account. This means that for high values the Reynolds number has less influence on the flow. It would be interesting to check this behaviour by a water tunnel simulation at the pump flow Reynolds number. In such case a refined mesh would be necessary.

9. Conclusion

The aim of our work was to quantify the blade roughness influence on the hydraulic performance of a mixed-flow pump. A previous study has shown that a significative head decrease can be correlated with an increase of the blade surface roughness. Our work consisted in a calibration of the roughness parameter (z_0) used for the previous study to a real roughness.

In this way the CREMHyG has performed an experiment in a hydrodynamic tunnel. The roughness effects on the boundary layer separation was quantified by surface conditions and velocities measurements. The experimental results were compared to the N3S simulations.

The results obtained with N3S in the smooth conditions are satisfactory in regard to the simple wall modelling (law of the wall). This means that N3S can predict a correct evolution of the separation abscissa. The deviation between the measured and calculated abscissa is about 15%.

Several difficulties were analysed before suggesting a calibration first in the water tunnel and second in the pump. The main difficulty was to define a calibration criterion in the hydrodynamic tunnel. The presence of a recirculation put the general law of rough wall in the wrong track. Therefore we had to consider the following hypothesis: the relative evolution between the experimental and the numerical results has to be respected for the different velocities. The second difficulty concerned the extrapolation of our suggested calibration to the pump (scale one) flow. Firstly, the Reynolds numbers between the pump and the hydrodynamic tunnel conditions are completly different (a ratio of 10). Secondly, more experimental results are needed to correctly take into account the Reynolds number effects: for example an intermediate case between 8 and 24 m/s.

Despite these difficulties, we suggested a calibration for the pump flow (scale 1) and the scale model.

In case of the scale model the roughness of 3μm can be linked to $z0=10^{-4}$(m). This calibration would be reviewed considering the last simulations results: the flow is extremely sensitive to the roughness in opposite to the flow in the water tunnel (at similar Reynolds number).

In case of the prototype pump the parameter $z_{02}=10^{-5}$ (m) would correspond to a roughness of 7 μm at the most and $z_{01}=10^{-6}$ (m) to 3μm. Therefore a roughness variation of 4μm could provoke a head decrease of 2%. The measured and calculated head evolution is given figure (5).

This study have shown the complexity of modelling the roughness effects by means of a wall law. Several supplementary analysis can be envisaged:
- a direct calibration for the scale model by comparisons between N3S computations and measurements,
- a finer comparison of the flow structures between the prototype pump and the scale model conditions (Reynolds number effects),
- an improvement of the rough wall laws.

10. References

[1]   I. GRIMBERT. *Dérive des charactéristiques hydrauliques d'une roue de pompe; premières analyse de l'écoulement interne.* Internal report EDF/DER HP-41/91.17

[2]   V.MOULIN. *Analyse complémentaire de l'écoulement dans une roue de pompe avec le code N3S.* Internal report EDF/DER.HP-41/94/007.

[3]   **C. REBATTET**. *Etude expérimentale de l'influence de la rugosité sur le comportement de la couche limite dans une veine*. Rapport CREMHyG 1994.

[4]   **J.P CHABARD**. *Projet N3S de Mécanique des Fluides, Manuel théorique de la version 3*. Internal report EDF/DER HE-41/91.30B.

[5]   **F.M. WHITE**. *Viscous Fluid Flow*. McGraw-Hill, Inc, Second Edition.

11.Figures

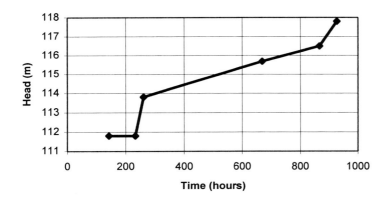

Figure 1 measured head evolution in time during the fifth test serie

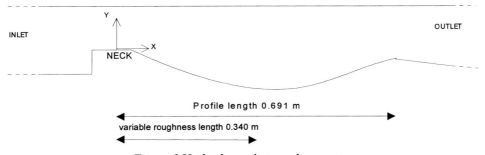

Figure 2 Hydrodynamic tunnel geometry

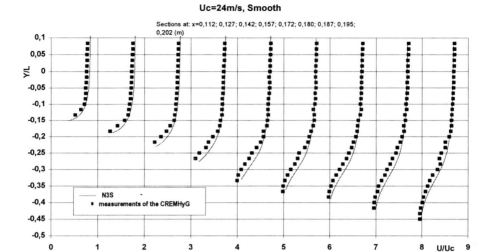

Figure 3 Comparison measured/calculated velocity profiles(Uc=24m/s)

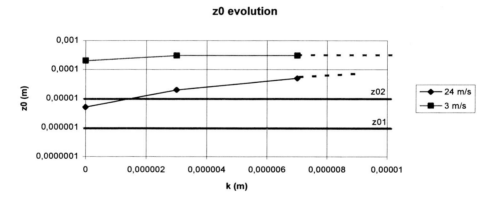

Figure 4 z0 evolution (water tunnel)

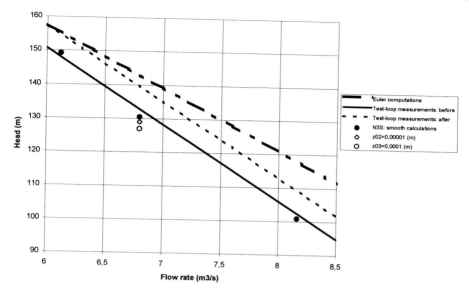

Figure 5 Head of the mixed-flow pump with the flow rate (mesaurements and computations)

IMPROVEMENT OF PERFORMANCE OF CENTRIFUGAL PUMPS BASED ON COMPUTATIONAL AND THEORETICAL METHODS AND EXPERIMENTAL DESIGN

A.F. Vinokurov,
(J-S-C "POMPA", Moscow region, Russia)
A.V. Volkov, G.M. Morgunov, S.N. Pankratov
(Moscow Power Engineering Institute, Moscow, Russia)

Introduction

The problems of quality performance improvement of centrifugal pumps are actual and long-range problems. This type of pumps is widely used in all branches of national economy. Improvement of various technological processes requires modernization of old centrifugal pumps or manufacture of new ones able to provide improved performance.

The problem of raising efficiency of a pump unit , same as improvement of its anticavitation characteristics, belong to the main tasks which attract attention of research workers and designers of pump equipment.Increase of pump efficiency by some per cent, and in other cases - by fractions per cent, considering high driving capacities or a big number of operating hours/year of the pump unit, results in significant energy saving. Improvement of anticavitation characteristics increases reliability of the pump, its ecological safety during operation.

Modernization of the pump, in contrast to making a new hydraulic machine,allows to lower very significantly the level of expenditures on development and use existing foundations, main pipelines, piping adjacent to the pumps, electrical equipment. Besides, necessity of modernizaton also arises in those cases when the pump operation does not agree with the characteristicof the network for which it functions. As a result, the hydraulic machine operates at non-optimum conditions. It means it is necessary to bring operating conditions in line with the optimum point of the pump operational characteristic. Fabrication of a new impeller only, in accordance with the conditions of the network required, makes it possible to get a considerable saving.

Sustantial Part

A successful settlement of the problems under consideration is achieved by using more precise computational methods of research on hydraulic machines allowing to carry out with a high degree of authenticity an analysis of work processes going on in the liquid passage part of the pump for a wide spectrum of integrated and local characteristics, and apply optimization methods to design phases.

E. Cabrera et al. (eds.), Hydraulic Machinery and Cavitation, 438–444.
© 1996 *Kluwer Academic Publishers. Printed in the Netherlands.*

inspection operations during impeller manufacture. The above mentioned improvements of the pump design made it possible to increase reliability of the pump unit operation and make the hydraulic machines produced by joint- stock company " POMPA " more competitive.

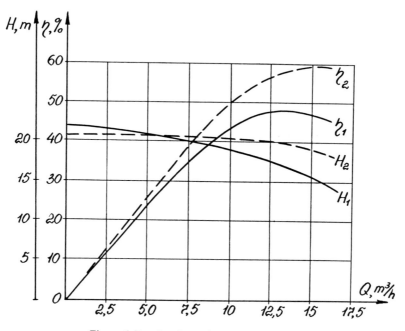

Figure 6. Results of experimental investigation.

References

1. Morgunov G.M.,Volkov A.V.,Gorban V.M. Combined 1D-3D cad method of hydromachine water pessages design..- "Mathematics, computer, control and investments",Int. Conf.,Moscow,15-19 february, 1993, p. 58.

2. Vinokurov A.F.,Volkov A.V. , Kuzmin S.V. One of combine of the methods designing flowing of parts
      centrifugal of pumps.- Moscow power engineering institute 1992. N 652. p.52-60. (in Russian)

3. Morgunov G.M. Calculation unseparetion of the flow spacing blades of systems with the account
      viscous. - Izv. AS USSR, Engineering and transport, 1985, N 1, p.117-126. (in Russian)

LIQUID-PARTICULATE TWO-PHASE FLOW
IN CENTRIFUGAL IMPELLER BY TURBULENT SIMULATION

WU Yu-lin, DAI Jing and MEI Zu-yan
Tsinghua University
Beijing, CHINA 100084

OBA Risaburo and IKOHAGI Toshiaki
Tohoku University
Sendai, JAPAN 980-77

Abstract

In the present work, turbulent liquid-particulate two-phase flows at the particle dilute concentration through a centrifugal impeller has been simulated by using the K-ε-Ap turbulence model. In the present numerical treatment, the finite volume method, based on the SIMPLEC algorithm, is applied in a body fitted coordinate system. The liquid-particle two-phase flow turbulence model used in this study can correctly predict essential features of this flow in centrifugal impellers at dilute concentrations.

1. Introduction

Sevaral theoretical and experimental techniques for investigation of multiphase flows are available. Basically, there are two fundamental theories for two-phase flows, namely, the macroscopic continuum mechanics theory and the microscopic kinetic theory. Marble (1963) was one of pioneer researchers to describe basic equations of two-phase flows. From the sixties to seventies, general governing equations of two phase flows for the two fluid model and the mixture model based on the continuum mechanics theory were set up by Drew (1971), Ishii (1975) and others by using different average methods. In the mixture model, it is assumed that there exists only one type of mixture in fluid-dispersed flows, and that the whole flow space is full of the mixture. The mixture flow is described by using a set of unique characters. Roco & Reinhart (1980) applied this model to calculate the liquid-particle mixture flow through centrifugal pump impellers .

In the two-fluid model, the dispersed phase is treated as a pseudo-fluid. In the Eulerian approach, the flow of the dispersed phase is described by conservation equations of mass, momentum and energy in continuum mechanics. In this model, there exists the slip of parameters between the carrier fluid and the dispersed phase. Based on the two-fluid model, increasing interest in the prediction of turbulent multi-

445

E. Cabrera et al. (eds.), Hydraulic Machinery and Cavitation, 445–454.
© *1996 Kluwer Academic Publishers. Printed in the Netherlands.*

phase flows has been noticed during the last twenty years, such as Danon et al (1974), Al Taweel & Landau (1977), Genchiev & Karpuzov (1980), Melville et al (1979), Sharma & Crowe (1978), Michaelides & Farmer (1984) and Shuen et al (1983). Danon et al (1974) tried to develop the one-equation turbulence model by adding a length scale determined experimentally to the turbulent kinetic energy equation for calculating two-phase round jet-turbulent-flows. Al Taweel & Landau (1977) deduced an energy spectrum equation at the transient frequency band, adding a supplement dissipation term to consider appearance of the dispersed phase. Calculated results show that the increase of volumetric density of the dispersed phase causes the turbulence intensity at the high frequency band decrease. Genchiev & Karpuzov (1980) proposed that a sink term has to be added to the single phase turbulent kinetic energy k equation to simulate the influence of the disperse phase. These authors applied the two-equation turbulence model to calculate two-phase turbulent flows. Two-equation turbulence models also have been proposed for dilute fluid-particle flows by Elghobashi & Abou-Arab (1983) and Crowder et al (1984). Algebraic and one equation turbulence models have been suggested for the dense liquid-solid interaction by Roco & Shook (1983). Bertodano et al (1990) applied the two-fluid model combined with the Reynolds stress turbulence model to analyze water-bubble flows. Dai & Wu (1993) have used the k- ε -a-kp liquid-particulate two-phase turbulence model to calculate this flow through centrifugal impellers at dilute concentration. Most of existing models are suitable only for relatively dilute mixtures, in which the particle-particle collision effect and the fluctuation energy interaction between the liquid and the particulate phases are negligible. During the last decade, there has been an interesting development in modeling dense solid-liquid flows, as shown, in References of Ma & Roco (1988), Ahmadi & Abu-Zaid (1990) and Gidaspow et al (1991). As mentioned above, Roco & Shook (1983) applied the macroscopic continuum theory and the one-equation turbulence model with double averaging to calculate dense liquid-particle flows. A probabilistic microscopic model for solving shear motion of spherical particles dominated by friction and lubrication force was proposed by Ma & Roco (1988) to slow granular flow and high dense coal slurry. Gidaspow et al (1991) have developed a computer code that solves a generalization of N-S equations for multiphase flows. Particulate phase viscosity and pressure were derived by using the mathematical techniques of the dense gas kinetic theory.

2. Governing Equations

For incompressible liquid-particle flows through a centrifugal impeller, the continuity-, the momentum- and the turbulent character-equations for both the liquid and the particulate phases in 2D Cartesian coordinate system, fixed on the impeller rotating with an angular velocity ω (rad/s), can be written as follows:

2-1 LIQUID PHASE

(a) Continuity equation

$$(u_j)_{xj} = 0 \tag{1}$$

(b) Momentum equations

$$\{\rho\, u_i u_j - \mu\, e[(u_j)_{xi} + (u_i)_{xj}]\}_{xj} = -p'_{xi} + \rho_p\, (u_{pi} - u_i)/\tau_{rp} - 2\rho\, e_{ilk}\, \omega_l\, u_k \tag{2}$$

where

$$\tau_{rp} = \rho_p d_p^2\, (1 + Re_p^{2/3}/6.0)^{-1}\, / 18\,\mu \qquad Re_p = |u - u_p|\, d_p / v$$

$$p' = p - \rho\, \omega^2\, r^2 / 2$$

The subscript xj represents the derivative with respect to xj. The term $2\rho\, e_{ilk}\, \omega_l\, u_k$ represents the Coriolis force.

(c) Equation of turbulent kinetic energy k

In order to calculate turbulent flows in a practical way, a turbulence model is necessary for simulating turbulence process in the flows. In this study, the two-phase turbulence model, particularly, the k-ε –Ap turbulence model is used for simulating the liquid-particulate turbulent flow in centrifugal impellers. The k-ε –Ap model is actually the k-ε turbulence model for liquid flows combined with an algebraic turbulence model for particulate phase flows. The equation of the turbulent kinetic energy k is indicated as follows:

$$\{\rho\, k u_j - (\mu_e / \sigma_k)\, k_{xj}\}_{xj} = G_k - \rho\,\varepsilon \tag{3}$$

where $\qquad G_k = \mu_e\, [(u_i)_{xj} + (u_j)_{xi}]\, (u_i)_{xj}$

(d) Equation of turbulent energy dissipation rate ε

$$\{\rho\,\varepsilon\, u_j - (\mu_e / \sigma_\varepsilon)\,\varepsilon_{xj}\}_{xj} = C_{\varepsilon 1}\,\varepsilon / k\, G_k\, (1 + C_{\varepsilon 3}\, R_f) - C_{\varepsilon 2}\,\rho\,\varepsilon^2 / k \tag{4}$$

where $Rf = (k/\varepsilon)^2\, (w/r^2)\, (wr)_r$ represents the effect of rotation on liquid turbulence, w is the circumferential velocity component.

(e) Boussinesq eddy viscosity μ_t

$$\mu_t = C_\mu\, k^2 / \varepsilon \tag{5}$$

Constants in the k-ε turbulence model are taken as:

$$C_\mu = 0.09 \quad \sigma_k = 1.0 \quad \sigma_\varepsilon = 1.3 \quad C_{\varepsilon 1} = 1.44 \quad C_{\varepsilon 2} = 1.92 \quad C_{\varepsilon 3} = 0.8$$

2.2 PARTICLE PHASE

(a) Continuity equation

$$[\rho_p u_{pj} - \nu_p / \sigma_p (\rho_p)_{xj}]_{xj} = 0 \tag{6a}$$

The continuity equation (6a) may be expressed in the form of the particle number n_p, that is,

$$[n_p u_{pj} - \nu_p / \sigma_p (n_p)_{xj}]_{xj} = 0 \tag{6b}$$

where $\rho_p = m_p n_p$, and m_p is the mass of a particle.

(b) Momentum equations

$$\{\rho_p u_{pi} u_{pj} - \mu_p [(u_{pi})_{xj} + (u_{pj})_{xi}]\}_{xj} = \nu_p / \sigma_p [u_{pi} (\rho_p)_{xj} + u_{pj} (\rho_p)_{xi}]_{xj} + \rho_p / \tau$$

$$rp(u_i - u_{pi}) \ -2 \rho_p e_{ilk} \omega_l u_{pk} +0.5(\rho_p \omega^2 r^2)_{xi} \tag{7}$$

(c) Eddy viscosity of particulate phase ν_p

The algebraic turbulence model of the particulate phase is used in this computation. So the eddy viscosity ν_p of the particulate phase can be expressed in the following Hinze-Tchen equation:

$$\nu_p / \nu_t = kp/k = (1 + \tau_{rp}' / \tau_t)^{-1} \tag{8}$$

where the particle dynamic response time τ_{rp}' and the fluctuation time of liquid turbulence τ_t are shown in the following forms:

$$\tau_{rp}' = \rho_p d_p^2 / (18 \mu) ; \qquad \tau_t = l/u' = (1.5)^{0.5} C_\mu^{3/2} k/\varepsilon$$

The governing equations of motion for the two-dimensional relative steady incompressible liquid-particle flow through such an impeller, expressed in the BFC system (ξ, η), can be represented by the model transport conservation equation (Wu et al 1995 b).

3. Boundary Conditions

The boundary conditions of liquid flows through the impeller for turbulence computation are the same as those used in the reference by Wu et al (1995 a). Only additional boundary conditions of particle turbulent flows are imposed.

The variables in the particulate phase also possess zero normal gradient specification everywhere except the inlet. At the inlet, particle velocity components have the same values as those in the liquid flow. And the bulk density in the particulate phase at the inlet can be easily determined according to the known volumetric concentration.

4. Solution Procedure

Discretization of the governing equations (1), (2) and so on in the form at the BFC system are performed, by using the same finite volume approximation as used in Ref. of WU et al (1995a). The second order central differencing is used for approximating the diffusion and source terms. For convective terms, the hybrid differencing scheme is employed. Solutions of incompressible turbulent flows are nonlinear and strongly coupled. The SIMPLEC algorithm is applied to drive the pressure field and the velocity field to be divergence free and convergence. The present numerical method for solving liquid-particulate flows contains the following steps:

(1) Estimate the initial velocity and the pressure field of a single liquid phase flow in the computational domain.

(2) Solve for the velocity, the pressure and the turbulent characters k and ε in the single phase flows, by using the SIMPLEC method, to get coarse convergence solution of the flow.

(3) Based on the single phase flow solution, we solve for the particulate turbulent flow field, by using the algebraic particulate phase turbulence model.

(4) Calculate the interaction terms between the liquid- and the particulate-phases to get some source terms in the control equations for the liquid flow.

(5) Solve again the momentum equations, the turbulent energy and its dissipation rate equation for single phase flow, but including the interaction terms.

(6) Set the new variables and return to step (3); repeat the process until it converges.

5. Calculated Results

Calculation of the liquid-particulate two-phase turbulent flow has been carried out for the same model slurry pump as that used for experiments in the references by Wu et al (1992, 1993), by using the Cray Super Computer in the Institute of Fluid Science of Tohoku University, JAPAN. The geometry and the inlet flow parameters of the impeller, as well as the grid system of computational domain have been shown in the references by Wu et al (1995 a and b). The reliability of the computational method and computer program in the present work can be obtained by comparison between the calculated results of the pure water flow through the same impeller and the experimental results by using the PIV technique (Wu et al 1995 a) and also by comparison between the calculated results and the experimental data for a rectangular region with a backward-facing step (Wu et al, 1995 b).

Figs. 1, 2 and 3 show the results at the following conditions:

Volumetric concentration of solid particle phase $C_v = 0.1$

Particle diameter	$d_s = 0.1$ mm
Particle density	$\rho_s = 2.65$ g/cm3
Inlet radial velocity	$u_{r1} = 6$ m/s
Rotation speed	$n = 1450$ rpm

Figs. 4, 5 and 6 show the results at the conditions:

Volumetric concentration of solid particle phase $C_v = 0.1$

Particle diameter $d_s = 1.0$ mm

Particle density $\rho_s = 2.65$ g/cm3

Inlet radial velocity $u_{r1} = 6$ m/s

Rotating speed $n = 1450$ rpm.

From these calculations, the following remarks can be obtained.

(1) According to the comparison between Figs. 2(a), 2(b) and 5(a), 5(b), as well as figures in the reference by Wu et al (1995 b), the existence of solid particles in the impeller has a little influence to the liquid flows at dilute solid concentrations, that is, C_v is smaller than 0.1. In the single liquid phase flow (Figs. 1 (a, b)), the flow recirculation occurs near the pressure side and by the trailing edge, resulting from the large expansion of transverse passage sections, caused by the large blade outlet angle $\beta_{b2} = 34^o$ and the small value of blade number, $N_b = 4$. This phenomenon was verified by the PIV observation results in the reference by Wu et al (1995 a).

But the existence of particle phase makes the maximum value of relative velocity at the leading edge on the pressure side lower slightly, e.g., at the single-phase flow $W_{max} = 13.27$ m/s (Ref. Wu et al 1995 b), but at the two-phase condition of $C_v = 0.1$ and $d_s = 0.1$ mm, $W_{max} = 12.24$ m/s (Fig. 2(a)), and at the condition $C_v = 0.1$ and $d_s = 1.0$ mm, $W_{max} = 12.84$ m/s (Fig. 5 (b)). Also the existence of particles makes the area of the recirculation and small velocity near the pressure side outlet different. For example, at a condition of $C_v = 0.1$ and $d_s = 0.1$ mm, that is, at a small particle condition (Fig. 2 (a)), the area is getting smaller a little than that at the single-phase flow. And at the condition of $C_v = 0.1$ and $d_s = 1.0$ mm, that is, at a large particle condition (Fig. 5 (a)), the area is slightly larger for the large diameter particles possessing greater velocity than that of liquid flows and making the liquid flow circulation more severely.

The existence of particles also affects to the pressure developed by the impeller. The larger the particle diameter is at the same concentration and the same density conditions, the lower the pressure in the liquid flow field is at the impeller outlet. For example, at the single-phase flow P_{max} is 4.99×10^5 Pa, in the two-phase flow with $d_s = 0.1$ mm, P_{max} is 4.85×10^5 Pa and at the condition with $d_s = 1.0$ mm, P_{max} is 4.68×10^5 Pa.

(2) The velocity field in the particulate phase is different from that of the liquid phase (Figs. 1(b), 2(b), 4(b) and 5(b)). The velocity difference between these two phases takes place mainly near the outlet of the impeller. That is, the relative particle velocity components at the outlet are higher than those in the liquid flow. The phenomenon agrees with the measurements by using LDV by Carder et al (1991). The results also agree with the observation of individual particle movement by Herbich & Christopher (1963) and by Wu et al (1992, 1993). That is, the particles are not appreciably slowed down even at the large cross-sectional area of the impeller discharge ring, and their absolute tangential velocities do not follow those of the transporting liquid.

(3) An increase in particle diameter results in a marked difference between the particle velocity and the liquid one. From comparison between Figs. 1(b) and 4(b), it is very clear that the particle relative velocity Ws near blade pressure side at the outlet under the condition of d_s = 1.0 mm (Fig. 4(b)) is either larger than that of the liquid flow (Fig. 4(a)), or larger than that of particles with d_s = 0.1 mm in Fig. 1 (b). Because the particle with a larger diameter as a pseudo-fluid in the present model keeps more inertia. This phenomenon also agrees with the observation of individual particles by Wu et al (1993). And an increase in the volumetric particle concentration results in the similar effects.

(4) The existence of particles in the liquid flow makes the turbulent kinetic energy k of the liquid phase lower. This trend will be severe with an increase in particle diameters at the same volumetric concentration. In the single-phase flow, the maximum value of k, i.e., k_{max} is 8.13 m2/s2, but in the two-phase flow with d_s=0.1 mm, k_{max} is 7.03 m2/s2 and at d_s=1.0 mm condition, k_{max} is 6.54 m2/s2 , which is much lower than that in single-phase flow.

6. Conclusions

(1) The liquid-particle two-phase flow turbulence model used in this study, the k- ε –Ap turbulence model, can correctly predict essential features of this flow in centrifugal impellers at dilute concentrations.

(2) The velocity field of particulate phase is different from that of the liquid phase in centrifugal impellers, especially near the outlet ring.

7. References

Ahmadi, G. and Abou-Zaid, S. (1990) A Rate-dependent thermodynamical model for rapid granular flows, *Int. J. Non-Newtonian Fluid Mechanics*, Vol. **35**, 15-3.

Al Taweel, A.M. and Landau, J. (1977) Turbulence modulation in two-phase jets, *Int. J. Multiphase Flow*, Vol. **3**, 341-351.

Bertodano, M.L., Lee S-J., Lahey, R.T. and Drew, D.A. (1990) The prediction of two-phase turbulence and phase distribution phenomena using a Reynolds stress model, *Trans. ASME, J. Fluid Engng.*, Vol. **112**, 107-113.

Carder, T., Masbernat, O. and Roco, M.C. (1991) Liquid-solid slip velocity around a centrifugal pump impeller, *ASME FED*-Vol. **118**, *Liquid-Solid Flows*, 101.

Crowder, R.S., Daily, J.W. and Humphrey, J.A.C. (1984) Numerical calculation of particle dispersion in a turbulent mixing layer flow, *J. Pipelines*, Vol. **4**, **3** , 159-170.

Dai, J. and Wu, Y.L. (1993) Numerical simulation of turbulent liquid-particle flows in centrifugal impeller, *Proc. Int. Symp. on Aerospace and Fluid Science*, Sendai, 404-410.

Danon, H., Wolfshtein, M. and Hetsroni, G. (1974) Numerical calculations of two-phase turbulent round Jets, *Int. J. Multiphase Flow*, Vol. **3**, 223-234.

Drew, D.A. and Segal, L.A. (1971) Averaged equation for two-phase flow, *Studies in Applied Math.*, Vol. **50**, 205-231

Elghobashi, S.E. and Abou-Arab, T.W. (1983) A two-equation turbulence closure for two-phase flows, *Phys. Fluids*, Vol. **26**, pp. 931-938.

Genchiev, Zh.D. and Karpuzov, D.S. (1980) Effects of motion of dust particles on turbulence transport equations, *J. Fluid Mech.*, Vol. **103**, 833-842.

Gidaspow, D., Jayaswal, U.K. and Ding, J. (1991) Navier-Stokes equation model for liquid solid flows using kinetic theory, *ASME* FED-Vol. 118, *Liquid-Solid Flows,* 165-172.

Herbich, J. B. and Christopher, R. J. (1963) Use of high speed photography to analyze particle motion in a model dredge pump, *Proc. IAHR Congress,* London, 89-98.

Ishii, M. (1975) *Thermo-fluid dynamic theory of two-phase flow,* Eyrolles, Paris.

Ma, D.N. and Roco, M.C. (1988) Probabilistic three-dimensional model for slow shearing particulate flow: wet friction, *ASME* FED-Vol. **55**, 53-60.

Marble, F.E. (1963) Dynamics of a gas containing small solid particles, *Proc. of 5th AGARD Combustion and Propulsion Colloquium,* Pergamon Press, London.

Melville, W.K. and Bray, K.N.C. (1979) A model of the two-phase turbulent jet, *Int. J. Heat Mass Transfer,* Vol. **22**, 647-656.

Michaelides, E.E. and Farmer, L.K. (1984) A model for slurry flows based on the equation of turbulence, *ASME* FED-Vol. **13**, *Liquid-Solid Flows and Erosion Wear in Industrial Equipment,* 27-32.

Roco, M.C. and Reinhart, E., (1980) Calculation of solid particles concentration in centrifugal pump impeller using finite element technique, *7th Hydrotransport,* 359-376.

Roco, M.C. and Shook C.A. (1983) Modeling slurry flow: The effect of particle size, *Canadian J. Chem. Engng.,* Vol. **61**, **4**, 494-504.

Sharma, M.P. and Crowe, C.T. (1978) A novel physico-computational model for quasi one-dimensional gas particle flows, *Trans. ASME, J. Fluid Engng.,* Vol. **100**, 343-349.

Shuen, J.S., Chen, L.D. and Faeth, G.M. (1983) Prediction of the structure of turbulent particle laden in rounded jet, *AIAA J.,* Vol. **21**, 1480-1483.

Wu, Y.L., Xu, H.Y. and Gao, Z.Q. (1992) Experimental study on particle motion in slurry pump impeller, *J. Tsinghua University,* Vol. **32**, **5**, 52-58, (in Chinese).

Wu, Y.L., Xu, H.Y., Dai, J. and Wang, L. (1993) Experiment on solid-liquid two-phase flow in centrifugal impeller, *Proc. Int. Symp. on Aerospace and Fluid Science,* Sendai, 379-387.

Wu Y.L., Sun Z.X., Oba, R., Ikohagi, T. (1995 a) Study on flows through centrifugal impeller by turbulent simulation, *ASME* FED-222, 63-68.

Wu, Y.L., Dai, J., Oba,R., Ikohagi, T. (1995 b) Turbulent flow simulation through centrifugal pump impeller at design and off-design conditions, *Proc. of 2nd ICPF,* Tsinghua University, Beijing, 155-167.

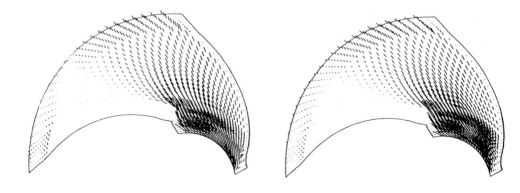

(a) Water Flow in Two-Phase Flow Condition   (b) Particle Phase Flow
Fig. 1 Velocity Vectors at Condition (u_{r1} =6m/s, n=1450rpm, C_v=0.1, d_s= 0.1mm and ρ_s=2.6 g/cm3)

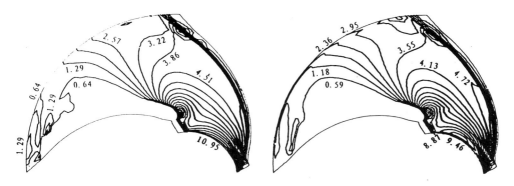

(a) Water Flow (W_{max} = 12.24 m/s)   (b) Particle Flow (W_{pmax} = 11.13 m/s)

Fig. 2   Relative Velocity at Condition (u_{r1} =6m/s, n=1450rpm, C_v=0.1, d_s= 0.1mm and ρ_s=2.6 g/cm3)

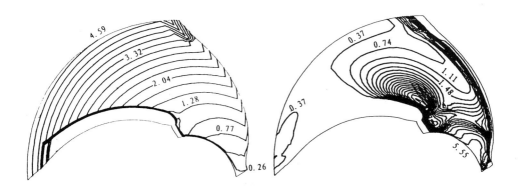

(a) Pressure Contour (P_{max} = 4.85 × 10^5 Pa)   (b) Turbulent Kinetic Energy Contour (k_{max} =7.03 m^2/s^2)

Fig. 3   Calculated Results at Condition (u_{r1} =6m/s, n=1450rpm, C_v=0.1, d_s= 0.1mm and ρ_s =2.6 g/cm^3)

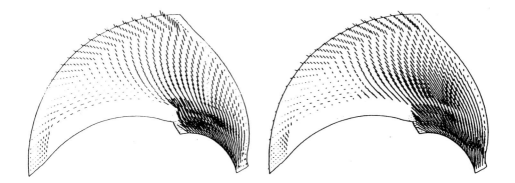

(a) Water Flow in Two-Phase Flow Condition (b) Particle Phase Flow

Fig. 4   Velocity Vectors at Condition (u_{r1} =6m/s, n=1450rpm, C_v=0.1, d_s=1.0mm and ρ_s =2.6 g/cm^3)

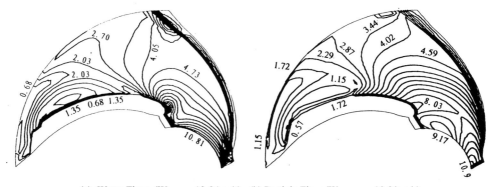

(a) Water Flow (W_{max} =12.84 m/s)   (b) Particle Flow (W_{pmax} = 10.90 m/s)

Fig. 5 Relative Velocity at Condition   (u_{r1} =6m/s, n=1450rpm, C_v=0.1, d_s= 1.0mm and ρ_s =2.6 g/cm^3)

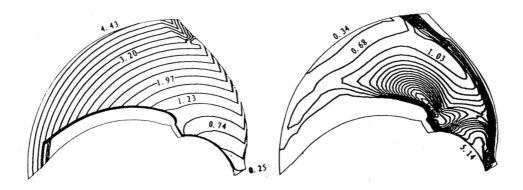

(a) Pressure Contour (P_{max} = 4.68 × 10^{5}Pa)   (b) Turbulent Kinetic Energy Contour (k_{max} = 6.51 m^2/s^2)

Fig. 6 Calculated Results at Condition (u_{r1} =6m/s, n=1450rpm, C_v=0.1, d_s= 1.0mm and ρ_s =2.6 g/cm^3)

APPLICATION OF THE METHOD OF KINETIC BALANCE FOR FLOW PASSAGES FORMING

M. BENIŠEK, S. ČANTRAK,
Faculty of Mechanical Engineering, Belgrade, Yugoslavia
B. IGNJATOVIĆ
"HPS Djerdap", Kladovo, Yugoslavia
D. POKRAJAC
Faculty of Civil Engineering, Belgrade, Yugoslavia

Definition of flow field boundaries is a common problem in hydraulic engineering. Their shape should enssure stable fluid flow, without separation and transient phenomena. Such problems are often solved using either the acquired experience or a large number of expensive experiments, with many trials, finally leading to the proper solution.

The inner fluid current is at hydrodynamic equilibrium. It contains the zones of "sound flow" and boundary layer zones. Besides them, there may be some zones of fluid at rest, or moving slowly, separated from the sound flow by the discontinuity surface. These zones are called "dead water". Although they interact with the "sound flow", their flow patterns are completely different. The "dead water" zones are formed if the fluid cannot follow the rigid flow boundaries. Separation happens either due to boundary layer thickening or because of significant fluid inertia. The former case can be avoided by boundary layer suction, but the latter one cannot. It was named "inertial separation" by Strscheletzky (1957). In the rotating fluid, close to the rotation axis, swirling core is formed, which can be treated as "dead water" compared to the "sound flow" between the core and the boundary layer at the pipe walls. In the "sound flow" region, the influence of viscosity can be neglected, flow can be assumed homogeneous and the fluid ideal and incompressible.

The problem of fluid boundaries shaping, which shall ensure the stable fluid flow with minimum undesirable phenomena was studied by Strscheletzky (1957) and (1969). He developed the method of kinetic balance, based

E. Cabrera et al. (eds.), Hydraulic Machinery and Cavitation, 455–463.
© *1996 Kluwer Academic Publishers. Printed in the Netherlands.*

on the Euler flow equation for ideal, incompressible fluid

$$\frac{d\vec{C}}{dt} = \vec{F} - \frac{1}{\rho}grad\,p. \tag{1}$$

Assuming conservative volume forces ($\vec{F} = grad\,U$), for the elementary mass $dm = \rho dV_i$, of the fluid, from (1) the momentum equation can be written as

$$\rho\frac{d\vec{C}}{dt}dV_i + \rho grad\,U\,dV_i + grad\,p\,dV_i = 0 \tag{2}$$

where $\vec{C}$ and p are local velocity and the pressure for the elementary volume dV_i.

Zone of fluid flow V_i has its characteristic flow mode ("sound flow" od "dead water"). The whole flow domain contains n elementary volumes V_i ($i = 1, 2, 3 \ldots n$). The "sound flow" and "dead water" zones are separated by the surfaces of zero, second or higher order of discontinuity, Strscheletzky (1969)

Virtual work of forces acting on the fluid in the volume V_i, at the moment t, for virtual displacement $\delta\vec{r}$ is

$$\int_{V_i} \rho(\frac{d\vec{C}}{dt}\delta\vec{r})dV_i + \int_{V_i} \rho(grad\,U\,\delta\vec{r})dV_i + \int_{V_i} (grad\,p\,\delta\vec{r})dV_i = 0. \tag{3}$$

For the whole flow domain V

$$\sum_{i=1}^{n}\left\{\int_{V_i} \rho(\frac{d\vec{C}}{dt}\delta\vec{r})dV_i + \int_{V_i} \rho(grad\,U\,\delta\vec{r})dV_i + \int_{V_i} (grad\,p\,\delta\vec{r})dV_i\right\} = 0. \tag{4}$$

Since the integrals are additive, and $\sum_{i=1}^{n} V_i = V$ from (4) it follows that

$$\int_{V} \rho(\frac{d\vec{C}}{dt}\delta\vec{r})dV_i + \int_{V} \rho(grad\,U\,\delta\vec{r})dV_i + \int_{V} (grad\,p\,\delta\vec{r})dV_i = 0. \tag{5}$$

The equation (5) states Lagrange's principle of virtual work. Flow equilibrium in the volume V, at the moment t, is achieved, when the sum of virtual works of the forces acting on the fluid equals zero.

Introducing kinetic energy dE_k and potential energy dE_p, of the elementary mass dm

$$dE_k = \frac{1}{2}c^2 \rho dV, \tag{6}$$

$$dE_p = U \, dm + p dV, \tag{7}$$

from the equation (5) it follows that

$$\int_{t_1}^{t_2} \delta \int_V (dE_k - dE_p) dt = 0. \tag{8}$$

The equation (8) states the balance condition for the fluid in motion. Non-viscous and incompressible fluid is in equilibrium if the difference between kinetic and potential energy is at minimum.

Introducing the realistic assumption that **total energy in the fluid stream does not change after virtual displacement** i. e. $\delta(dE_k + dE_p) = \delta(dE_t) = 0$, the equilibrium condition (8) is expressed as the variation of the sum of **integrals of action** I_i) formed for the characteristic flow domain zones V_i

$$\delta I = \sum_{i=1}^{n} \delta I_i = \sum_{i=1}^{n} \delta \int_{s_1}^{s_2} \rho \int_{V(1)} \vec{c} d\vec{s} dV_i = 0, \tag{9}$$

where: V_i - i-th fluid flow region, $\vec{c}$ - local flow velocity, dV_i - elementary volume, bounded by the inflow A_{ei} and outflow A_{oi} control surfaces (fig. 1) as well as by the given boundaries, s_1, s_2 - respective positions of the fluid particle at the moments t_1 and t_2 with $d\vec{s} = \vec{c}dt$

The equation (9) also states the equilibrium condition for the fluid in motion. It is more convenient for the solution of practical problems than (8). If one of the flow region boundaries changes, the action integral changes as well. The most convenient flow domain boundary, from the series of trial shapes, is the one which has minimum value of action integral.

In many practical problems the inner flow consists of only one main sound flow region, and one or more closed secondary flow regions, which are separated from the main flow by the free boundaries. For the ideal, non-viscous fluid flow, this boundaries are the discontinuity surfaces i.e. vorticity dissipative layers in the real fluid. In "dead water" zones the fluid is at rest or moves very slowly. Variational conditions can be applied to the sound flow regions only (action integral for the "dead water" equals zero). In the region V_1 having total discharge Q, from the equation (9) it follows that

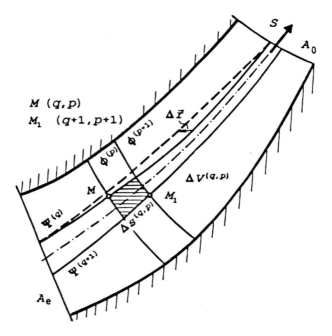

Figure 1: General view of flow passage and virtual streamline displacement

$$\delta I = \delta \int_{s_1}^{s_2} \int_{V_1} \rho \vec{c} dV \, d\vec{s} = 0. \qquad (10)$$

Analytical solution of the equation (10) exists only for the special cases. For that reason, the grapho-analytical or numerical solution is used. The equation (10) is applied to the small, but finite elements $\Delta V^{(q,p)} = \Delta V(x^{(q,p)}, y^{(q,p)}, z^{(q,p)})$ of q-th stream tube $(q = 1, 2, \ldots, m)$. Each $\Delta V^{(q,p)}$ is divided into p $(p = 1, 2, \ldots, k)$ elementary volumes. The distance between the two respective positions $s^{(q,p)}$ and $s^{(q,p+1)}$, along the mean streamline of the q-th stream tube (fig. 1). The action integral I is approximated as:

$$I = \rho \sum_{q=1}^{m} \sum_{p=1}^{k} \vec{c}(q, p) \Delta V^{(q,p)} \Delta s^{(q,p)} \qquad (11)$$

where $\vec{c}(q, p)$ is local flow velocity corresponding to the mean streamline of the q-th stream tube divided into k parts.

Practically, shaping of flow field boundaries reduces to one boundary variation, while the others are fixed, as well as the control surfaces. For each case, the action integral is computed, and the one with minimum integral is adopted as the final solution.

To illustrate kinetic balance method application, some flow field boundaries forming examples are shown.

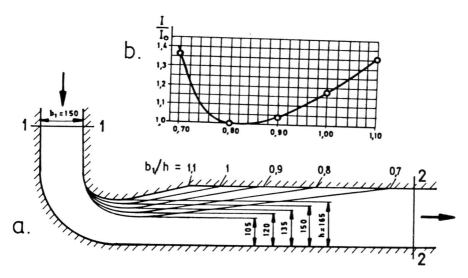

Figure 2: Difusor with parallel lateral walls

Example 1 **Forming of the inner (convex) curved contour of the diffusor with parallel lateral walls** is presented in Strscheletzky (1957). The form of diffusor model is shown in figure 2a. The inner (convex) boundary form is varied as defined by the ratio $h/b_1 = 1.1; 1.0; 0.9; 0.8; 0.7$. The other flow passage contours are fixed. For every flow passage form (h/b_1) the equipotential lines and streamlines are determined, and the action integrals calculated, using the expression (11). The change of dimensionless action integral values I/I_0 is shown in figure 3.b . For the value $h/b_1 = 0.815$, the function I/I_0 has minimum, so this contour was found optimum. The experiments presented in Strscheletzky (1957) have confirmed such a theoretical result.

Example 2   **Forming of the aeration duct of the bottom outlets of
the Haditha dam in Iraq**, Benišek (1982). The shape of the aeration
duct is shown in figure 3.a. The inner boundary forms and positions are
defined by the value $r/b_u = 1.57; 1.37; 1.14$ The other contours are fixed.
The calculation results of the dimensionless integral I/I_0 are shown in figure
3.b. For the value $r/b_u = 1.37$ (case II) the function I/I_0 has minimum.

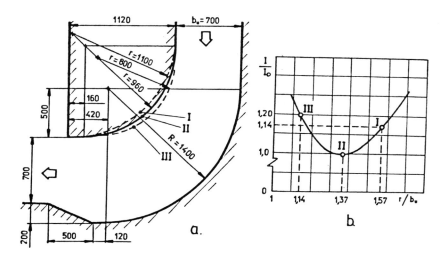

Figure 3: Aeration duct of the bottom outlets

Example 3.   **The analytical determination of the intake structure
lower contour of the additional hydro-turbine plant on the spillway of the "HEPS Djerdap I"**, Benišek (1993), Ignjatović, (1994) The
lower contour of the intake structure is determined by given method. The
considered cases are shown in figure 4.a, denoted by numbers 4 to 10, for
various head water levels $\nabla 69.5; \nabla 65.0; \nabla 63.0$. The computation results
are shown in figure 4.b. The minimums of dimensionless action integral
I/I_0, for various head water levels, are denoted by the arrows. The satisfactory contours are between 6 and 8. These results were confirmed by the

vibrations have equivalent amplitudes, some units present vibration in one direction as approximately half of the other direction, indicating a significant static load in these cases.

1.2   ESTIMATION OF OIL FILM STIFFNESS

Oil film stiffness of some guide bearings were estimated using data achieved on the rotor balancing at the commissioning time. For some units the coefficients obtained in such way are very close to the values estimated from the design. However, in other hydrogenerators were detected large discrepancies between experimental and theoretical values.

1.3   EXCITATION OF NATURAL FREQUENCIES

When some units are in a narrow band near to 80% of the nominal load, therefore above the range of low load vortex in the draft tube, there were observed excessive shaft vibrations at the turbine guide bearing (up to 750 μmpp). Such vibrations occurs always in two different frequencies, in the range of three to four times the nominal rotating speed. It is most interesting to observe that both frequencies may change up to 20% from a test to another. Such a phenomenon is still being investigated, but these vibrations are very probably due to the excitation of rotor natural frequencies.

1.4   CHANGES IN BEARING CLEARANCES

The clearances adjusted during assembling vary due to several factors, mainly influenced by thermal expansions. The difference between the journal and pad support ring temperatures can reach some degrees centigrades, which will cause expressive changes in the clearance and in the hydrogenerator bearing dynamic coefficients. Tests have shown that thermal expansions are strongly affected by the bearing cooling water temperature. As this water is taken directly at the spiral case, its temperature changes up to 15°C from summer to winter. Therefore the units present seasonal variations in bearing clearances.

1.5   SEASONAL VARIATIONS ON SHAFT VIBRATION

Shaft vibration and bearing temperatures monitoring has shown that there are significant periodic variations on such magnitudes, with seasonal characteristics. This behavior is originated by the seasonal changing on the bearing cooling water temperature, as described above.

1.6   DYNAMIC BEHAVIOR AFTER A START UP

Shaft vibration monitoring on a generating unit, since its start up at environment temperature until its thermal stabilization seven hours later, demonstrated that vibration severity decreases exponentially to less than the half of the initial value. Such a behavior is due to a significant reduction on the bearing clearances.

1.7   BEARING BRACKET DEFORMATIONS

Significant discrepancies among clearances were detected, caused by deformations in the bearing. Such deformations are originated mainly by localized heating of the bearing bracket near to the generator phases and neutral outputs, by the action of eddy currents. This behavior was confirmed by a Finite Elements Method - FEM analysis on the bearing bracket.

1.8   LUBRICANT VISCOSITY

It was verified that the guide bearings pads temperatures, and consequently the temperatures of their oil films, may differ up to $20\,^{\circ}C$ from a bearing side to another. The oil film dynamic coefficients are directly proportional to the lubricant dynamic viscosity, which mainly depends on its temperature. Therefore, it is very important to take into account the actual oil film temperatures when determining dynamic coefficients.

2.   Theoretical Aspects

Mathematical models are essential tools for today's design of rotating machinery. A well elaborated numerical description of intended devices allows a previous knowledge of static and dynamic behavior of the later installed machinery. By this, the design time and costs can be reduced and safety aspects can be improved at an early stage of planning.

Likewise, there is a need for modeling already installed mechanical structures to determine and optimize operational parameters, as well as trade with maintenance aspects. The maintenance optimization extends the operation intervals to increase productivity and prolongs machine life.

The ideas of on-line monitoring and incipient failures diagnostic through correlating vibration signals to a base line or simulated data, is based on the philosophy named predictive maintenance approach, which is already in use in less complex industrial processes. To do that it is desirable a mathematical model which describes the points of interest as good as possible.

The main goal in creating a numerical image of the hydrogenerator dynamic behavior is to help the maintenance team in detect and diagnostic incipient failures, through vibration monitoring. This image contains a huge number of dynamic influences like a structure with natural frequencies and resonances depending on the rotation speed, the gyroscopic effect, electric excitation at the generator and hydrodynamic forces at the turbine, as well as the interactions at turbine seals and hydrodynamic bearings. For such reasons, as well as to find more important influence factors for future modeling, an experimental identification is necessary to validate modelling and understand and estimate the machine dynamics.

This paper describes a procedure to determine stiffness and damping coefficients by means of a simplified calculation, derived from an analytical study of Reynolds equation. Despite being simplified, this study represents adequately the hydrodynamic bearings commonly used in large hydrogenerators: segmented type, with more than six concentric pivoted pads. For this bearing type the pad radius is approximately equal to the sum of journal radius and the bearing clearance, which allows to represent the oil film thickness along the bearing length by a linear function.

3   Dynamic characteristics of oil film

When a journal supported by hydrodynamic bearings vibrates under the influence of external forces, the oil film pressure oscillates. This oscillation gives rise to stiffness and damping forces in the oil film, which influence the critical velocities and vibration amplitudes of the rotor. The coefficients that represent such stiffness and damping effects are defined from forces resulting from the oil film pressure, derived from the Reynolds equation.

Let $F_X = F_X(X, X', Y, Y')$ and $F_Y = F_Y(X, X', Y, Y')$ be the components of the resulting force from oil film hydrodynamic pressure between the journal and pads, in X and Y directions respectively. The velocities of the displacements of the journal centre, X and Y, are $X' = \partial X/\partial t$ and $Y' = \partial Y/\partial t$.

The force $F = \{F_X, F_Y\}^T$ is non-linear for each variable. However, it can be approximated around the point of the journal static equilibrium with adequate precision, using the Taylor Series, if the vibration displacements and velocities are sufficiently small. The vectorial expression for the oil film force, disregarding terms of power equal or greater than two, is: $F = F_0 + \Delta F$, where $\Delta F = k \{\Delta X, \Delta Y\}^T + c \{\Delta X', \Delta Y'\}^T$ and k and c are given by Equation 1.

$$
k = \begin{bmatrix} \dfrac{\partial F_X}{\partial X} & \dfrac{\partial F_X}{\partial Y} \\[2mm] \dfrac{\partial F_Y}{\partial X} & \dfrac{\partial F_Y}{\partial Y} \end{bmatrix} = \begin{bmatrix} k_{XX} & k_{XY} \\ k_{YX} & k_{YY} \end{bmatrix} \quad , \quad c = \begin{bmatrix} \dfrac{\partial F_X}{\partial X'} & \dfrac{\partial F_X}{\partial Y'} \\[2mm] \dfrac{\partial F_Y}{\partial X'} & \dfrac{\partial F_Y}{\partial Y'} \end{bmatrix} = \begin{bmatrix} c_{XX} & c_{XY} \\ c_{YX} & c_{YY} \end{bmatrix} \tag{1}
$$

The subscript 0 on F_0 denotes the force at the point of the bearing static equilibrium, i.e., the force originated only by the oil wedge pressure, caused by the static load. The symbol Δ indicates that the dynamic force and vibration amplitudes are small. The coefficients k_{ij} and c_{ij} are respectively the oil film stiffness and damping coefficients in the directions $i, j = X, Y$.

To calculate the coefficients defined above, it is necessary to differentiate the force acting on the oil film with relation to the displacements and velocities of the relative vibration between journal and pads, in order to determine stiffness and damping coefficients, respectively.

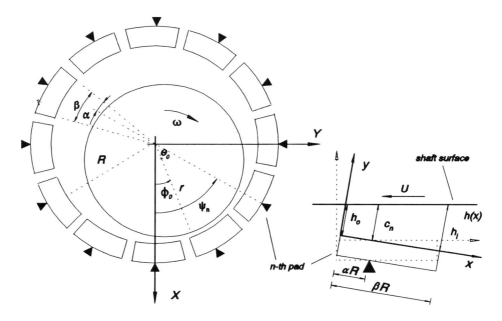

Figure 1. Sketch of a large hydrogenerator guide bearing

4.   Oil Film Pressure: Unidimensional Analysis

The Reynolds equation for hydrodynamic bearings can be derived from the movement equation of an oil film element in a laminar flow between parallel plates, in a steady state condition. It is given by:

$$\frac{\partial}{\partial x}(\frac{h^3}{\mu}\frac{\partial p}{\partial x}) + \frac{\partial}{\partial z}(\frac{h^3}{\mu}\frac{\partial p}{\partial z}) = -6\,U\frac{\partial h}{\partial x} - 12\,X'\cos\theta - 12\,Y'\sin\theta \qquad (2)$$

Disregarding the axial losses ($\partial p/\partial z = 0$), the pressure distribution over the surface of the n-th pad is determined analytically by integrating the Reynolds equation twice with respect to x. The integration constants are obtained by the boundary conditions for the pressure ($x=0$, $p(x)=0$) and ($x=\beta R, p(x)=0$).

The hydrodynamic pressure bends the pivoted pad, assuming an equilibrium position defined by $\eta = h_i/h_o$, the relationship between the oil film thickness at the leading edge (h_i) and at the tailing edge (h_o) of the pad. To determine such position, it should be remembered that the torque originated by the hydrodynamic pressure at the left and at the right side of the pivot point must be equal in this condition. By this, the following equation for η is obtained [1]:

$$2 \ln\eta \ [(1+2\eta)+\frac{\alpha}{\beta}(\eta^2-1)] + (\eta-1)(\eta+5) - 4\frac{\alpha}{\beta}(\eta-1)^2 = 0 \qquad (3)$$

By the previous equation can be noted that η is a constant which depends exclusively on the pad geometry. Once the pad length βR and pivot point αR are determined, η is obtained by solving Equation 3 and it is valid to all bearing pads.

5. Oil Film Pressure: Two Dimensional Analysis

In a finite bearing there are axial losses, which can not be disregarded ($\partial p/\partial z \neq 0$). The effect of this flow in z, the width direction, is considered by using an approximated solution for the complete Reynolds equation. This solution is derived from the specific case where the bearing pad width is much smaller than the bearing length. For such a case the pressure variation in x direction can be disregarded ($\partial p/\partial x = 0$) and the Reynolds equation can be solved analytically again. It can be shown that the pressure has a parabolic distribution in the z direction. The proposed solution has the following form [1]:

$$p(x,z) = \frac{-6\mu U}{\beta R} \frac{(\eta-1)}{(\eta+1)} \frac{x\,(x-\beta R)}{h^2(x)} \ [-|\frac{2z}{L}|^m + 1] \qquad (4)$$

where coefficient m depends on the pad geometry. For an infinitely wide pad, m is infinite, which means a constant pressure distribuition along the pad width. For the infinitely short pad, m is equal 2 and the pressure distribution in the z direction is parabolic. In fact, such idea was used before in cylindric bearings [4], where m is determined in relation of bearing dimensions, by an empirical function.

In the present paper another procedure to determine m is used for a finite bearing. The basic idea is to put Equation 4 into the complete Reynolds equation. If $p(x,z)$ actually were the solution to that equation, this procedure would result in a null equality. However, in this case there will be a residue, a function of m and bearing dimensions. The coefficient m is determined [1] by minimizing the total residue, obtained by integrating the residue over the pad surface. This procedure leads m to:

$$m = -\frac{1}{2} + \frac{1}{2} \sqrt{1 + 12\,(\frac{L}{\beta R})^2} \qquad (5)$$

6   Oil Film Dynamic Coefficients

The stiffness coefficients are determined by differentiating the components of the resulting force (F_X and F_Y) with respect to the displacements X and Y. This forces are obtained by the integration of the pressure over pad surface. Thereby, the following equations are reached [1]:

$$k_{XX} = \frac{\Gamma_0}{c_0} \sum_{n=1}^{n_r} \frac{1 + \cos2\psi_n}{\chi_n^3} \qquad\qquad k_{XY} = \frac{\Gamma_0}{c_0} \sum_{n=1}^{n_r} \frac{\sin2\psi_n}{\chi_n^3}$$

$$k_{YX} = \frac{\Gamma_0}{c_0} \sum_{n=1}^{n_r} \frac{\sin2\psi_n}{\chi_n^3} \qquad\qquad k_{YY} = \frac{\Gamma_0}{c_0} \sum_{n=1}^{n_r} \frac{1 - \cos2\psi_n}{\chi_n^3} \qquad (6)$$

where:

$$\Gamma_0 = \frac{m}{m+1} \frac{6\mu\, UL[(\eta-1)\alpha+\beta]^2}{(\eta-1)^2} [-\ln\eta + \frac{2(\eta-1)}{(\eta+1)}] \left(\frac{R}{c_0}\right)^2 \qquad (7)$$

The damping coefficients are determined by differentiating F_X and F_Y with respect to the velocities X' and Y'. Thereby, the following equations are obtained [1]:

$$c_{XX} = \Lambda_0 \sum_{n=1}^{n_r} \frac{[C_{\beta\eta}\cos\theta_n + S_{\beta\eta}\sin\theta_n]\cos\psi_n}{\chi_n^3}$$

$$c_{YX} = \Lambda_0 \sum_{n=1}^{n_r} \frac{[C_{\beta\eta}\cos\theta_n + S_{\beta\eta}\sin\theta_n]\sin\psi_n}{\chi_n^3}$$

$$c_{XY} = \Lambda_0 \sum_{n=1}^{n_r} \frac{[C_{\beta\eta}\sin\theta_n - S_{\beta\eta}\cos\theta_n]\cos\psi_n}{\chi_n^3}$$

$$c_{YY} = \Lambda_0 \sum_{n=1}^{n_r} \frac{[C_{\beta\eta}\sin\theta_n - S_{\beta\eta}\cos\theta_n]\sin\psi_n}{\chi_n^3} \qquad (8)$$

where $\theta_n = \psi_n - \alpha$ and the constants Λ_0, $C_{\beta\eta}$ and $S_{\beta\eta}$ are determined as follows:

$$\Lambda_0 = \frac{m}{m+1} \frac{6\mu\, L[(\eta-1)\alpha+\beta]^3}{(\eta-1)^2} \left(\frac{R}{c_0}\right)^3 \qquad (9)$$

$$C_{\beta\eta} = \frac{-\beta\eta-\beta\eta\cos\beta-2\eta\sin\beta}{\beta\eta\,(\eta+1)} + [\ \frac{-(\eta^2+1)\,\delta^2}{(\eta+1)} + \frac{2\ln\eta}{(\eta-1)}\ ]\cos\delta\ +$$

$$+[\ \frac{-(\eta^4-2\eta^3+2\eta-1)\,\delta^3}{6\,(\eta^2-1)} + 3\delta - \frac{2\eta\,\delta\,\ln\eta}{(\eta^2-1)}\ ]\sin\delta \qquad (10)$$

$$S_{\beta\eta} = \frac{2\eta+\beta\eta\sin\beta-2\eta\cos\beta}{\beta\eta\,(\eta+1)} + [\ \frac{-(\eta^2+1)\,\delta^2}{(\eta+1)} + \frac{2\ln\eta}{(\eta-1)}\ ]\sin\delta\ +$$

$$+[\ \frac{(\eta^4-2\eta^3+2\eta-1)\,\delta^3}{6\,(\eta^2-1)} - 3\delta + \frac{2\eta\,\delta\,\ln\eta}{(\eta^2-1)}\ ]\cos\delta \qquad (11)$$

7   Procedure Validation

The dynamic coefficients of the Itaipu generating units upper guide bearing were calculated using two methods, i.e., by using Equations 6 and 8 and the Dynko software. Dynko was elaborated by the 'Arbeitsgruppe Machinendynamik' at Kaiserlautern University - Germany, in order to determine oil film dynamic coefficients using the FDM - Finite Difference Method.

In Figure 2 the results are presented in a graphical form, where the dimensionless dynamic coefficients ($K_{ij} = c_0\,k_{ij}\,/\,W$ and $C_{ij} = \omega\,c_0\,c_{ij}\,/\,W$) are plotted against the Sommerfeld Number ($So = (\mu\ U\ L\ r^2)\ /\ (\pi\ W\ c_0^2)$). It can be observed that the results of both calculation methods have a good agreement. Moreover, it was verified that there is a constant for a given bearing that matchs the stiffness curves perfectly.

8   Conclusions

The experimental observation has shown that the guide bearings oil film forces have a strong influence on large hydrogenerators dynamic behavior. Therefore, such forces must be considered to get a reliable response with the mathematical models of those machines.

The expressions to calculate the oil film dynamic coefficients present a good agreement with the traditional time consuming method. Despite simplified, such expressions can be used to study several important phenomena that occur in large hydrogenerators (bearing deformations, clearances variations, different viscosities), which are usually disregarded. It is important to notice that the influences given by those phenomena are much more significant than those obtained by the simplification.

A future works will improve the description of coefficient m, by defining it as a function of bearing length. It will be also studied the bearing of non concentric type, looking for a procedure to linearize the oil film thickness function.

9   References

1.   Brito, G.C. and Weber, H.I. (1996) *Dynamic Behavior of Large Hydrogenerators Guide Bearings* - Manuscript of M.Sc. Thesis (in Portuguese) - Unicamp - Campinas State University - Brazil

2.   Fuerst, A.G.A. (1993) *Modellierung einer vertikalen Wasserkraftmaschine in Xavantes - Brasilien* - Diplomarbeit - Universität Kaiserslautern - Germany

3.   Ohashi, H. (1991) *Vibration and Oscillation of Hydraulic Machinery* - Hydraulic Machinery Book Series - Avebury Technical - England

4.   Varga, Z.E. (1971) *Wellenbewegung, Reibung und Oeldurchsatz beim segmentierten Radialgleitlager von beliebiger Spaltform unter Konstanten und Zeitlich veränderlicher Belastung* - Dissertation Nr. 4734 - ETH - Juris Druck + Verlag - Zurich

11.   List of notation

c_n	Clearance at the n-th pad	α	Angle between the pad pivot point and the origin of (x,y) co-ordinates
c_0	Nominal clearance		
e_0	Journal eccentricity	β	Angle between pad extremities
h	Oil film thickness ($h=h(x)$)	δ	Auxiliary variable $\delta=\beta/(\eta\text{-}1)$
L	Pad width	ϕ_0	Journal attitude angle
n	Pad numbering ($n=1..n_s$)	η	Relationship between the oil film thickness at the input (h_i) and the output (h_o) of the pad ($\eta=h_i/h_o$)
n_s	Number of bearing pads		
p	Oil film pressure ($p=p(x,z)$)		
r	Journal radius	μ	Dynamic or absolute viscosity
R	Pad radius	ψ_n	Angular position of the n-th pad pivot point
t	Time		
X,Y	Journal center displacements	θ	Angular position of a point on the pad surface ($\theta=\theta_n+x/R$, $\theta_n=\psi_n\text{-}\alpha$)
X',Y'	Journal center velocities		
U	Journal surface velocity	χ_n	Dimensionless clearance $\chi_n=c_n/c_0$
W	Static load acting on journal	ω	Journal rotating speed

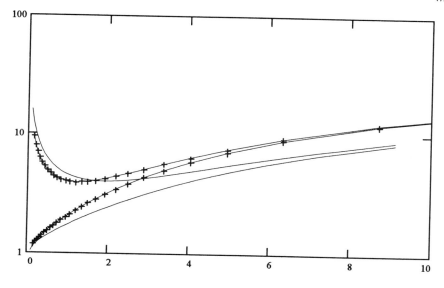

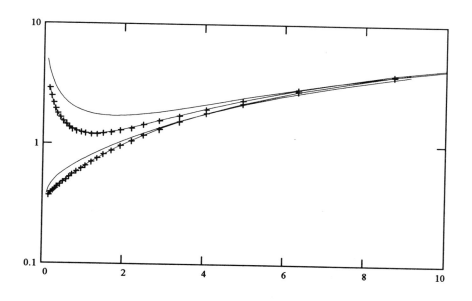

Figure 2.   Comparing stiffness (upper) and damping (lower) coefficients: Equations 6 and 8 - crossed lines versus Dynko Program - solid lines.

$$(\ K_{XX} > K_{YY} \ , \ \ C_{XX} > C_{YY} \ , \ \ K_{XY} = K_{YX} = 0 \ , \ \ C_{XY} = C_{YX} = 0 \ )$$

INSTABILITIES IN A FLOW-CONTROL VALVE

CIGADA A.(*), GUADAGNINI A.(**) and ORSI E.(**)
(*) Dip. Meccanica (**) D.I.I.A.R.
Politecnico di Milano, P.zza L. da Vinci, 32, 20133 Milano (Italy)

1. Abstract

This paper is about some instabilities that have been observed during the testing procedure of a flow-control valve. It was observed that the complete closure of the valve was very often reached after a transient state with oscillating behaviour of its mobile element whose displacement was recorded by means of proximity sensors, facing on opposite sides the moving body of the valve. The outputs have been collected by two signal conditioning devices and sent to a dual channel spectrum analyser. Measurements have also been performed through a servo-accelerometer which has been fixed to the pipe. The influence of the pipe length on the behaviour of the system was studied. The oscillations recorded by the proximity sensors are fairly well reproduced by a series of square waves, with a number of low amplitude, strongly damped, higher frequency contents. During a single test the frequency of the oscillations is not a constant but increases in its central part. The phenomenon cannot be related to one of the natural frequencies of the piping system, that are much higher. The observed frequency depends on the length of the adduction pipe.

2. Introduction

Some devices, even though very simple, in a complex piping system may cause noticeable instabilities (e.g. Brunone *et al.*, 1995; Fanelli, 1976; Föllmer and Zeller, 1980; Green and Wood, 1980; Orsi, 1994; Tanda and Zampaglione, 1991). A very careful analysis has to be performed to fully identify the dynamics of these phenomena and to propose an interpretation which has to be checked by experimental data. We will be concerned with the testing procedure of a flow control valve, that is designed to stop the flow in a pipe when the flow rate reaches a given value. These elements are of great interest especially for safety reasons in industrial processes. The structure of the valve, as reported in *Figure* 1, is very simple. It was mounted on a steel pipe line, with 50 mm internal diameter and some preliminary tests allowed to obtain characteristic curves as a function of the orifice diameter δ, of the initial opening h and of the spring preload F_S. In standard operating conditions the device is like an orifice in the main pipe of diameter D with transversal section s = (π d δ), d being the stem displacement

E. Cabrera et al. (eds.), Hydraulic Machinery and Cavitation, 474–483.

$(0 \leq d \leq h)$. The dynamic force of the fluid is opposed to the preload of the spring. When a fixed calibration limit is trespassed, the flow rate should instantaneously go to zero. We have performed a laboratory study of the behaviour of such a device. When the flow rate is progressively incremented and the conditions at the valve calibration limits are reached, some instabilities arise: the complete closure of the valve is very often reached after a transient state with oscillations of its mobile element. Therefore, the laboratory tests have been focused on the analysis of this transient state.

The displacement of the mobile element was recorded by means of proximity sensors, facing on opposite sides the moving body of the valve. The outputs have been collected by means of two signal conditioning devices and then sent to a dual channel spectrum analyser capable of directly summing the two inputs. Measurements have also been performed through a servo-accelerometer which has been fixed to the pipe. The servo-accelerometer output has sometimes been recorded together with the one coming from one of the proximity sensors to eventually get the relation between the two. During this first phase the influence of the pipe length, L_P, on the observed behaviour was studied: tests were performed for $L_P = 2$ and 400 m, the sampling rate f_S was varied between 50 and 1280 Hz and the phenomenon was recorded until the complete closure of the valve was obtained.

3. Experimental set - up

The way to approach the problem passed through the measuring of the movements of the valve head inside its body. An adequate accuracy level had to be achieved, but the main worries were about the eventual disturbances on the oncoming flow due to the measuring transducers, which could affect the phenomenon to be studied to such an extent to make it disappear. The choice had to take into account the fact that the adopted sensors had to work in water, too. The most attractive solution looked the one which employed some non-contact eddy current sensors. However, as it was not possible to fix one of these sensors in front of the plate of the valve head, due to the already mentioned disturbing fears, some alternative solution had to be looked for. This solution is shown in *Figure 1*: a cone has been screwed and glued to the valve stem, allowing for a radial positioning of the eddy current pick-up, which has been fixed drilling a hole in the tube carrying the valve. The conicity has been chosen in such a way to adjust the head stroke to the probe measuring field. The sensor has been put downstream the valve throat and it is "hidden" by the guide plate beading, to limit the effects due to its presence to a very small consequence. To prevent the plate from rotating and therefore leaving the probe exposed to the free stream, a steady pin has been inserted between the valve body and the plate. During the first laboratory tests to check the efficiency of the whole system, it was observed that one probe alone was not enough to get a satisfactory accuracy. In fact, as a certain backlash is observed between the guide and the stem, a movement of the head, relative to the valve seat, is always allowed. Therefore, the single probe output can't distinguish between the axial and the radial movements of the stem. A suitable way for separating these two effects is to fix another probe facing the cone at the opposite side of the same diameter where the first probe was. The two measurements combined together give the possibility of detecting the needed part of the signal out of the global measurement. The outputs of the two probes, properly conditioned, have been sent to a dual-channel spectrum analyser, for both data storing and analysis. A further step was the measuring system calibration. Calibration curves have been obtained by means of a slip table moved by a micrometric screw allowing for an accuracy of 1/100 mm. The combination of component errors in the overall system accuracy had to be pointed out. Calibration has been performed following two different procedures, therefore giving the needed values to perform an uncertainty analysis (e.g. Doeblin, 1990). The first way is to simply establish a link between the displacement superimposed to the valve head and the average change of the probe outputs, with respect to the initial rest position. This gives the calibration curve of *Figure* 2, expressing the along-stream stem motion as a function of the average voltage readings. The second way passes through the eddy current probe calibration and the cone angle according to the following:

$$d = \frac{(\Delta V_1 + \Delta V_2)}{2k_P} \frac{1}{tg\,\alpha} = \frac{\Delta V}{k_P} \frac{1}{tg\,\alpha} \tag{1}$$

where ΔV_i are the probe outputs [V], ΔV is the average of the probe outputs [V], k_P is the eddy current calibration factor [V/mm], and $tg\,\alpha$ is the cone angle. Uncertainties are bound to each term of the equation. The eddy current probes have a fixed gain and their calibration is given by the manufacturer for a certain number of materials. A common calibration value is 7.874 V/mm (200 mV/mils) with a resolution of a few μm for AISI 4140: the cone screwed to the stem has therefore been made up of this steel; anyway, as the probe faced a not-flat surface and as water was between the probe and the cone, some doubts arose about the possibility of a significant calibration change. The cone angle has been built so that $tg\,\alpha = (1.5/12 \pm 0.1/12)$ mm/mm. As several calibration curves were available, it has been decided to carry out an estimation of both k_P and $tg\,\alpha$ by means of a least square method minimising the difference between eq. (1) and the superimposed stem displacements, to get reliable values of the uncertainty bound to both k_P and $tg\,\alpha$.

The identified values are $tg\,\alpha = 1.6 / 12$ mm/mm, and $k_P = 8.635$ V/mm, thus allowing for an estimation of the uncertainties related to $tg\,\alpha$ and k_P (which have been fixed close to the difference between the known theoretical values and the identified ones). The $tg\,\alpha$ value is coherent with the a-priori supposed uncertainty, the estimation of k_P suggests a conservative uncertainty limit of ± 1 V.

The following uncertainties have been therefore assumed: $unc(\Delta V) = \pm 0.01$ V, $unc(k_P) = \pm 1$ V/mm, $unc(tg\,\alpha) = \pm 0.1/12$ mm/mm. A sensitivity analysis has been carried out to point out the weight of the calibration error $unc(k_P)$ as compared to the others. This leads to the calculation of the following derivatives coming from eq. (1):

$$\frac{\partial d}{\partial \Delta V} = \frac{1}{k_p\,tg\,\alpha} \tag{2}$$

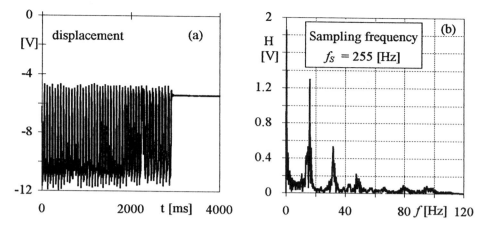

Figure 5. (a) Example of output from proximity sensor (L_P = 2 m; total sampling time T = 4000 ms) and (b) associated spectrum.

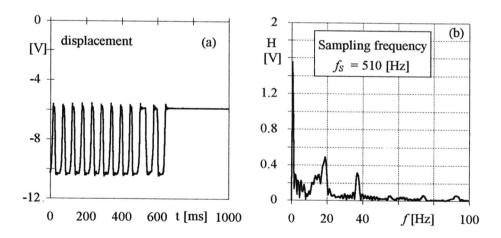

Figure 6. (a) Example of output from proximity sensor (L_P = 2 m; total sampling time T = 2000 ms) and (b) associated spectrum.

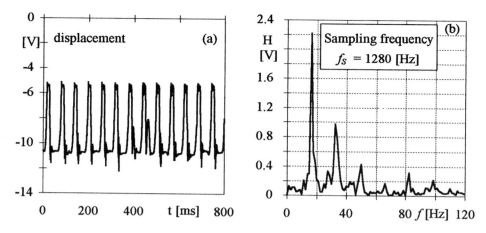

Figure 7. (a) Example of output from proximity sensor (L_P = 2 m; total sampling time T = 800 ms) and (b) associated spectrum.

From the readings of the accelerometer it is clear that the phenomenon cannot be related to one of the natural frequencies of the piping system, that are higher.
From a hydrodynamic point of view two tentative interpretations of the observed instabilities may be proposed. The first one is related to some analogies with a water hammer process : this may cause the motion of the mobile element of the valve, as it is subjected to the opposing hydrodynamic and spring forces. The second one is related to the behaviour of the system as the valve is reaching its closing position: as the flow section narrows, the fluid velocity increases and the local pressure drops. This action adds to the flux momentum. When the valve is closed, the spring force is opposed only to the force given by the fluid column in the upstream pipe, so that it might be possible that the valve opens again and an instability is generated. This observations are somehow consistent with the experimental evidences reported by Föllmer and Zeller (1980). At the present it is not possible to fully accept or cast aside any of these two schemes. Further experiments are needed, with different pipe lengths ; it is also important to properly measure the pressure values immediately upstream and downstream the valve.

5. Conclusions

A series of laboratory tests have been performed to study the oscillating behaviour that has been detected during the testing procedure of a flow-control valve. This transient phenomenon may assume a remarkable importance as it may be caused and maintained by several reasons which have to be thoroughly investigated. The displacement of the mobile element was recorded by means of proximity sensors ; moreover, a servo-accelerometer has been fixed to the pipe. The sampling rate was varied between 50 and

500 Hz and the phenomenon was recorded until the complete closure of the valve was obtained. The influence of the pipe length, L_P, on the observed behaviour was analysed. The time length of the transient conditions varies; in one case a persistence of the phenomenon was observed to last several minutes. During a single test the frequency of the oscillations is not a constant but it tends to increase in its central phase. The observed frequency, f, depends on the length of the adduction pipe: it ranges between 16 and 18 Hz for $L_P = 2$ m and it is about 1 Hz for $L_P = 400$ m. This behaviour suggests some analogies with a water hammer process and this problem is currently under investigation, also with the recording of the pressure values upstream and downstream the valve.

6. References

Brunone, B., Golia, U.M., and Greco, M. (1995) Effects of two-dimensionality on pipe transients modeling, *J. of Hydr. Eng.* ASCE, **121**(12), 906-912.

Doeblin, E. O. (1990) *Measurement Systems: Application and Design*, Mc Graw-Hill, New York.

Fanelli, M. (1976) Bibliographie raisonnee sur : "Les resonances hydrauliques dans les circuits industriels", ENEL - CRIS, Milano.

Föllmer, B. and Zeller, H. (1980) The influence of pressure surges on the functioning of safety valves, *Proc. of the Third Int. Conf. on pressure surges*, Canterbury, England, paper J2, 429-444.

Green, W.L. and Wood., G.D. (1980) The stability of direct acting spring loaded relief valves taking into account the upstream conditions, *Proc. of the Third Int. Conf. on pressure surges*, Canterbury, England, paper G4, 45-62.

Orsi, E. (1994) Su un fenomeno di instabilità in valvole a limitazione flusso (in Italian), *Proc. of the Seminar* "Moto vario nei sistemi acquedottistici", Bari, 72-75.

Press, W., Flannery, B., Teukolsky, S. and Vetterling, W. (1992) *Numerical Recipes: The Art of Scientific Computing* (Second Edition). Cambridge University Press, New York.

Tanda, M.G., and Zampaglione, D. (1991) Analisi di fenomeni di risonanza in un sistema distributore al termine di un'adduttrice a gravità: primi risultati (in Italian), Giornata di studio per la celebrazione del centenario della nascita di Girolamo Ippolito. Napoli.

STUDY OF STAYVANE VIBRATION BY HYDROELASTIC MODEL

Jean-Loup DENIAU
GEC-ALSTHOM NEYRPIC
82,Avenue Léon Blum BP 75, 38041 Grenoble Cedex, FRANCE

A method using hydroelastic model of stayvane, completed by mechanical calculations is presented. Comparison with measured stress value on a prototype is made. Some results obtained with different profiles are given.

1. Introduction

Usual method described to avoid stay vane cracking in hydraulic turbines consists in checking that frequency of vortex shedding is inferior to the first natural frequency of the vane.

Shedding frequency is a function of boundary layer characteristics near the separation points. Many workers have developed a Universal Strouhal Number Sh, based on the spacing h between the separation streamlines, and the velocity V along them.

$$\boxed{fk = Sh \times \frac{V}{h}} \qquad (1)$$

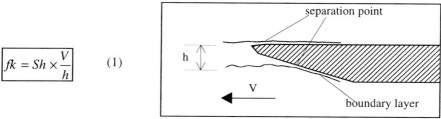

FIG. 1. Illustration of wake thickness

In this formula, Strouhal number Sh is independent of geometry.

Sh is difficult to evaluate, and different values were proposed. But from published works, and from our experimental studies, we can conclude in all cases :

$$Sh < 0.28$$

Evaluation of velocity V and wake thickness h is made by hydraulic calculations in stayring.

Then, the method of justification is to check :

E. Cabrera et al. (eds.), Hydraulic Machinery and Cavitation, 484–493.
© *1996 Kluwer Academic Publishers. Printed in the Netherlands.*

• For reduction factor of frequency induced by added mass, tests give a excellent concordance between mesured and calculated values (see appendice). Example for precedent profile 2 :

TABLE 2. Reduction factor of frequency from air to water (profile 2)

	bending mode 1	torsion mode 1
measurement	0.729	0.838
calculation	0.731	0.838

• Tests show a good concordance between peak to peak value of measured stresses in attachment of a prototype stayvane, and in the correspondent hydroelastic model (profile 1) :

In the prototype.

Stresses measured at AGUA-VERMEHLA (before modification) in 1979 : Smax = 82 MPa peak to peak for V estimated between 11 and 13 m/s at the trailing edge.

In the hydroelastic model.

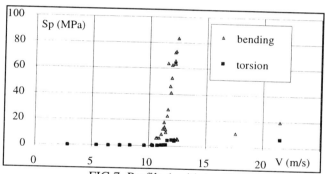

FIG.7. Profile 1 with i = 0°

7. Some results obtained with different other profiles

Following figures give, at the same scale, peak to peak stress values obtained during hydroelastic tests :

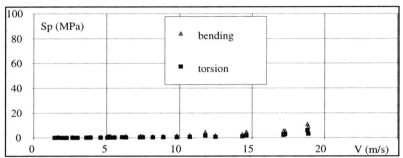

FIG.8. Profile 2 with i = 14°

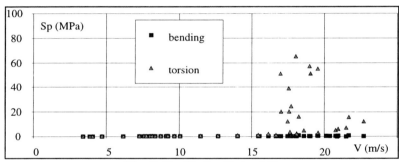

Fig. 9. Profile 3 with i = 14°

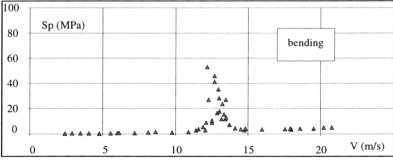

FIG. 10. Profile 4 with i = 0°

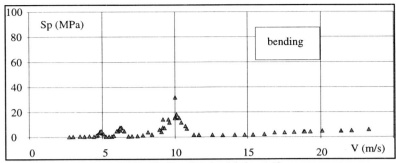

FIG. 10. Profile 4 with i = 14°

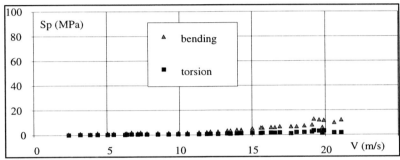

FIG. 11. Profile 5 with i = 14°

During resonance, stress amplification factor can reach values between 50 and 100.

During tests at i=14° with profile n°4, a strange phenomenon occurs : three successive resonance at the first bending mode appear. Only the third resonance at flow speed of 10 m/s is forecasted by classical design method (relation (1) in part 1.).
Following figures illustre this phenomena :

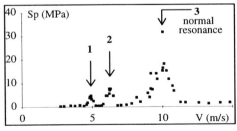

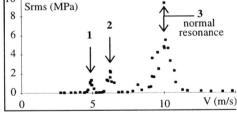

FIG. 12. Profil 4 : Peak to peak stress value in bending mode

FIG. 13. Profil 4 : Efficace stress value in bending mode at first natural frequency (1100 Hz)

8. Conclusions

Tests described in this paper have permitted to verify possibility of a experimental approach in the field of mechanical structural response at vortex shedding.

For each stayvane profile to study, these tests have to be completed by hydraulic calculations, to check that wake caracteristics behind the trailing edge are very similar in model and in stayring. Calculations have to take into account the effect of double cascade (stayvane and guidevane).

Some strange phenomena occured with a symetrical trailing edge, which exhibit three succcessive resonance at the first natural frequency.

Appendice : analytical calculation of natural frequencies

Natural frequencies of vibration can be calculated assuming stay vane are beams fixed at two ends.
For example, the frequency of the first bending mode is calculated by :

in air : $\quad f_b = \dfrac{22,35}{2 \times \pi \times H^2} \times \sqrt{\dfrac{E \times Im}{\rho \times S}}$

$$\begin{aligned}
\text{where :} \quad & H && \text{height of stay vane} \\
& Im && \text{minimal moment of inertia} \\
& S && \text{area of cross section} \\
& \rho && \text{steel density} \\
& E && \text{modulus of elasticity}
\end{aligned}$$

in water : f_b water $= \dfrac{1}{\sqrt{1 + \dfrac{\pi}{4} \times \dfrac{\rho \text{water}}{\rho} \times \dfrac{l^2}{S}}} \times f_b$ air

where $\quad l =$ length of cross section
$\rho_{water} =$ water density

For the natural frequency of the first torsion mode :

in air : $\quad f_t = \dfrac{1}{2 \times H} \times \sqrt{\dfrac{G \times J}{\rho \times Ip}} \times \dfrac{1}{\sqrt{1 - \dfrac{2}{\beta \times d} \operatorname{th}\left(\dfrac{\beta \times d}{2}\right)}}$

where : $\quad J = \dfrac{S^4}{4 \times \pi^2 \times Ip}$

$Ip =$ polar moment of inertia

$$\beta \times d = 2,148 \times \dfrac{H}{\left(2 \times \dfrac{S}{e} - 1\right)}$$

Precedent formula takes into account no warping of beam.

In water : $\quad f_t$ water $= \dfrac{1}{\sqrt{1 + \dfrac{\pi}{128} \times \dfrac{\rho \text{water}}{\rho} \times \dfrac{l^4}{Ip}}} \times f_t$ air

STUDY OF DYNAMIC BEHAVIOUR OF NON-RETURN VALVES

P. FRANÇOIS

Industrial Hydraulic

Technical Center for Mechanical Industries

BP 7617

F - 44076

NANTES Cedex 03

1 - Introduction

Although fluid carrying industrial installations have long been insured against non-return valves malfunctions, attention must be drawn to the risk of operating failures involved by the use of valves not well suited to the installation they are fitted to.

As a result of CETIM's research in this field, non-return valves operation can be characterized which facilitates the selection of a non-return valve for a given installation.

2 - Purpose of study

Non-return valves are essential oil pipe systems protecting devices. Their "one-way" function prevents many damages such as the draining of an installation in case of supply pressure decrease, back flows, unexpected liquid mixtures, liquid losses, etc .

E. Cabrera et al. (eds.), Hydraulic Machinery and Cavitation, 494–503.

However, they must be suitably designed for the pipe system they are fitted to because damages may happen in transient conditions when the stable fluid flow rate decreases suddenly and reverses on valve closure.

Subsequent pulsations influenced by the valve characteristics may then cause destructive water hammers (upstream and downstream pressure surges) likely to damage the valves and fittings.

Working in close collaboration with the valve manufacturers and the AFIR (Valves Industrial French Association), the CETIM's industrial valve committee has set out a method aimed at characterizing non-return valves operation during that stage. Current tests are required to provide an approved non-return valve closure characterization procedure giving results transposable from one installation to another.

3 - Test equipment on an experimental schakle

The test rig allows to simulate a pump shutdown with back pressure development in downstream pipework.

The final test arrangement and components are shown in Figure 1.
The test line is composed of the following elements (from upstream to downstream) :

- a pump shutdown which can be synchronized with the operation of a number of
    valves,
- a quick opening discharge valve which is mounted next to the test valve.

Its opening can be simulated when a shutdown signal is sent to the pump, thus eliminating the influence of the motor pump set and pipework inertia between the pump and the valve during the tests.

Then, there is :

- the test valve,

- a dynamic pressure transducer, pressure range 100 bar, frequency range 1 kHz,

- a fast acting electromagnetic flowmeter working in direct and reverse flow, frequency range 100Hz.

Finally, the test line ends by the operation of the quick-acting valve at the tank outlet which can be synchronized with the pump shutdown and the opening of the upstream valve.

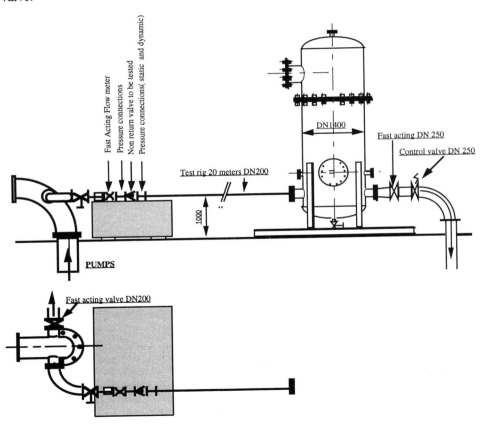

Figure 1 : Non return valves test rig diagram

In addition, the tests are aimed at validating the model built.

Then, it is necessary to evaluate the maximum pressure allowable in the installation in case water hammers are generated as it will be used in equation (1) to determine the reverse flow rate maximum velocity Vrmax.

The selection of the best suited non-return valve shall rely on the study of a variety of valves characteristic curves. The final choice shall not necessarily be the highest performing valve but certainly the one that is best suited to the installation under consideration.

In addition to pulsation-induced water hammers, a certain amount of fluid is lost upstream the installation which may be a more stringent criterion in some specific cases.

6 - Setting of test configuration

During the tests, the validation of Joukowski's equation has always been our preoccupation, because this is the first calcul for the use of dynamic characteristics.

We have noticed a good connection between the Vr measured speed and the pressure in the case that the test line is homogenous, that is to say when diameters of flowmeter, non return valve and the installation are equal.

When the reverse flow is too little, we use this relation (1) to determinate the reverse velocity.

As it was difficult to measure the twist speed with the relation, because of a too weak. We have decided to assert ourselves some conditions : the test line and the non return valve will have the same diameters.

Moreover, we dispose 3 flowmeters with 100, 200 and 250 mm diameters. We limit to a 250 mm non return valve for diameters which don't enter in the flowmeters range, we realise test with two different flowmeters.

From this point, we have noticed the divergent and convergent influence on the dynamic characteristic, which can be met in industrial using (Figure 5).

In this case, it is primordial to define the installation. The non return valve dynamic characteristic is single, but the junction of a divergent or convergent pipe leads to an another non return valve with different characteristics.

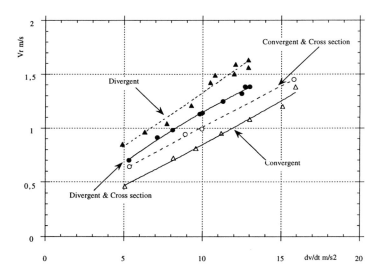

Figure 5 : Valves dynamic characteristics, a duo-check split disk DN 150

7 - Conclusion

For a given installation, i.e. deceleration $\dfrac{"dv"}{dt}$, pressure, wave propagation velocity "a" and maximum allowable pressure surge are defined, one can select the non-return valve best suited to these parameters, provided its operating characteristics are known.

CETIM's Industrial Hydraulics Departement test rig is designed to test pipes up to 250 mm, to deliver flow rates up to 4000 m^3/h for a 4 bar pressure and to create deceleration from 10 m/s^2 to 20 m/s^2 against tried diameters.

8 - References

1 - PROVOST GA (1982) The dynamic Characteristics of non-return valves, 11e Symposium AIRH AMSTERDAM, 1-7

2 - THORLEY A-R-D (1983) Dynamic response of check valves, Procedings 4e int. pressure Surges, 231-242.

OPTIMUM HYDRAULIC DESIGN OF TWO-WAY INLET CONDUIT
OF WANGYUHE PUMPING STATION

LU LINGUANG, ZHOU JIREN
Agricultural College,Yangzhou University, China

ZHANG RENTIAN
Jiangsu Surveying And Design Institute of Water Conservancy, China

ABSTRACT

The numerical analysis of 3-D turbulence flow through the two-way inlet conduit of is executed. In the light of the analysis the conduit is finally optimized by considering all geometrical parameters which influence the conduit shape. The hydraulic characteristics of two pump systems with different inlet conduit are compared with model tests, which show that the better characteristics are achieved for that with the optimized conduit.

1. INTRODUCTION

Wangyuhe Pumping Station will be built between the Yangtze River and the Tai Lake in China, where 9 axial flow pump sets are going to be installed, the diameter of each pump being 2.5 meters. The station is required to pump water in two directions: one for drainage and the another for irrigation. The alteration of the directions can be brought about only by regulating the four gates: two located in the inlet conduit and two in outlet conduit, respectively, as shown in Fig.1. Thus several hydraulic constructions could be saved which are otherwise necessary for the direction controlling.

A two-way inlet conduit is employed by the pumping station (See Fig.1).

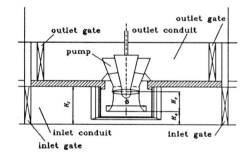

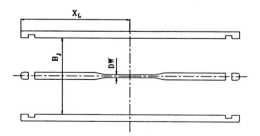

Fig.1 Configuration of two-way conduit

504

Due to the absence of study for the hydraulic design of the two-way inlet conduit before, two problems could be occurred to the inlet conduit. First, the flow field provided for the pump by the conduit may be non-uniform, which can reduce the efficiency of the pump system. Second, there probably be eddies in the non-working side of the inlet conduit, which may induce vortex band leading to severe oscillation of the pump.

In order to ensure the pump to be operated with safety and high efficiency it is necessary to optimize the hydraulic design of the conduit. The numerical analysis of 3-D turbulence flow through the two-way inlet conduit of Wangyuhe Pumping Station is executed, based on which the effects of all geometrical parameters of the conduit on the hydraulic characteristics are investigated one by one. Finally, the comparisons of model tests are conducted between two pump systems with the same model pump and outlet conduit but different inlet conduit. The calculated results of flow pattern in the inlet conduit and the hydraulic performance of the systems are experimentally validated.

2. MATHEMATICAL DESCRIPTION OF THE FLOW FIELD

2.1 GOVERNING EQUATIONS

The steady, Reynolds averaged Navier-Stokes equations for flows through the two-way inlet conduit, along with the closure standard k-ε model, can be written in Cartesian coordinates as:

(1) Continuity equation

$$\frac{\partial u_i}{\partial x_i} = 0 \tag{1}$$

(2) Momentum equation

$$u_j \frac{\partial u_i}{\partial x_j} = f_i - \frac{1}{\rho} \frac{\partial p}{\partial x_i} + \frac{\partial}{\partial x_j}[(\nu + \nu_t)(\frac{\partial u_i}{\partial x_j} + \frac{\partial u_j}{\partial x_i})] \tag{2}$$

(3) k-equation

$$\frac{\partial u_i k}{\partial x_i} - \frac{\partial}{\partial x_i}[(\nu + \frac{\nu_t}{\sigma_k})\frac{\partial k}{\partial x_i}] = P_r - \varepsilon \tag{3}$$

(4) ε-equation

$$\frac{\partial u_i \varepsilon}{\partial x_i} - \frac{\partial}{\partial x_i}[(\nu + \frac{\nu_t}{\sigma_\varepsilon})\frac{\partial \varepsilon}{\partial x_i}] = \frac{(C_{\varepsilon 1} \cdot \varepsilon \cdot P_r - C_{\varepsilon 2} \cdot \varepsilon^2)}{k} \tag{4}$$

where x_i, u_i and f_i, with the suffix i=1,2,3, represents the Cartesian coordinates, the Cartesian velocity components and the Cartesian force components, respectively. P_r, the generation ratio of the turbulent energy, may be mathematically expressed by

$$P_r = v_t (\frac{\partial u_i}{\partial x_j} + \frac{\partial u_j}{\partial x_i}) \frac{\partial u_i}{\partial x_j}$$ (5)

in which v_t stands for the dimensionless turbulent viscosity which is associated to k and ε by the relation:

$$v_t = C_\mu \frac{k^2}{\varepsilon}$$ (6)

The values of the empirical constants appearing in the above equations is given as follows[1]:

$C_\mu = 0.09$, $\sigma_k = 1.0$, $\sigma_\varepsilon = 1.3$, $C_{\varepsilon 1} = 1.44$, $C_{\varepsilon 2} = 1.92$

2.2 BOUNDARY CONDITIONS

2.2.1 *Upstream B.C.*
The upstream boundary is set at the inlet section of the suction sump, where the approach flow could considered as having a logarithmic velocity profile:

$$u = \frac{1}{\kappa} u_* \ln(\frac{z}{d_0})$$ (7)

where κ is the Von Kármán constant, u_* is the friction velocity, z is the vertical distance of the centre of the grid element to the bottom of the suction bay and d_0 is the roughness length of the bottom. The friction velocity may be obtained from:

$$u_* = \frac{\kappa q}{d(\ln\frac{d}{d_0} - 1)}$$ (8)

where q is the design discharge per unit width and d is the water depth in the suction sump.
The imposed profile of k and ε are of the forms:

$$k = \frac{u_*^2}{\sqrt{C_\mu}}(1 - \frac{z}{d})$$ (9)

$$\varepsilon = \frac{u_*^3}{\kappa}(\frac{1}{z} - \frac{z}{d})$$ (10)

2.2.2 *Downstream B.C.*
The outlet boundary is set at the inlet of the impeller chamber of the pump, where an absence of flow circulation over the cross section and a zero velocity gradient normal to the section are given[2,3,4].

2.2.3 *Solid Wall B.C.*

According to 'the law of the wall'[5], the following relation can be applied:

*b89Wâα Ç

$$u_* = \frac{\kappa u_w}{\ln(\frac{z_w}{d_w})}$$

(11)

in which z_w and u_w are the distance from the concerned element to the wall and the total velocity parallel to the wall in that element, respectively. The values of k and ε in the element are calculated via:

$$k = \frac{u_*^2}{\sqrt{C_\mu}}$$

(12)

$$\varepsilon = \frac{u_*^3}{\kappa Z_w}$$

(13)

2.2.4 *Free Surface B.C.*

The free surface may be considered as a symmetry plane for the velocity components and for the turbulence energy when wind shear is neglected. For the dissipation, ε, the used boundary conditions reads:

$$\varepsilon = \frac{(\sqrt{C_\mu} k_s)^{1.5}}{\kappa (Z_s + C_{BE} d)}$$

(14)

where z_s and k_s are the distance from the centre of the concerned element to the free surface and the turbulence energy in that element, respectively[5], and $C_{BE} = 0.07$ is an empirical constant[6].

3. OBJECTIVE FUNCTIONS

According to the design conditions for the impeller of an axial pump, it is necessary for the pump to be operated with maximum efficiency that the velocity distribution over the inlet section of the impeller chamber should be uniform and the velocities should have no any radial component. Correspondingly, the two objective functions[7] are employed for the following optimum hydraulic computations.

4. BASIC FLOW PATTERN IN THE CONDUIT

The basic flow pattern calculated is shown in Fig.2, from which it can be seen that the pump intakes water through the bellmouth. The flows entering the pump are converged to the bellmouth from all the directions: the front, the two sides and the rear of the bellmouth. Obviously, the velocities from different directions are not uniform: that from

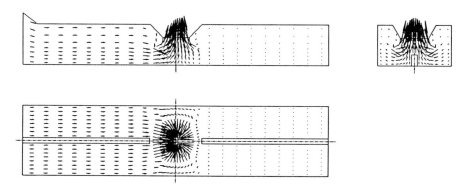

Fig.2 Basic flow pattern in the two-way inlet conduit

the front are the largest, while that from the rear are the smallest. This is the most important feature of the intake form in the two-way inlet conduit. It reveals that to obtain full uniform velocity distribution for the pump is difficult or even impossible. What we can do is to equalize the flow flux in all the directions as far as possible by hydraulically optimizing the design of the conduit.

The velocity distribution is clearly shown in the model test by the colour wires stuck on the inlet section of the bellmouth and the floor under the bellmouth. The flow pattern observed is essentially consistent with that calculated.

5. RESULTS OF THE OPTIMUM COMPUTATIONS

5.1 THE FIRST PART OF THE COMPUTATIONS

In the light of the numerical analysis, the hydraulically optimum computations can be executed. The undetermined parameters for the first part of the computation, in which the bellmouth is made of cast iron, so its outer shape is the the same as its inner one, are: clearance of bottom floor, H_B, diameter and height of the bellmouth, D_L and H_c, thickness of the baffle plate, DW, height, length and width of the inlet conduit, H_j, X_L and B_j (see Fig.1). The single-factor comparison method is applied in the optimum calculations. The schemes with different geometry parameters are principally taken in order according to the influence extent of the parameters on the flow field in the inlet conduit. The number and the relative geometry parameters of computational schemes in the first part are listed in Table 1, where scheme 11 is the original inlet conduit. The effect of the parameters upon the objective functions may found in Table 2 ~ Table 8. Following comments could be made based on the results:

(1) The clearance of bottom floor effect the degree of velocity uniformity obviously (See Table 2). The original value seems to be too small, so the centre of the pump impeller has to be raised to make the clearance increase to 0.84D (D is the diameter of the impeller).

(2) The diameter of the bellmouth has notable effect on the objective functions. From Table 3 it can be seen that the original one 1.44D is suitable.
(3) The height of the bellmouth influences markedly on the objective functions, too. Table 4 shows the larger the height, the better the inlet flow field.

Table 1 Number and relative geometry parameter of computation schemes

No.	X_L	H_j	B_j	H_B	H_c	ϕ	DW
11	5.40D	1.48D	2.60D	0.68D	0.56D	1.44D	0.5m
12	5.40D	1.48D	2.30D	0.68D	0.56D	1.44D	0.5m
13	5.40D	1.28D	2.60D	0.84D	0.56D	1.60D	0.2m
14	5.40D	1.28D	2.75D	0.84D	0.56D	1.44D	0.2m
15	5.40D	1.48D	2.60D	0.84D	0.56D	1.44D	0.5m
16	5.40D	1.28D	2.60D	0.84D	0.56D	1.44D	0.5m
17	5.40D	1.28D	2.60D	0.84D	0.56D	1.20D	0.2m
18	5.40D	1.28D	2.60D	0.84D	0.56D	1.44D	0.2m
19	5.00D	1.28D	2.60D	0.84D	0.56D	1.44D	0.2m
20	5.40D	1.28D	2.60D	0.84D	0.46D	1.44D	0.2m

Table 2 Effect of clearance of bottom floor

No.	H_B	u_{max}(m/s)	u_{min}(m/s)	$\bar{u}$(m/s)	V_u(%)	$\bar{\theta}_{max}$(°)	$\bar{\theta}_{min}$(°)	$\bar{\theta}$(°)
11	0.68D	5.52	2.56	3.98	83.60	90.0	60.0	78.0
15	0.84D	5.47	2.64	3.99	85.23	90.0	64.0	78.4

Table 3 Effect of diameter of bellmouth

No.	ϕ	u_{max}(m/s)	u_{min}(m/s)	$\bar{u}$(m/s)	V_u(%)	$\bar{\theta}_{max}$(°)	$\bar{\theta}_{min}$(°)	$\bar{\theta}$(°)
16	1.44D	5.34	2.04	3.98	85.77	90.0	65.7	78.1
13	1.60D	5.09	1.47	3.93	86.25	90.0	59.0	74.9
17	1.20D	5.11	0.33	3.93	75.60	90.0	64.1	76.4

Table 4 Effect of height of bellmouth

No.	H_c	u_{max}(m/s)	u_{min}(m/s)	$\bar{u}$(m/s)	V_u(%)	$\bar{\theta}_{max}$(°)	$\bar{\theta}_{min}$(°)	$\bar{\theta}$(°)
18	0.56D	5.10	1.77	4.02	87.79	90.0	64.6	77.4
20	0.46D	5.06	0.33	3.80	84.33	90.0	52.7	71.1

Table 5   Effect of thickness of baffle plate

No.	DW	u_{max}(m/s)	u_{min}(m/s)	$\bar{u}$(m/s)	V_u(%)	$\bar{\theta}_{max}$(°)	$\bar{\theta}_{min}$(°)	$\bar{\theta}$(°)
16	0.5m	5.34	2.04	3.98	85.77	90.0	65.7	78.1
18	0.2m	5.10	1.77	4.02	87.79	90.0	64.6	77.4

Table 6   Effect of width of inlet conduit

No.	B_j	u_{max}(m/s)	u_{min}(m/s)	$\bar{u}$(m/s)	V_u(%)	$\bar{\theta}_{max}$(°)	$\bar{\theta}_{min}$(°)	$\bar{\theta}$(°)
18	2.60D	5.10	1.77	4.02	87.79	90.0	64.6	77.4
14	2.75D	5.37	2.64	3.99	88.14	90.0	63.9	77.1

Table 7   Effect of height of inlet conduit

No.	H_j	u_{max}(m/s)	u_{min}(m/s)	$\bar{u}$(m/s)	V_u(%)	$\bar{\theta}_{max}$(°)	$\bar{\theta}_{min}$(°)	$\bar{\theta}$(°)
15	1.48D	5.47	2.64	3.99	85.23	90.0	64.0	78.4
16	1.28D	5.34	2.04	3.98	85.77	90.0	65.7	78.1

Table 8   Effect of length of inlet conduit

No.	X_L	u_{max}(m/s)	u_{min}(m/s)	$\bar{u}$(m/s)	V_u(%)	$\bar{\theta}_{max}$(°)	$\bar{\theta}_{min}$(°)	$\bar{\theta}$(°)
18	5.40D	5.10	1.77	4.02	87.79	90.0	64.6	77.4
19	5.00D	5.24	1.82	4.02	88.04	90.0	64.6	77.1

(4) From the results in Table 5, the thickness of baffle plate also has obvious effect on the functions and should be as thin as possible. In fact, the reason to found a baffle plate under the bellmouth is to prevent vortex bands from forming in the inlet conduit, and there is no requirement for thickness of the plate. Considering the smallest thickness of concrete pouring in the construction site, 0.2m is acceptable.

(5) In a certain range, the width, height and length of the inlet conduit have a little effect on the functions (See Table 6, Table 7 and Table 8). The values of the three basic dimensions of the pumping station greatly influence on the capital investment, so they had better to be determined more considering other factors. For example, the length may be decreased only if the arrangement of the upper structure of the pumping station is permissible.

5.2 THE SECOND PART OF THE COMPUTATIONS

To further reduce the velocities near the bellmouth the conduit width has been widened from the original value 2.6D to 2.8D in the second part of the computations. This change, however, does not mean any increasing of capital expenditure, because the widened part of the width is realized by thinning width of the supporting pier from 1.1 m to 0.6 m and remaining the width of 1.1 m in the vicinity of the gate grooves.

In order to avoid any possible vibration of the bellmouth the it is decided to be made from cast iron to concrete. In this part of the computations the purpose is mainly to compare two types of outer shape of the bellmouth: bell-like and cylinder outer shape. The calculated flow fields in the two inlet conduit are shown in Fig.3 and Fig.4, respectively. The calculated flow field reveals that the flow in the conduit with bellmouth of bell-like outer shape is easy to be induce large eddies with inclined axis in the non-working side of the conduit (See Fig.5(a)). These eddies complicate the flow between the bellmouth and the non-working side. Here flows entering into the pump and the eddies twist with each other, which makes it be possible for a vortex band to be produced. The flow in the non-working side of the conduit with bellmouth of cylinder outer shape, however, is quiet: no eddy can be found there (See Fig.5(b)), which is validated by the model test of the pump system[9].

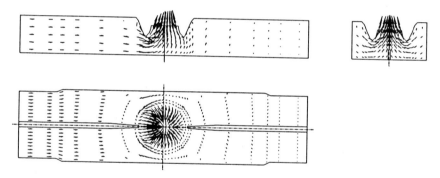

Fig.3 Flow field in the conduit with bell-like outer shape of bellmouth

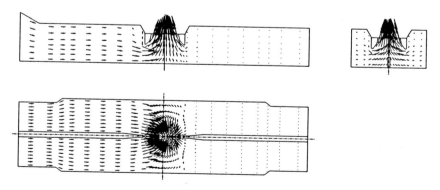

Fig.4 Flow field in the conduit with cylinder outer shape of bellmouth

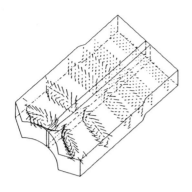   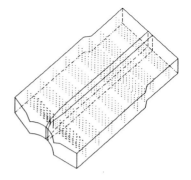

(a) Bell-like outer shape   (b) Cylinder outer shape

Fig.5   Flow field in the non-working side of the two-way conduit

Difference of hydraulic characteristics between the two pump systems with different bellmouth of outer shape may be compared according to the computation results listed in Table 9. The outer shape of bellmouth hardly effect the objective functions.

Table 9   Effect of outer shape of bellmouth

outer shape of bellmouth	u_{max}(m/s)	u_{min}(m/s)	$\bar{u}$(m/s)	$V_u(\%)$	$\bar{\theta}_{max}(°)$	$\bar{\theta}_{min}(°)$	$\bar{\theta}(°)$
bell-like	3.96	2.66	3.45	90.48	90.0	63.58	78.11
cylinder	3.98	2.73	3.45	90.78	90.0	66.50	77.76

6. EXPERIMENT COMPARISONS

Model tests with the two pump systems have been finished. The model pump and the outlet conduit are the same in the pump systems, while the inlet conduit is different: one is original and another is optimized. A brief comparison between the hydraulic characteristics of the two pump systems at the optimal operating point is given in Table

Table 10   Comparison between hydraulic characteristics of the two pump systems

	inlet conduit	Q(m³/s)	H(m)	P(kW)	$\eta(\%)$	n_s	$\beta(°)$	reference
1	original	0.299	2.463	11.78	61.34	1269	+2	[8]
	optimized	0.2986	2.03	9.36	63.51	1466	0	[9]
2	original	0.311	2.848	15.04	57.76	1161	+4	[8]
	optimized	0.3164	2.21	11.08	61.85	1416	+2	[9]

10, where H, Q, P, η, n_s and β is the head, discharge, power, efficiency, specific speed of the pump system and the blade angle of the pump, respectively.

Measured results in Table 10 show that under the condition of the discharge Q being equal the hydraulic characteristics of the pump system for one with the optimized inlet conduit are improved as indicated below:

(1) The head H at the optimal operating point decreases 0.46m for the first group of data and 0.64m for the second group of data, respectively, and the specific speed n_s increases correspondingly from 1269 to 1466 for the first group of data and from 1161 to 1416 for the second group of data, which enable the new pump system to be more suitable to the feature of the special low head of the pumping station.

(2) The efficiency of the pump system at its optimal operating point increases 2.17% for the first group of data and 4.09% for the second group of data.

7. CONCLUSIONS

By the help of the hydraulic optimization, the efficiency of the pump system at the optimal operating point is increased by 2% ~ 4% while the head is decreased by 0.46m to 0.64m.

The geometry parameters clearer to the inlet section of the pump, such as the hight and diameter of the bellmouth, the clearance of the bottom floor, the thickness of the baffle plate, have more notable effect on the hydraulic characteristics of the pump system, so they need more carefully to be optimized.

The outer shape of the bellmouth greatly influence the flow field in the non-working side of the conduit. The outer shape of cylinder is much better than that of bell-like which is easy to lead eddies.

REFERENCES

1. Jin.,Z. (1989) Numerical Solution to the Navier-Stokes Equations and Turbulence Models, *Hohai University Press*, pp54.

2. Song,C.C.S., He,J.M. and Chen,X.Y.(1991) Calculation of Turbulence Flow Through a Francis Turbine Runner and an Elbow Draft Tube, *International Power Generation Conference*, San Diego, CA.

3. Vu,T.C. and Shyy,W.(1990) Navier-Stokes Flow Analysis for Hydraulic Turbine Draft Tubes, *Journal of Fluid Engineering*, Vol.112, pp199-204.

4. Vu,T.C. and Shyy,W.(1990) A Comparative Study of Three Dimensional Viscous Flows In Semi and Full Spiral Casings, *IAHR Symposium*, Belgrade Yugoslavia.

5. W.Rodi,W.(1980) Turbulence models and their application in hydraulics, *IAHR Section on Fundamentals of Division II: Experimental and Mathematical Fluid Dynamics*, Delft, The netherlands, pp44-46.

6. Hossain,M.S.(1980) Mathematical Modelling of turbulent buoyancy flows, Ph.D.Thesis, University of Karlsruhe.

7. Lu,L. and Booij,R.(1994) Numerical Determination of A Hydraulically Optimized Suction Box of A Large Pumping Station, *IAHR Symposium*, Beijing China.

8. Jiangsu University of Science and Technology (1993) Report of Model Test for Pump System of Wangyuhe Pumping Station (1), pp8.

9. Jiangsu University of Science and Technology (1996) Report of Model Test for Pump System of Wangyuhe Pumping Station (2), pp10.

FLOW ANALYSIS FOR THE INTAKE OF LOW-HEAD HYDRO POWER PLANTS

A. RUPRECHT, M. MAIHÖFER, E. GÖDE

Institute for Fluid Mechanics and Hydraulic Machinery
University of Stuttgart
Pfaffenwaldring 10, 70550 Stuttgart, Germany

Abstract

In low-head hydro power plants often severe flow problems arise at the inlet region. Since model tests are mostly too expensive, especially for small power plants, numerical flow analysis is introduced. In this paper the numerical analysis of the flow behavior is shown for the intake region of two small power plants. For the first power plant the problem was that due to a disturbed flow in the trash rack, causing severe head losses, the power output was too low. By suggesting a relative simple modification of the geometry the problem could be cured. In the second case investigated there were oscillation problems initiated by a vortex shedding at the separation pier. Again, by changing the shape of the pier based on the results of the flow analysis, this problem could be avoided.

1. Introduction

Hydro turbines today obtain a very high hydraulic efficiency. Great effort is focused on all parts of the machine (stay vanes, wicket gates, runner, draft tube) in order to get the best power output. This is also the case for low-head turbines, which are frequently installed in rivers. In contrast to the high level turbine design, the shape of the intake region often is often not optimized in terms of flow behavior. This may have different reasons:

- the power plant is only of secondary importance, since the geometry is dominated e. g. by ship movements or flood resistance,
- the power plant is integrated into existing civil work arrangements or
- simply in order to achieve a cheap construction.

A poor design, however, can either result in severe energy losses, which can be caused by flow separations or by an oblique inflow to the trash rack. Or it can cause operational problems e. g. by unsteady vortex structures. Often minor changes in the geome-

E. Cabrera et al. (eds.), Hydraulic Machinery and Cavitation, 514–523.

try can increase either the power output of the plant considerably or it can lead to smooth turbine operation. Therefore it is necessary to pay the same attention to the intake as to the other components.

Hydraulic model tests of the entire intake region are very expensive and time consuming as well and are usually not applicable for smaller power plants. Therefore a numerical method is developed for the analysis of theses types of flow. In this paper the intake geometries of two small low head power plants are investigated using this numerical method. The existing flow problems were analyzed and modifications of the geometry are examined to cure the problems.

2. Basic equations and numerical methods

A viscous incompressible flow is assumed, either two-dimensional or three-dimensional and either steady or unsteady, depending of the problem. The calculations are based on the Reynolds-averaged Navier-Stokes equations combined with the eddy viscosity assumption of Boussinesq. The momentum equations are obtained to

$$\frac{\partial U_i}{\partial t} + U_j \frac{\partial U_i}{\partial x_j} + \frac{1}{\rho} \frac{\partial P}{\partial x_i} - \frac{\partial}{\partial x_j}\left[(v + v_t)\left(\frac{\partial U_i}{\partial x_j} + \frac{\partial U_j}{\partial x_i}\right)\right] = 0 \qquad (1)$$

all equations are written in tensor notation assuming the summation convention of Einstein (summation on repeated indices). The continuity equation is given by

$$\frac{\partial U_i}{\partial x_i} = 0 \qquad (2)$$

The turbulent viscosity v_t is calculated from a Prandtl mixing length formulation

$$v_t = |\omega| \cdot l_m^2 \qquad (3)$$

where $|\omega|$ is the amount of vorticity and l_m is the mixing length.

The mixing length formulation is preferred instead of the frequently used k-ε model, since it is relatively simple and robust and therefore reduces the computational effort. In addition the k-ε model in unsteady flows often shows too high dissipation and some times suppresses vortex shedding. Especially the standard k-ε model with logarithmic wall functions causes severe problems, since the separation of the boundary layer is not predicted accurate enough by wall functions.

For the flow predictions it is assumed that the free surface is of minor influence to the global flow structure and therefore is assumed to be constant. The trash rack cannot be considered exactly in the calculation, because its resolution by a computational grid would cause an extreme computational effort. Therefore the trash rack is introduced in form of a "porous" medium with an additional local pressure loss. The pressure loss is calculated from the empirical formulas of Kirschmer [1] (straight inflow) and Spangler [2] (oblique inflow).

The simulation is performed by the finite-element code FENFLOSS, which is developed at the university of Stuttgart. FENFLOSS uses a segregated solution algorithm for the momentum equations combined with a modified Uzawa type pressure correction. The linear systems of equations are solved by an ILU-preconditioned conjugate gradient algorithm (BICGSTAB2). For details the reader is referred to [3,4].

3. Applications

3.1. VERIFICATION

Since it is complicated, time consuming and therefore very expensive to do field measurements and because the calculations are often used to predict the flow behavior before the power plant is built, a simple model test rig has been constructed in order to compare the calculation results with some flow visualizations in the experiment. The test rig, shown in fig. 1, consists of 3 inlet gates and 3 turbines, which can be opened or closed in order to obtain different flow situations. In the test rig no trash rack has been considered. A two-dimensional calculation has been applied.

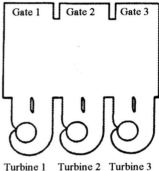

Fig. 1: Geometry of the test rig

In fig. 2 the streamlines of the calculation and the experiment are compared for a straight inflow. In this case the inflow enters through the middle gate and all three turbines are open. One can observe, that there are two recirculation regions behind the closed gates. The flow to the middle turbine is nearly undisturbed, whereas the flow to the left and to the right turbine show a strong turn on the piers.

In Fig. 3 the behavior for strong oblique inflow conditions are presented. The flow enters the test rig by the right gate and the right turbine is closed. Therefore a vortex occurs behind the closed gates and in front of the closed turbine. The flow to middle and the left turbines is highly disturbed and flow separations at the piers can be observed. These flow separations can cause severe losses and unsteady phenomena within the turbine and should be avoided if possible.

The comparison between the experiment and the calculations shows a good agreement. The structure of the flow can be predicted quite accurately. Therefore the computational procedure can be applied for the prediction of field conditions and to search for solution of the arising problems.

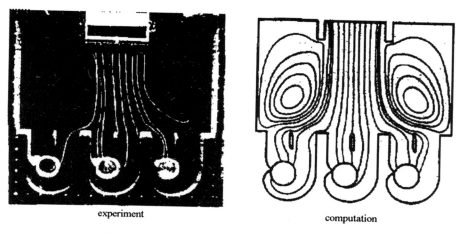

experiment   computation

Fig. 2: Comparison of experiment and calculation for straight inflow conditions

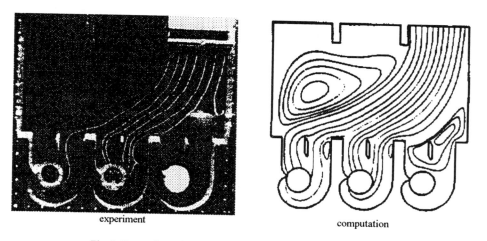

experiment   computation

Fig. 3: Comparison of experiment and calculation for oblique inflow conditions

3.2. APPLICATION 1

As a first example a small hydro power plant, schematically shown in fig. 4, is presented. Next to an existing weir a new power plant has been build. After commissioning it was found that the power output did not reach the estimated level. It was assumed that the problem is caused by the shape of the inlet. It could be observed that the flow separates at the sharp corner when going into the channel to the turbine, leading to a disturbed flow field in the trash rack cross-section.

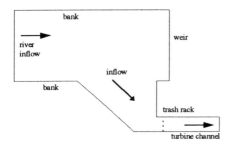

Fig. 4: Geometry of application 1

Since the water depth is nearly constant the investigation is carried out two-dimensional. In order to assure the results and to show the influence of secondary motion also a three-dimensional calculation has been carried out.

The flow behavior described above has already been found by the two-dimensional flow analysis. The streamlines are shown in fig. 5. It clearly can be observed that there exists a flow separation which extends into the trash rack. This leads to a locally higher velocity in the trash rack cross-section, fig. 6, and therefore to higher losses. Looking to the flow angle of attack, fig. 7, it can be seen that the flow enters the trash rack with an up to 10° wrong inflow angle. This also causes an increase of flow losses. It has to be pointed out, that the trash rack itself makes the flow more uniform. A simulation without trash rack leads to a flow angle deviation up to 20° and a more disturbed velocity profile. But this equalization of the flow causes additional losses.

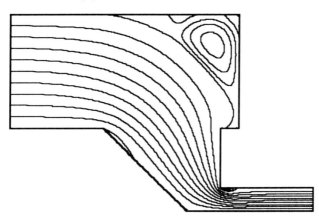

Due to the disturbed flow in the trash rack cross-section the losses in the trash rack are approximately 3 times higher than for a uni-form flow. Since the velocity in the channel is rather high anyhow, losses in the trash rack are severe and cause the drop in thepower output.

Fig. 5: Streamline plot

In order to be sure that there are no mayor three-dimensional effects dominating the flow behavior a 3D flow analysis has been carried out. In fig. 8 the velocity profile in the trash rack cross-section is shown. Again it can be observed that there is also the recirculation region initiated by the sharp corner. Except the influence of the boundary layer at the bottom the flow shows a two-dimensional behavior. Therefore a 2D calculations is sufficiently accurate.

The high energy losses in the trash rack mean, that if it is possible by changing the inflow geometry in such a way, that the velocity profile in the trash rack is quite uni-

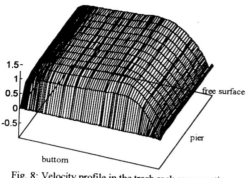

Fig. 8: Velocity profile in the trash rack cross-section

form, a considerable increase of power output can be achieved. Since it would be very expensive to reconstruct the power plant, it was tried to cure the flow situation only by minor changes of the geometry. The shifting of the trash rack cross-section further down-stream had to be avoided, because it is too expensive to move the trash rack inclusive the automatic cleaning devices.

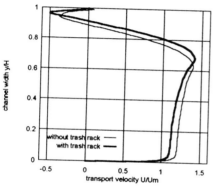

Fig. 6: Velocity distribution in the trash rack cross-section

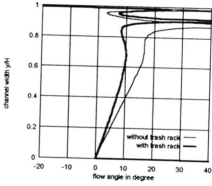

Fig. 7: Flow angle in the trash rack cross-section

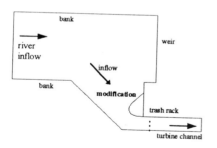

Fig. 9: Modified geometry

By giving the inflow section a round shape, see fig. 9, the flow behavior could be improved. Fig. 10 shows the streamlines for the modified contour. It can be seen, that the flow enters the trash rack quite undisturbed. There is no longer a flow separation at the inlet to the turbine channel. This can be seen more clearly in fig. 11 and fig. 12, where the velocity distribution and the flow angle in the trash rack cross-section are shown. One can observe that the velocity profile is quite uniform and the maximum angle of attack at the trash rack is lower than 2°. Therefore the head losses in the trash rack are reduced remarkably.

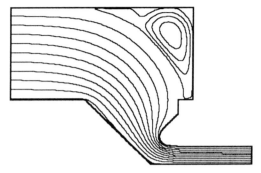

Fig. 10: Streamline plot of the modified geometry

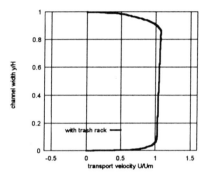
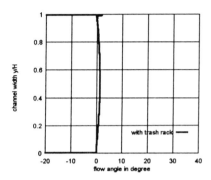

Fig. 11: Velocity distribution in the trash rack cross-section,   Fig. 12: Flow angle in the trash rack cross-section ,
              modified geometry   modified geometry

3.3. APPLICATION 2

As a second example another small hydro power plant with two turbines, shown in fig.13, is presented. In this plant one turbine shows strong oscillations during "steady" operation, which causes severe problems to the bearing (water lubricated bearings). These oscillations are initiated by vortex shedding from the separation pier. Since the water depth nearly remains constant in the whole intake region and strong secondary flow behavior is not expected, a two-dimensional simulation is applied.

The calculations show, that the flow is unstable. A steady flow simulation did not converge. Therefore, an unsteady flow analysis has been carried out. The flow patterns can be seen in fig. 14, where the streamlines for different time steps are presented.

It can be observed that vortices are shedding from the separation pier and move towards the turbines. When reaching the turbine inlet channel on the inner side the vortices disappear but the strong velocity changes causes the role up of smaller vortices in the channel. These vortices move down-stream to the turbine. The unsteady behavior can also be seen in fig. 15. The variation in time of the velocity is shown for three different locations of the intake is shown. It can be seen, that the flow is periodic with an oscillation time of about 1 minute.

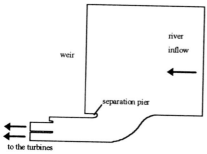

Fig. 13: Geometry of application 2

In order the cure the problem, the geometry of the pier was modified. In fig. 16 the streamlines for the new contour is shown. With this modification the flow remains stable and no oscillations occur anymore. After this modification have been built by the end of last year it was observed that the oscillation problems do no longer exist and the turbines are running smoothly.

4. Conclusions

Numerical flow analysis has been applied in order to evaluate the flow in the inlet region of different low-head power plants. By comparing the calculation with a flow visualization on a small test in the laboratory it was shown that the characteristics of the flow can be predicted with sufficient accuracy.

The application of the numerical procedure to two different existing power plants is shown. In both of the plants severe operating problems exist concerning the flow in the intake region. In the first case the energy losses in the trash rack were extremely high due to a highly disturbed velocity field. In the second plant oscillation problems existed caused by vortex shedding at a separation pier. In both cases the flow problems could be analyzed and understood and appropriate changes in the geometry could be found to cure the problems.

It has been shown that the numerical simulation can be used as a tool for the design of the inlet region of hydraulic turbines. It can predict the characteristics of steady state flows as well as unsteady vortex shedding. By applying this investigation in advance of the construction many operational problems could be avoided. Since the effort of the numerical flow simulation is rather low compared to model tests, it should be justifiable to analyze the flow behavior even of very small power plant in advance.

References

[1] Kirschmer, O., "Untersuchungen über den Gefälleverlust an Rechen", Mitteilungen des Hydraulik Instituts der TH München, 1-5 (1926-1932).

[2] Spangler, J., "Untersuchungen über den Verlust an Rechen bei schräger Zuströmung", Mitteilungen des Hydraulik Instituts der TH München, 2, 1928.

[3] Ruprecht, A. et al. "Einsatz der numerischen Strömungsmechanik in der Entwicklung hydraulischer Strömungsmaschinen", Mitteilungen Nr. 9, Inst. f. Strömungsmechanik und Hydraulische Strömungsmaschinen, Universität Stuttgart, 1994.

[4] Ruprecht, A., "Finite Elemente zur Berechnung dreidimensionaler, turbulenter Strömungen in komplexen Geometrien", Mitteilungen Nr. 3, Inst. f. Strömungsmechanik und Hydraulische Strömungsmaschinen, Universität Stuttgart, 1989.

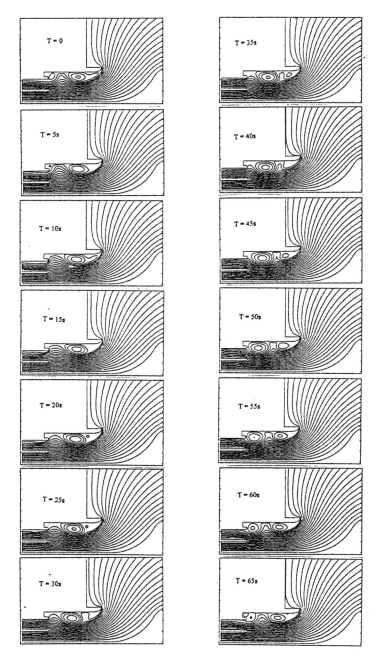

Fig. 14: Streamline distribution for different time steps

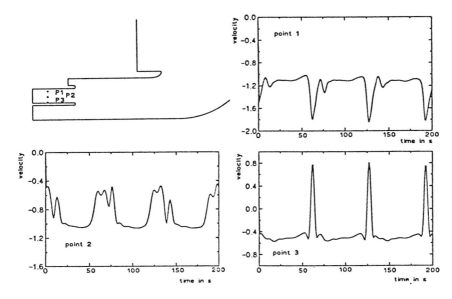

Fig. 15: Velocity variation in time for three different locations

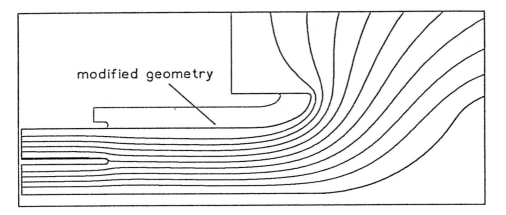

Fig. 16: Streamline distribution for the modified geometry

EFFICIENCY ALTERATION OF FRANCIS TURBINES BY TRAVELLING BUBBLE CAVITATION

Experimental and theoretical study

CH. ARN, PH. DUPONT AND F. AVELLAN
IMHEF-EPFL
33, av. de Cour, CH–1007 Lausanne, Switzerland
Email: christophe.arn@imhef.dgm.epfl.ch

Abstract.
The setting level of a hydraulic machine, specially for low head machines, is decided with respect to the possible alteration of the efficiency due to the cavity development. This alteration can easily be noticed by following the evolution of the efficiency η as a function of the Thoma number σ leading to the so-called $\eta - \sigma$ cavitation curves. Observation of the cavity extent in the flow passage of the runner allows to associate the drop of efficiency with a particular type of cavity development.

However, depending on the type of cavities this drop cannot be very easily explained. Obviously, for a leading edge attached cavity corresponding to high head operating points, the presence of the vapour phase on the blade suction side limits the pressure at the vapour tensile strength value which causes the flow alteration. In the case of travelling bubble cavitation, corresponding to the outlet cavitation at the nominal head, previous experiments with a 2-D NACA profile show that the modification of the mean pressure field is mainly due to the bubble dynamics. The aim of this paper is to present the results and the analysis of two experiments intending to explain the influence of the nuclei content on the mean pressure field correction due to the bubble dynamics.

1. Introduction

Travelling bubble cavitation takes place for the design value of the head, at the throat of the Francis runner flow passage, close to the outlet and corresponds to a low flow angle of attack. The development of this type of cavitation, responsible of an efficiency alteration of the machine is very sensitive to the content of cavitation nuclei and to the value of the setting level. For this reason, this setting level is determined with respect to this type of cavitation. The physical phenomenon underlying such an efficiency

E. Cabrera et al. (eds.), Hydraulic Machinery and Cavitation, 524–533.

alteration of the turbine is not yet very clear. Previous experiments [9] performed with a NACA009 profile mounted on a hydrodynamic balance show us the hydrodynamic loads to be dependent on the cavitation coefficient σ and on the content of the cavitation nuclei. We can so deduct that these two parameters are responsible for that pressure modification on the blade as the pressure distribution is generating the lift.

An approach to study this influence is to consider that the pressure field modification is mainly due to the bubble dynamics [8] [10] [7]. If we consider the travelling bubble like a non moving sphere in expansion in a still fluid, the computation of the pressure field around the bubble is possible. Indeed, the relative velocity between the bubble and the fluid is very low. The potential theory of incompressible flows allows to class this case as the determination of the pressure generated by a potential $\phi = -\dot{R}R^2 r^{-1}$ where R is the radius of the sphere. The expansion of this sphere generates a radial pressure field around. The Bernoulli equation leads to the expression of the pressure field p. Thus, the pressure on the blade can be obtained by superposing the two bubble potential fields according to the image superposition technique, the wall being a flow surface. The Rayleigh-Plesset model allows us to determine the evolution of the bubbles radius, which is necessary to compute the radial pressure field. In the case of the NACA 009 bidimensional blade whose pressure field is computed by a Navier-Stokes finite elements code, the evolution of the bubbles along streamlines is determined solving the Rayleigh-Plesset equation using a fifth order Runge-Kutta method. One can see an example on Figure 1 with a value of the cavitation coefficient σ equal to 0.43. The evolution of the pressure computed in a point of the wall is also reported.

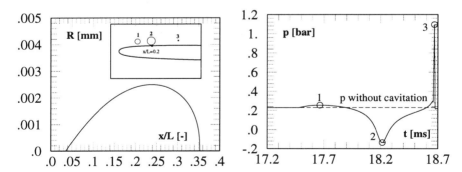

Figure 1. Bubble radius evolution along a streamline in a NACA009 blade and pressure generated at $x/L = 0.2$. $\sigma = 0.43$.

Obviously, this method is valid for the case where the bubbles remain relatively spherical which corresponds to high values of the cavitation coefficient σ. Moreover, the amplitude of the pressure are higher as the measured one [3]. Thus, this model allows qualitatively a good approach of the physical phenomenon. For lower values of σ, the size of the bubbles is greater than the layer containing the activated nuclei. Thus, the shape of the bubbles have a trend to be flattened up to form an hemisphere [1]. This trend depends strongly to the Weber number. In this case, the pressure on the wall under the bubble is naturally the vapour pressure value p_v. If the nuclei content is great enough to obtain the cavitation saturation, the pressure becomes constant at the value of p_v. Then we can define three different zones of travelling bubble cavitation influence. The first one is the part of the blade where the bubbles are spherical. The pressure modification is there mainly due to the bubble dynamics. The second zone is the region where the bubbles becomes hemispherical and the pressure under the bubbles are limited by the vapour tensile strength. The last one is the part of the blade where a saturation of the development of the bubbles is obtained and where the pressure on the blade is constant and equal to the value of p_v. The aim of this paper is then to present the results of an experiment intending to confirm these different points by measuring the pressure distribution on a two dimensional blade and on a Francis runner with travelling bubble cavitation.

2. Experimental set-up

2.1. MEASUREMENTS ON THE NACA 009 BLADE

The test are carried out in the IMHEF high speed cavitation tunnel [5]. The experimental hydro-foil is a 2D NACA 009, 100 mm long and 150 mm wide, truncated at 90 % of its chord length. 18 piezoresistive absolute pressure transducers are distributed on the suction side of the blade, as shown on Figure 2. The measurement range covers 0 to 100 bar. Each transducer is supplied by an independent current source and its output pressure signal is separately amplified and band-pass filtered. Data acquisition is performed with the help of two digital transient recorders with a 12 bits resolution per sample. The first one (Lecroy 8212a) allows simultaneous data acquisition of 32 signals at a maximum sampling frequency of 5 kHz whereas the second one ensures simultaneous data acquisition of 12 channels at a maximum sampling rate of 1 MHz (3 Lecroy 6810 Modules) the static calibration of the pressure transducers is performed with the blade mounted in the test section by varying the static pressure in the tunnel from 0.3 to 10 bar. To achieve a dynamic calibration of the pressure transducers, a special technique is developed to generate a pressure impulse in the test section [6]. The transducers output are compared to the output of a 601 Kistler pressure transducer mounted in the test section too. The main result is a good concordance up to 15-20 kHz.

A control of the nuclei content is performed during all the experiment. Indeed, the travelling bubble cavitation is impossible without the injection of cavitation nuclei. The nuclei are generated by an expansion of air-saturated water in a series of injection modules [2]. By varying the number of these modules, one can obtain the required quantity of cavitation nuclei

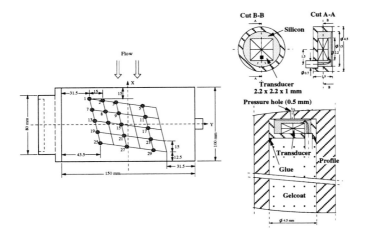

Figure 2.   NACA 009 hydro-foil equipped with 15 transient pressure transducers. Pressure transducers mounting

in the test section. The distribution of these nuclei is measured by a cavitation nuclei counter [4] which use the Venturi effect for the detection of the nuclei explosive growth.

2.2. MEASUREMENTS ON THE FRANCIS RUNNER

The experiment is conducted with a Francis runner model which has a specific speed $\nu = 0.33$. It represents a standard case of a Francis turbine design. The tests of the model are performed on the high performance research test rig of IMHEF. This closed loop test rig covers a range of flow rates up to 1.5 cubic meters per second, with a maximum net head of 60 meters. The primary quantities as the torque, the flow rate, the head, the angular velocity and the water temperature are continuously recorded to determine the power and the efficiency of the machine with an overall accuracy better than 0.1 percent.

Two blades of the runner are equipped with each five absolute transient pressure transducers similar to these mounted on the 2D blade. The position of the transducers is described on Figure 3. The acquisition set-up is the same as the other experiment except the conditioning electronics connected to the on-board transducers. This electronics is placed in the head of the runner, it has an amplifier with a remotely variable gain and a multiplexer that allows the scanning of all the transducers. Signals are converted to frequency to avoid electro-magnetic disturbances. They are brought out from the rotating part through a slip ring collector and finally reverted to voltage signals which are digitized on the waveform recorders.

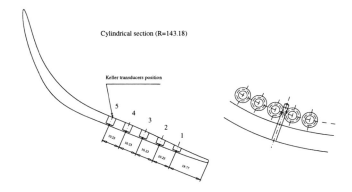

Figure 3.   Transducers position on the two blades of the Francis runner

3. Results and discussion

3.1. NACA 009 BLADE

For the case of the 2D blade, the cavitation coefficient $\sigma = \frac{p_{ref}-p_v}{1/2\rho C^2}$ is defined by a relation similar as the definition of the local cavitation factor $\chi_E = \frac{p_{ref}-p_v}{\rho E}$ for the turbomachines where C is the velocity and E the energy. A previous experiment conducted with the same blade mounted on a five components hydrodynamic balance give us the evolution of the lift coefficient c_z as a function of this cavitation coefficient σ [3]. We can observe an increase of the lift from the inception of the travelling bubble cavitation. Then, after a maximum value corresponding to $\sigma = 0.32$, the lift falls. These results are reported on the Figure 4.

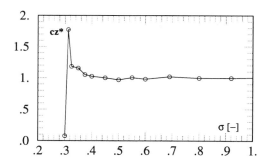

Figure 4.   Measured lift coefficient as a function of the cavitation coefficient σ.

An approach to explain this lift drop phenomenon is to measure the pressure distribution at the same operating points, for an upstream velocity of 20 m/s and with cavitation nuclei injection in order to obtain the saturation condition. These distributions are reported on Figure 5 and compared with the result of a Navier-Stokes computation. The frames corresponding to each σ are too illustrated on this Figure. One can observe a good agreement between the pressure distribution measured for a high value of σ and the computed distribution. This provides a certain confidence in the results. From the inception of travelling bubble cavitation, a depression appears in the zone where the size of the bubbles is maximum ($\sigma = 0.4$ and $x/L = 0.3$). In this condition, the bubbles stay relatively spherical. For a value of $\sigma = 0.375$, the bubbles become hemispherical and the depression extends over the suction side but doesn't go below the vapour pressure value. For lower values of σ, the measured pressures correspond in all positions to the vapour pressure. Indeed, the travelling bubble cavitation saturation is reached as shown in the different frames on Figure 5. The main noticing is that the maximum lift corresponds to the situation where the value of $-\sigma$ reaches the plateau of the pressure coefficient distribution. In fact, this is mainly the position of the point where the pressure is minimum which determine the behavior of the hydraulic characteristic in function of the σ. In the case where the $c_{p,min}$ value is near the trailing edge of the blade, which corresponds to low flow angle of attack, only a lift drop can be appear since the vapour pressure goes to limit the suction of the blade. In the other case where the $c_{p,min}$ is reached near the leading edge, an increase of the lift is generated by the development of travelling bubble cavitation in the zone where the pressure is higher as the vapour pressure. The drop appears only when the vapour pressure is reached on the suction side.

3.2. THE FRANCIS RUNNER

Generally, the pressure distribution in a Francis runner at the best operating point corresponds to the first case that we described in the previous section. The flow angle of attack is low and the minimum pressure is localized on the outlet of the blade. The tests are conducted at this operating point for three different heads: 10 m, 15 m and 20m. The results of a flow computation in the case of a head of 15 m are presented on Figure 6. One can observe on this figure that the minimum of the pressure is localized in the outlet of the blade for each computation mesh line except in the zone close to the runner band (k=19). Based on the conclusion of the previous described experiment, the efficiency must directly decrease with the cavitation coefficient. All the more so since the whirl has no effect at the best operating point. The cavitation coefficient usually used in hydraulic turbomachines is the Thoma number $\sigma = NPSE/E$.

The cavitation curves obtained are illustrated on Figure 7. The values of the efficiency are related to the cavitation free values. We can determine the value of σ_0 which is the highest value of the Thoma number where the efficiency is modified. This value is the same for all the tests with nuclei injection and is equal to 0.068. The values of χ_E corresponding to this Thoma number are equal to 0.024, 0.022 and 0.02 for the three test heads. The com-

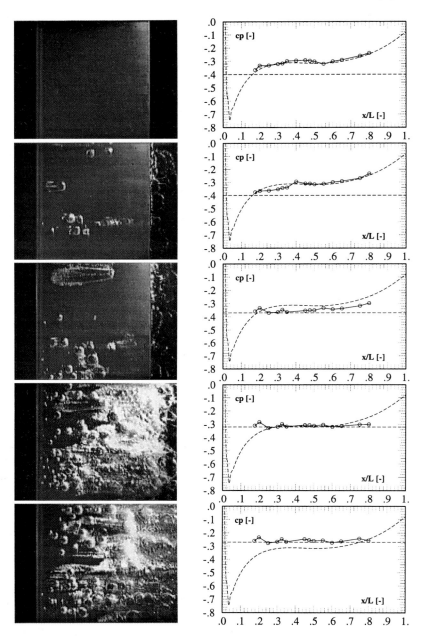

Figure 5.   Measured and computed pressure distribution on the NACA 009 blade for the following values of σ: 1.0, 0.4, 0.375, 0.325, 0.275. (From top to bottom). The horizontal dashed line represents the $-\sigma$ value

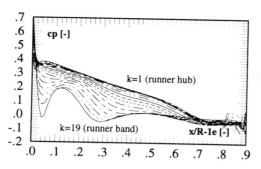

Figure 6. Results of a Navier-Stokes flow computation at the best operating point. The reference pressure is the computation domain outlet pressure. H=15 m

parison with the pressure coefficient computed by the Navier-Stokes code shows us that the beginning of the efficiency drop appears when the all the values of the minimum pressure coefficient are lower as $-\chi_{E,0} = -0.022$. At this state, it corresponds to an important development of the outlet cavitation. For the case corresponding to the computed one, the first bubbles are developed for a value of the Thoma number of 0.085. Moreover, based on the computed results, we can determine the maximum value of the Thoma number where the vapour pressure is reached on the blade. The minimum pressure coefficient on the blade provides this value of σ which is 0.11. Thus, we have a difference of the σ from 0.04 between the point where the vapour pressure is reached on the blades and the beginning of the efficiency drop. The development of travelling bubble cavitation must be then relatively established before the efficiency falls.

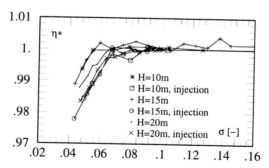

Figure 7. Normalized $\sigma - \eta$ curves for the Francis runner tests.

As in the case of the 2D blade, we can plot the results of the pressure measurements with the computed distribution at the corresponding computation mesh line. The Figure 8 presents these results where we can observe that the correspondence between the measurements and the computation is very good for the case without cavitation. For lower values of the Thoma number, the pressure distribution is modified by the development of travelling bubble cavitation. The different pressure measurements show us a decrease of the pressure until the vapour pressure value. The development of the bubbles is due to the upstream pressure drop localized close to the runner band. However, the main performance of the runner decrease since the development of the travelling bubble cavitation take effect mainly at the outlet of the blade.

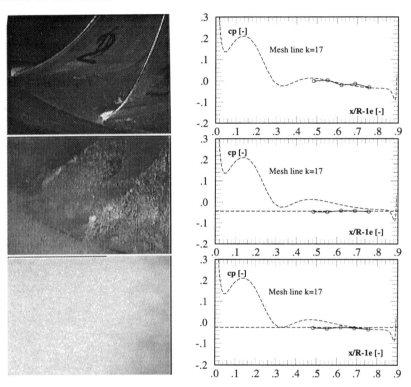

Figure 8. Measured and computed pressure distribution on the Francis runner blade for the following values of σ: 0.43, 0.07, 0.06. H=15 m. (From top to bottom). The horizontal dashed line represents the $-\chi_E$ value.

4. Conclusion

Based on these two experiments, the effects of the travelling bubble cavitation on the turbomachines performances are better described. The efficiency modification appears mainly when the saturation of the cavitation is reached. The saturation causes a modification of the pressure distribution by limiting the pressure field to the vapour pressure value. The part of the machine where the shape of the bubbles stay spherical is not really influenced in term of performances. The acoustic emission caused by the bubbles growth or collapses is an other problem which justify the study of the acoustic pressure field around the active bubbles. However, in order to obtain a modeling of the pressure distribution modified by travelling bubble cavitation, it is necessary to determine the different zones of the blade where the bubbles are spherical, hemispherical and the cavitation saturation is reached. The size and the position of these different regions depends naturally to the cavitation nuclei concentration and to the operating point of the machine. This is the reason why the study of the bubbles evolution close to a wall in a variable pressure field is important for the characterization of the performances alteration in a hydraulic turbomachine. The way to this characterization in the case of the efficiency alteration by travelling bubble cavitation is certainly the study of a reliable model of the bubble evolution in a runner, and then the modification of its shape including the case of the saturation. The effects of the bubbles on a pressure field being now relatively known and described, this is the knowledge of the distribution of these different travelling bubble cavitation development on the runner that will be make possible a prediction of the efficiency drop in the Francis turbines due to this type of cavitation.

5. Acknowledgment

The authors are particularly grateful to the members of the IMHEF Cavitation research group and to its technical staff. This research is financially supported by the PSEL "Fonds Suisses pour Projets et Etudes de l'Economie Electrique".

References

1. Y. Kuhn de Chizelle et al. Observations and scaling of travelling bubble cavitation. *J. Fluid. Mech.*, 293:99–126, 1995.
2. C. Brand et al. The imhef system for cavitation nuclei injection. Sao Paolo, 1992. AIRH.
3. Ch. Arn et al. Experimental and theoretical study of the 2d blade lift alteration by traveling bubble cavitation. ASME, August 1995.
4. F. Avellan et al. Theoretical and experimental study of the inlet and outlet cavitation in a model of francis turbine. pages 38–55, Stirling, August 1984.
5. F. Avellan et al. A new high speed cavitation tunnel for cavitation studies in hydraulic machinery. volume 57, pages 49–60. ASME, 1987.
6. F. Pereira et al. Dynamic calibration of transient sensors by spark generatd cavity. Symposium of Bubble Dynamic and Interface Phenomena, September 1993.
7. J.P. Franc et al. *La Cavitation, Mécanismes physiques et aspect industriels.* Presses Universitaires de Grenoble, 1995.
8. S. Kumar et al. A study of pressure pulses generated by traveling bubble cavitation. *J. Fluid. Mech.*, 225:541–564, 1993.
9. B. Gindroz. *Lois de Similitude dans les essais dans les Essais de Cavitation des Turbines Francis.* PhD thesis, EPFL, 1991.
10. J.T. Daily R.T. Knapp and F.G. Hammit. *Cavitation.* Mac GRaw Hill, New York, 1970.

CAVITATION EROSION PREDICTION ON FRANCIS TURBINES-PART 1 MEASUREMENTS ON THE PROTOTYPE

P. BOURDON, M.FARHAT, R. SIMONEAU
Hydro-Québec
1800 boul. Lionel Boulet, Varennes, Québec, Canada, J3X 1S1
F. PEREIRA, P. DUPONT., F. AVELLAN
IMHEF/EPFL
33 av. de Cour, CH 1007 Lausanne, Switzerland
J.-M. DOREY
Electricité de France
6, Quai Watier, 78401 CHATOU Cédex, France

1. Abstract

In the process of developing tools for cavitation erosion prediction of prototypes from model tests, 4 on board aggressiveness evaluation methods were tested on a severely eroded blade of a 266 MW Francis turbine. These are pressure, pit counting, DECER electrochemical and vibration measurements. All methods provided coherent results on the blade mounted measurements. The test program provided understanding of the heterogeneous erosion distribution of the prototype blades and quantitative data for comparison in subsequent tests on the model of the machine.

2. Introduction

The prediction of cavitation erosion of a prototype turbine from model tests requires that measurement tools be available to characterize the aggressiveness of the cavitating flow in both scales. To develop such tools, IMHEF, Electricité de France and Hydro-Québec pooled their resources in an ambitious research program involving various measurement techniques both on the prototype and on the model of a 266 MW Francis turbine with a well documented cavitation erosion history. Preliminary measurements on this prototype (1) had proven to be incomplete but also very encouraging. A more ambitious test program was conceived with improvements in the array of sensors utilized, sample mounting methods, data acquisition systems and hydraulic test conditions. The test program was divided in two parts, the first with pressure sensors mounted on the suction side of blade #4, a well eroded blade, to identify the type of cavitation present and its aggressiveness in terms of pressure pulses. In the second, polished metallic samples were mounted in place of the sensors for pitting studies. This experimentation program took place in June 1995.

E. Cabrera et al. (eds.), Hydraulic Machinery and Cavitation, 534–543.

3. Test set-up

3.1 MECHANICAL INSTALLATIONS

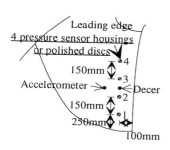

Figure 1. Blade 4 sensor and disc positions

This machine has the particularity that the erosion is much worse on about half of the blades and is localized towards the trailing edge near the blade to band fillet. Blade #4 was retained for testing as it had been used in the preceding tests because of its severe erosion history. Consequently, six 45 mm diameter holes were counterbored in this blade as shown in Figure 1 on the low pressure side. An intricate drilling and shimming scheme was developed by IMHEF to insure perpendicularity of the hole axis with the complex blade surface and optimal flush mounting of the various sensor housings. All housings or sample mounting supports were securely bolted down to the blade from the pressure side. Figure 2 shows the magnetic drill mounting base positioned in place as well as some of the already bored holes.

Holes were also bored above the draft tube access door to mount a dynamic pressure sensor, pass cables for 5 underwater accelerometers and cables to immersed spark plugs for the dynamic calibration of the set-up in watered conditions. A cylinder

Figure 2. Drill mounting base

housing the custom designed onboard data acquisition electronics was welded in place in the runner nose cone. Cables were run from the sensors to this unit through a stainless steel tube welded to the pressure side of bade 4 and covered with an hydraulically smooth buildup of resin to minimize turbulence. Signal and control cables were then brought out to the stationary world through the hollow turbine-alternator shaft using the air injection path and a double slip ring arrangement fixed to the end of the alternator exciter shaft. A threaded hole was also machined at the lower turbine guide bearing to receive a piezoelectric force exciter for sinusoidal sweep calibration of the transmissibility function between the guide bearing and 5 underwater accelerometers mounted on the runner blades.

3.2 SENSORS

For the pressure tests, 4 of the blade holes (# 1-4) received stainless steel sensor housings bearing each 4 piezoresistive pressure sensors allowing to measure both static and dynamic pressures. Each housing contained 2 400 bar and 2 1000 bar Keller sensors. These housings were replaced with optically polished 316L stainless steel discs for the pitting tests. The two remaining holes were occupied respectively by a DECER electrochemical cavitation erosion sensor and a 4kHz bandwidth damped piezoresistive accelerometer. These were used in both the pressure and pitting tests.

In addition to the onboard sensors mentioned above, four high frequency accelerometers monitored the cavitation impacts at the lower guide bearing while another was installed above the draft tube access door along with a Kistler 701 wide band dynamic pressure sensor. A tachometric probe allowed to synchronize data acquisition on multichannel wide band data acquisition systems. The test program included calibration of the setup through the measurement of transmissibility functions both in air and in water between the blades and measurement points at the lower guide bearing. This step is required to determine absolute cavitation aggressiveness levels from remotely measured vibration or pressure data. For this purpose, 5 accelerometers were mounted on blades 3, 4 ,7 10 and 13 on the suction side at a position corresponding to the location of the DECER probe on blade 4 shown in Figure 1. Direct and reciprocity impact and direct spark generated bubble calibration techniques were utilized. Two underwater spark plugs were used to generate bubbles at two locations close to the suction side of blade 4. A sine sweep technique was used in a reciprocity mode by exciting the structure with controlled force at the lower guide bearing.

3.3   DATA ACQUISITION SYSTEMS

For the purpose of this program, IMHEF designed an onboard data acquisition system with two variable gain 100kHz bandwidth data channels and two front end remotely controlled channel multiplexers. This system allowed to scan the 16 blade mounted pressure sensors as well as the 4kHz bandwidth accelerometer. Signals were converted from voltage to frequency for transmission through the dual slip ring arrangement to avoid noise pickup in the alternator

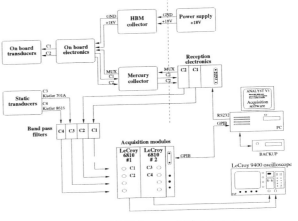

Figure 3. IMHEF data acquisition system.

environment. The signals were then converted back to voltage with a frequency to voltage converter. Extra tracks on the slip rings were used to carry an RS 232 control link, the DECER electrochemical work and auxiliary potentials and the power supply current for the onboard electronics.

Data acquisition was assured by two separate systems provided by IMHEF and Hydro-Québec. The first system is based on 2 LeCroy 6810 acquisition modules configured to acquire each 2 input signals with 12 bit resolution at sampling frequencies up to 5MHz. A 1 Mbyte memory in each module is shared between the two input channels. The acquisition is controlled via a GPIB port by a PC running Analyst software developed at IMHEF in the ASYST environment. On line monitoring of signals from this system is performed with a LeCroy DO9400 oscilloscope. Signals are first passed through antialiasing filters before acquisition. This system is illustrated in Figure 3.

The second includes a 9 channel HP3565S 13 bit 100kHz bandwidth data acquisition system controlled by LMS Fourier Monitor software, a Nicolet 500 4 channel 12 bit 10 Mhz digitizer with a 1 megasample memory on each channel and a Sony PC216A 16 bit resolution digital tape recorder used in an 8 channel 10 kHz bandwidth configuration.

Both systems were utilized for the dynamic calibration of the test setup and for acquisition during the tests. The instrumentation was completed with a bank of programmable bandwidth and gain analog filters and envelope detectors. A potentiostat circuit conditioned the DECER signal which was recorded on an analog chart recorder as well as on the SONY tape recorder. An audio amplifier and loudspeaker combination allowed to assess audibly on site the impulsive nature and the intensity of the cavitation signals perceived by the blade mounted accelerometer.

4. Test conditions

Conditions for the pressure and pit counting tests appear in the following table.

TABLE 1. Test conditions

Test Identification	Guide vane opening (%)	Power (MW)	Downstream level (m)	Duration (h.min)
Pressure tests				
1	78	228	205.7	6.0
2	90	259	205.8	3.0
3	78	228	207.4	4.0
4	90	255	207.6	2.15
Pitting tests				
M1	78	227	206.2	.41
M2	78	223	207.4	.41
M3	90	253	207.5	.40
M4-1	85	247	205.7	.17
M4-2	80	234	205.6	.16
M4-3	75	219	205.5	.15
M4-4	70	201	205.5	.15
M5	72	206	205.9	.40

In order to evaluate the effect of the variation of the cavitation index on the cavitation development, tailwater levels were raised in the downstream reservoir over a two week period to allow testing of the machine with high downstream levels with all available machines operating. These conditions had not been possible in the preliminary program (1). Low tailwater testing was done by operating only the test unit in the powerhouse.

5. Test results

5.1 CALIBRATION

Before running the operating tests on the prototype, an extensive calibration program was performed to characterize the transmissibility function (ratio of output acceleration power spectrum to input force power spectrum) between the attacked areas on the blades and the monitoring points either on the blades or at the lower guide bearing. Measurements were performed with the runner in air and then in water. The linearity

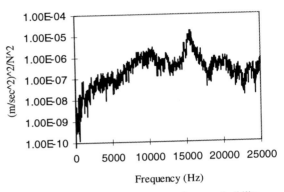

Figure 4. Lower guide bearing to blade transmissibility.

of the structure was verified by reciprocity measurements using instrumented hammer excitation techniques with the machine in air. The average transmissibility function with the runner under water was then evaluated by impacting at the lower guide bearing monitoring point at 90 degrees from upstream on the spiral case upstream side and measuring simultaneously the response on 5 blades. The result of this measurement is shown in Figure 4. This function was then used to infer acting forces on the blades from acceleration measurements during subsequent testing.

5.2 PRESSURE TESTS

The first pressure test performed with a low tailwater level produced cavitation conditions on blade 4 such that violent implosions were localized in the area of the two downstream pressure sensor housings. This was later confirmed in pitting test M1 under similar conditions and by the severe marking of the stainless steel sensor housings after the pressure tests. As a result, the 8 sensors in the most downstream positions and 2 more on the next upstream housing were progressively put out of service as the time exposure increased. The remaining sensors exhibited a main

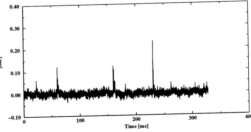

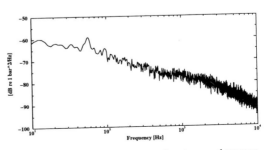

Figure 5. Typical pressure sensor time trace and average Power Density Spectrum

component in their power spectral density spectra around 60 Hz which corresponds to the guide vane passing frequency for a rotating observer. An example is given in Figure 5. This result suggests the existence of a cavitation cloud whose pulsation is modulated by the pressure variations generated by passing in front of the guide vanes. This in turn commands forced vortex shedding in the closure area of this cloud and impacts on blade 4 at this rate. This is confirmed by the envelope analysis of the high frequency acceleration of the blade which shows a strong component at or near this frequency. In pressure test #3, this hypothesis was supported by the remaining upstream sensors which were still operational. As the downstream level was raised, the cavitation cloud moved in the upstream direction in such a way that sensors on. housing 4 measured vapor pressure while those on housing 3 showed intense activity with enormous fluctuations which occasionally saturated the frequency to voltage converter. The pitting tests confirmed these indications as disc 3 was the most severely marked under the same hydraulic conditions.

5.3 PITTING TESTS

5.3.1 Pit counting

The 1993 test program had seen the most exposed glued discs torn away in the flow and the impossibility of testing with high downstream levels to evaluate the effect of this parameter on the cavitation development. As mentioned earlier, these two elements were corrected in this program. After the pressure tests, the pressure sensor housings were replaced with polished disc holders bolted down to the blade and the discs were replaced after each of the roughly 40 minute exposure periods. An exception to this was made in test M4 which consisted of 4 successive 15 minute exposures at 85, 80, 75 and 70% guide vane opening in one sequence with a low downstream level. Three of the pressure test conditions were repeated, the 90% low downstream level test was skipped as it had been performed in 1993, the sweep test was added and was followed by a final point at 72% guide vane opening.

The results of these tests are summarized in Figure 6 which shows all of the disc samples along with a 10X micrograph of typical pitted areas on those discs that were marked by the cavitation impacts. The size of the pits can be appreciated by the 1 mm references which are shown below samples 2M1 and 3M2. The first digit in the sample number corresponds to its position on the blade (see Fig. 1). Downstream is left and upstream right. At the left is indicated the guide vane opening and downstream level for each test. From this Figure it can be clearly seen that the largest pits are observed near the trailing edge of the blade in tests M1 and M4 with the low downstream level. Smaller pits occur on discs 3 or 2 when the downstream level is raised as in test M2 or the guide vane opening is reduced as in test M5. Opening the guide vanes to 90% has for effect to move the pitting area to disc 4 towards the trailing edge or even beyond as no pits are observed on upstream discs 2 to 4. The sweep test shows the worst pitting on disc 2, the result of the combined marking during the 80 and 75% guide vane opening exposure.

78 %
206.2 m

78 %
207.4 m

90 %
207.5 m

85-80
75-70 %
205.6 m

72 %
205.9 m

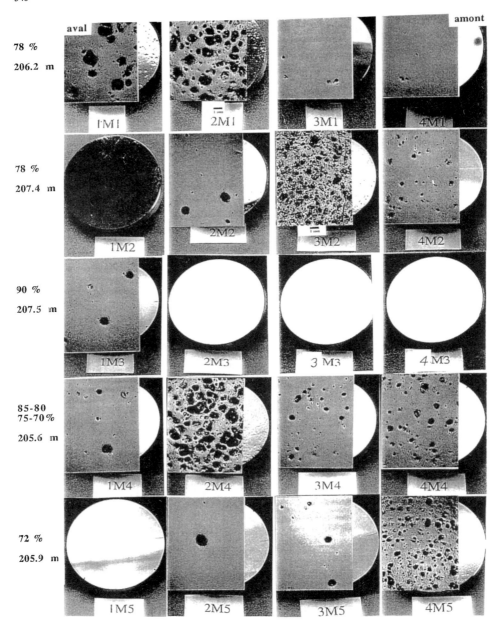

Figure 6. Pitted discs and 10X micrographs of pitted areas.

These results are consistent with those of the previous campaign and show the combined effect of the increasing velocity with larger guide vane opening which promotes a more voluminous cavitation development and the adverse effect of a higher downstream level which contracts the cavitation development and forces implosions on the blade in the closure area of the cavitation cloud in the pressure recovery region.

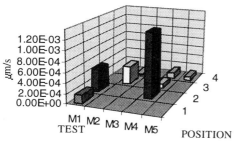

Figure 7. Volume pitting rate vs test and disc location

Since larger developments generate bigger shedded vortices, larger pits are found closer to the trailing edge of the blade. These observations are summarized in Figures 7 and 8 which show respectively the Volume pitting rate V_d (total volume of pits per unit area and time) and the characteristic radius R_v (average of pit radius weighted by V_i, the pit volume). The data for these figures are obtained by scanning the pitted samples with an UBM Laser profilometer and processing the output files with the "Adresse" software developed for EDF by CREMHyG (2).

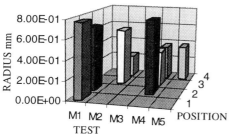

Figure 8. Characteristic radius Rv of the pits vs test and disc position on blade 4.

These Figures clearly show that the largest Volume pitting rates are recorded in tests M1 and M4, the sweep test, in position 2 closer to the trailing edge. This is due to conditions at 80 and 75% guide vane opening which probably both contribute to pitting at this location. Individual pit diameters approaching 2 mm, the largest we have ever observed, were found in test M1. These pits are found downstream where the larger and more energetic vortices implode. This is consistent with the notion of erosive power (3) where the potential energy of each cavity scales with its volume. In test M3 at 90% guide vane opening, no pits were recorded on samples 2 to 4 and very few on sample 1. Under these conditions implosions occur on the blade towards the trailing edge or in the flow beyond.

5.3.2 DECER electrochemical erosion detection

The DECER erosion detector located between samples 2 and 3 produced a localized maximum erosion rate at 75% guide vane opening in sweep test M4 with a low downstream level and a next higher value at 78% with a high downstream level in test M2. This is illustrated in Figure 9 which shows the DECER erosion current in μa in relation to the guide vane opening and downstream level. This current is proportional to the actual erosion rate on the surface of the titanium sensor.

The initial pressure tests, the pit counting tests and the DECER measurements all indicate a variable cavitation development on blade 4 affected both by guide vane opening and downstream level with maximum aggressiveness in the 75 to 80% span of guide vane opening.

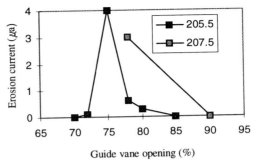

Figure 9. DECER erosion current vs guide vane opening

5.3.3 Vibration measurements

Vibration and pressure measurements were made during all tests. The blade mounted accelerometer was retained for these latter tests. The results are summarized in Figure 10. The three curves represent data normalized to the maximum observed value of each parameter during the low downstream level pressure and pit counting tests. The free symbols represent similar data for the high downstream level tests of both types.

Impacts were found to occur on the blade at or near a 60 Hz rate which corresponds to the guide vane passing frequency (20 guide vanes and 3 Hz rotation frequency) as indicated by the blade mounted accelerometer high frequency amplitude modulation envelope. The envelope of high frequency acceleration band pass filtered in a third octave bandwidth centered on 4 kHz was analyzed. Similar

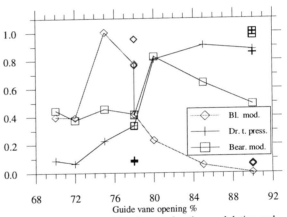

Figure 10. MSV of blade and bearing acceleration modulation and draft tube pressure around the guide vane passing frequency.

indications were obtained with the 0 degree upstream accelerometer at the lower guide bearing. In this case the 15 to 30 kHz bandwidth was retained for analysis. These results agree with those of the 1993 tests. The three curves present respectively with diamonds, squares and crosses the Mean Square Value of modulation of the high frequency acceleration on the blade and at the lower guide bearing and the MSV of pressure pulsation above the draft tube access door. These parameters were calculated at 20 +/- 1 and 2 times the runner rotation frequency over a quarter of a Hz bandwidth at each of these 5 frequencies.

These curves clearly show that maximum aggressiveness on blade 4 occurs around 75 to 80% G.V. opening depending on downstream level. A higher downstream level displaces the high intensity points towards the greater G.V. openings. The indications at the lower guide bearing are different with maximum values occurring at 80% G.V.

opening for low downstream levels and 90% for high downstream levels. The guide bearing sensor modulation envelope is in good agreement with the indications of the draft tube pressure sensor. As opposed to the blade mounted accelerometer which is affected essentially by local events, the response of this pressure sensor, like the guide bearing accelerometer, is determined by the overall runner cavitation performance.

6. Discussion and conclusions

The availability of tools to characterize the aggressiveness of cavitation at the prototype level has been demonstrated in this test program. Localized pressure, pitting, DECER electrochemical erosion detection and vibration measurements have all concurred to characterize the cavitation behavior of blade 4 of this Francis prototype. Increasing guide vane openings or lower downstream levels produce a larger cavitation development and greater aggressiveness inasmuch as the implosions do not occur beyond the trailing edge of the blade. Overall vibration measurements at the guide bearing and pressure measurements at the draft tube reveal maximum aggressiveness at the maximum tested guide vane opening and downstream level. This suggests that all blades do not behave identically under the same hydraulic conditions from a cavitation point of view and that maximum overall aggressiveness occurs under high flow velocity and downstream levels which produce the best conditions for high erosive power of the cavitating flow. A set of blades with high erosion, of which blade 4 is typical, go into cavitation early and are followed later by the others when hydraulic conditions are more severe. This explains the heterogeneous blade erosions that have historically been observed on this machine. Because of this behavior, the transposition of these results to the model level may not be simple. However, the tools required to characterize quantitatively cavitation aggressiveness at the prototype level have been demonstrated and data is available for comparisons with those obtained in the model test program (4). In closing, the authors would like to thank Louis Bezençon and Georges Jotterand of IMHEF and Pierre Lavigne and Jacques Larouche of IREQ for their fundamental contributions to the success of this program.

7. References

1. Bourdon, P., Simoneau R., and Dorey, J-M., "Accelerometer and Pit Counting Detection of Cavitation Erosion on a Laboratory Jet and a Large Francis Turbine", Proceedings of the *XVII IAHR Symposium*, Beijing, China, 1994, Vol. 2, pp. 599-615.
2. Fortes -Patella R and Reboud, J.L., "Analysis of cavitation erosion by numerical simulation of solid damage", Proceedings of the *XVI IAHR Symposium*, Sao Paulo, Brazil, 1992, Vol. 2, pp.617-626.
3. Farhat, M., Pereira, F. and Avellan F., "Cavitation Erosion Power as a Scaling Factor for Cavitation Erosion in Hydraulic Machines", Proceedings of the *ASME Symposium on Bubble Noise and Cavitation Erosion in Fluid Systems*, ASME Fed-Vol. 176, pp. 95-104, New-Orleans, USA, Dec. 1993.
4. Caron, J.-F., et al., "Cavitation Erosion Prediction on Francis Turbines, Part II: Model Tests and Flow Analysis", paper to be presented at the XVIII IAHR Symposium, Section on Hydraulic Machinery and Cavitation, Valencia, Espania, September 1996.

DETERMINATION OF CRITICAL CAVITATION LIMIT IN THE PRESSURE CONTROL DEVICES

A. CASTORANI[1], G. DE MARTINO[2], U. FRATINO[1]

[1] *Water Eng. Dep. - Polytechnic of Bari (Italy)*
[2] *Hydraulic and Environmental Eng. Dep. - University of Naples (Italy)*

Abstract

This paper describes a method for predicting the cavitation intensity of hydraulic devices by means of the analysis of recorded pressure signal. Infact, it seems possible to get in detail the behaviour of the observed variable through the determination of some frequencies of the pressure fluctuations produced by flow turbulence and by the cavitation impulses. It has been found that it is possible to define the numerical value of the critical cavitation limit by means of the determination of a global index as the root mean square of the pressure fluctuations and in which way the spectral signal analysis proves this approach.

1. Introduction

The determination of the performances of the pressure control devices must necessarily consider a careful analysis of the cavitation characteristics to assure their correct operations in every working conditions [4].
Therefore, in view of the modern guidances, the evalutation of the risks involved in the development of the cavitation phenomena has become a crucial moment of the design analysis. From this point of view, using the results of previous experiments, the laboratory procedures for their determination has been partly standardized, as proved by the standards used in several countries [1].
The used methodologies are based on the quantification of several secondary effects produced by the cavitation and, as a rule, these approachs are based on the measurement and evalutation, in a relative way, of the vibrational and acoustic parameters produced by the cavitation. Moreover, a method suitable to correlate the cavitation with the hydraulic variables that identify the fluid motion seems closer to a lagrangian problem approach.
Therefore, during an experimental research designed to point out the characteristics of several methods for predicting the cavitation intensity and carried out both on a Larner Johnson ported valve of 100 mm diameter and some 200 mm diameter orifices of several shapes, it seemed useful to examine a new approach based on pressure recordings.

E. Cabrera et al. (eds.), Hydraulic Machinery and Cavitation, 544–553.
© 1996 *Kluwer Academic Publishers. Printed in the Netherlands.*

2. The purpose of the research

Using some experimental results obtained from conventional analysis techniques, the temporal behaviours of the recorded pressure in several points both upstream and downstream the device location have been analized to characterize the pressure signal so that informations on the cavitation characteristics have been obtained[1].

The choice has been confirmed, in a preliminary study [5], by dimensional considerations that make possible an approach based on the evalutation of the pressure fluctuations produced on the fluid mass. Infact the description, through a dimensional analysis, of the phenomenon produced by the fluid flowing in a control section, allows to define the variables by the following relation:

$$\phi_1(V, \Delta P, P_m, \bar{r}, \rho, \mu, \tau, \varepsilon, \Delta\gamma, C_p, \Theta, P_w, T_r, D, \psi, shape) = 0 \tag{1}$$

where, the indipendent variables ρ, V, D and Θ being chosen and applying the π theorem, it leads to:

$$\phi_2(Eu, Ne, Ne_r, Re, We, Fr_d, Ma, Ne_w, Ec, St, \psi, shape) = 0 \tag{2}$$

that is the general formulation of the relation describing the fluid motion through a pressure control device.

The mathematical shape of the equation (2) can be modified to show the influence of the dimensionless cavitation parameter σ. Infact, σ being a pressure forces ratio as:

$$\sigma = \frac{P_m - P_w}{P_m - P_v} \tag{3}$$

it can be written as:

$$\sigma = \frac{(Ne - Ne_w)}{Eu} \tag{4}$$

from which it seems correct to use σ instead of Ne_w, the first one being just a linear combination of the other one.

Downstream of this mathematical position and the dimensionless parameters arised by hydraulic and bubble dynamics considerations, suitable to a lagrangian approach, being neglected we have:

$$\phi_3(Eu, Ne, Ne_r, Re, Ma, \sigma, St, \psi, shape) = 0 \tag{5}$$

expression in which the dimensionless parameters have a different weight according to the boundary conditions.

In the case that the fluid motion, in absolute turbulence conditions, produces the development of cavitation phenomena and modifying the equation so that the cavitation index σ becomes the dependent variable, it is possible to obtain:

$$\sigma = \chi(Eu, Ne, Ne_r, Re, Ma, St, \psi, shape) \tag{6}$$

that shows that σ is function of loss coefficient through the Euler number (Eu), of the mean system pressure and therefore of its relation with the vapor pressure of the water (Ne), of the pressure fluctuations around the mean value (Ne_r), of the Reynolds number (Re), of the Mach number (Ma), of the bubble permanency time in low pressure zones (St), of the opening degree (ψ) and of the valve shape.

---

[1] Several researchers previously have been performed many tests to correlate the pressure signal to the cavitation phenomena induced by the hydraulic devices and the results of these researches have been very useful [2, 6, 9].

In view of the previous relations an experimental research in a hydraulic device subjected to cavitation phenomena and for which it is required the design of whole similitude model, must necessarily consider the necessity to respect simultaneously all the analogies of the dimensionless parameters in the equation *(5)*, including the one of the cavitation index σ.

In the real situation, this requirement prevents to operate in a full respect of the similitude, determining the impossibility to use a different scale for the model, unless we accept several simplifying assumptions.

Therefore, it is necessary to neglect some dimensionless parameters defined in the equation *(5)* and to respect, in addition to σ, only the conditions imposed by the device geometry (the opening degree ψ and the shape) and by the Euler number.

In view of these restrictions, it is very critical to suggest experimental researches using scale models that bring unavoidably to introduce some distorsions between prototype and model and that determine those uncertainties, known as like scale effects, which many researcher have studied for a long time [*1, 5, 7, 8*].

3. Experimental apparatus

The experimental set-up is composed by two parallel 100 mm and 200 mm diameter pipelines on which the hydraulic devices have been installed, supplied by a electropump assuring a 0,9 MPa pressure value and discharges over 100 l/s, *(fig. 1)*.

The valve used in the tests is a control Larner Johnson ported valve which presents, at its end, a moving steel cylinder with sixteen parabolic-shaped openings that produce, in full opening position, a ψ value equal to 0,92. The four, 200 mm diameter and 19 mm thickness, steel sharp-edged orifices have 4, 5, 9 and 13 radial symmetry holes respectively. The 4 and 9 holes orifices have a ψ value equal to 0,40, whereas the 5 and 13 hole orifices have a ψ value equal to 0,16.

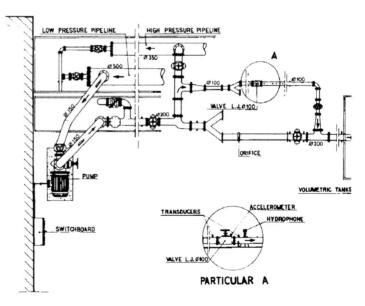

Fig. 1 - Experimental set-up

Preliminarily to the development of the cavitation research, on the 100 mm and on the 200 mm diameter pipelines, several tests have been performed to evaluate the Darcy-Weisbach friction factor in order to subtract the calculated friction loss between the two pressure taps from the measured gross pressure drop.

The pressure taps used for evaluating the gross pressure drop are located one diameter upstream and ten diameters downstream the device location and the pressure drop has been measured by a mercury differential manometer and by Bourdon tube pressure gauges. The discharge value has been valutated by a measuring orifice plate located on the delivery pipe and by some volumetric tanks.

In the laboratory circuit have been located several parallel slide gates, upstream and downstream the device location, for settling the desired pressure and discharge values in every test condition.

Therefore, during the laboratory tests, a pressure value equal to 0,5 MPa has been fixed upstream the valve and a pressure value equal to 25 Kpa downstream the orifices for making the experimental results indipendent from possible uncertainties due to the pressure scale effects that have been evaluated in previous tests [5]. In a preliminary study, several tests in steady condition have been performed to define, for the valve, the curve of the loss factor versus the opening degree and, for the orifices, the loss coefficient values [5].

The evalutation of cavitation intensity has been performed using several sensor types: a Larson & Davis 820 sound level meter and a Bruel & Kjaer 4384 accelerometer for the vibration measurements. For measuring the pressure fluctuations, two pressure transducers, having 200 Hz maximum frequency response, have been used. The pressure measurements have been performed in several points along the pipeline, located at 1, 3, 5 and 10 diameters downstream the device location.

These tests, in addition to provide informations about the cavition intensity, allowed to define the downstream section in which the pressure recovery occurs and the normal hydraulic gradeline is re-established.

4. Analysis of results

The experimental research, performed using the traditional cavitation techniques, provided the results in agreement to the conclusions of other researchers, proving the validity of these approaches [3, 5].

The reliability and the suitability of the test results has been proved, evaluating the mechanical and acoustic parameters produced by cavitation phenomena, but the limitation to use the laboratory procedures has been partially overcome when the problem has been studied through a detailed analysis of pressure signal. In fact, this approach seems to identify a typical behaviour of the pressure signal when the critical cavitation limit is reached. This circumstance is very meaningful because σ_c represents the universal design limit for a cavitation free device [1].

The recorded signal processing has been carried out according to two different ways: the first one by means of a global evaluation of the pressure signal and a second one in which the components and the frequencies involved in the cavitation inception have been defined for getting a new analysis procedure.

4.1. GLOBAL ANALYSIS

In this first phase, the evaluation of the signal characteristic has been carried out through a global analysis, verifying preliminarily the correctness of the study by means of the invariability of the mean and root mean square values versus the acquisition times. The root mean square value of the pressure signal, evaluated just upstream and downstream the device location, shows however that, for a fixed constant value of the upstream pressure, this quantity has a different behaviour starting from a certain σ value. An inspection, made to recognize the σ values for which the disjunction occurs, brings, for all investigated opening degrees, to identify it in a position close to the one conventionally defined, in the researches performed by accelerometer and sound level meter, as critical cavitation index $\sigma_c{}^2$. So, this remark suggests a physical interpretation of this index, through a dimensionless variable known as Newton number, connected with the turbulence conditions of the flow and defined by:

$$Ne_r = \frac{\sqrt{\dfrac{\sum\limits_{i=1}^{N}(P_i - \overline{P})^2}{N-1}}}{\rho V^2} \qquad (7)$$

All the graphs, plotting Ne_r versus σ, are similar to that in *fig. 2*, that is referred to the valve. From the figure, infact, it is possible to point out, close to the σ_c value, a clear divergence of the two curves. Hence, this behaviour of Ne_r confirms the role of the dynamics of the turbulence on the cavitation characteristics of the flow, as shown in *(5)*.

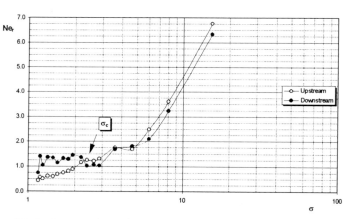

Fig. 2 - Dimensionless index Ne_r versus σ in the 100 mm valve (C=25%)

This behaviour has been found again, with the same characteristics, when the analysis has been extended to the 200 mm orifices, and particularly to the 5 and 13 holes ones characterized by a smaller ψ ratio. The orifices, on which many tests have been carried out in different flow conditions, show a strong tendency to cavitate and the Ne_r-σ curves, determined at the same pressure taps location, have no further doubts about the possibility that this occurence is somehow due to the shape of the streamlines in the

---

[2] In the technical literature [2, 3, 4, 5], the cavitation intensity is defined by means of four conventional levels called incipient cavitation, critical cavitation, incipient damage cavitation and chocking cavitation that represent the boundary marks between cavitation intensities associated to different risks for the hydraulic system.

outlet section of the device *(fig. 3)*[3]. The confirmation of this behaviour, obviously, defines a new methodological approach to the problem, expecially for its versatility and, moreover, permits to join a physical significance to a fundamental index in the evaluation of cavitation performance of a device, indipendently from a graphical settlement.

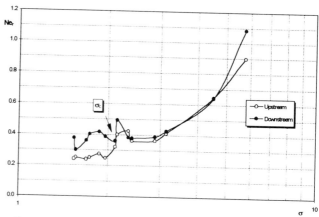

Fig. 3 - *Dimensionless index Ne, versus σ (13 holes 200 mm orifice)*

4.2. SPECTRAL ANALYSIS

Before describing the signal processing methods, it is important to underline that the frequency analysis of the recorded signal points out that the acquisition time was sufficient to estimate the phenomenon [5]. Settling the correctness of the pressure measurements, the problem is to identify the frequencies involved in the flow turbulence and in the inception of the cavitation phenomena. From this point of view, in the preliminary phase of the investigation, in absence of cavitation phenomena, a comparison between the pressure spectra in different sections was carried out, determining an essential agreement between the spectra relating to the upstream and downstream sections of the device. Hence, the turbulent nature of the flow doesn't appear to bring any changes during the passage through the section in which there is the local pressure drop, expecially if this one is of slight amount, so the spectral peculiarities of the pressure signal, mainly due to the pump and system characteristics are virtually the same. This investigation has been extended to the measurements performed along the downstream pipeline too, showing a similar behaviour.

However this situation changes clearly when, decreasing σ, the cavitation inception occurs with the development and subsequent implosion of vapor cavities. The amplitude spectra, at the same σ, at the section just downstream the valve location have a different trend from the ones defined at the upstream section. This trend is shown in *fig. 4*, where are reported the spectra for a σ value, equal to 2,77, less than σ$_c$. The two spectra appear different and the downstream one has components higher for frequencies less than 20 Hz, particulary in the ranges [6-12 Hz] and [16-20 Hz]. The component values in these frequency ranges, infact, are much higher than those defined, at the same frequencies, upstream. From this analysis, later improved by other measurements, it appears that, in presence of cavitation, some changes in the spectral

---

[3] The cavitation tests on the orifices carried out using the traditional techniques by accelerometer and by sound level meter show σ$_c$ values in agreement with those defined by this plots.

structure of the pressure fluctuations components in the downstream section occur. On the other hand, it is likely that the same character of the phenomen, by means of the number and the power of the induced implosions, gives rise to a singular sequence of time of the frequency intervals that, from time to time, are modified.

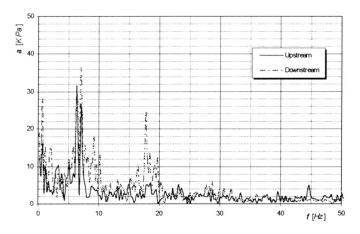

Fig. 4 - *Amplitude spectra upstream and downstream the 100 mm valve (C=40%)*

In a general way, if the upstream pressure measurements define the turbulence of flow and the vibrations originated by the pump, it is possible to observe that the flow turbulence, when the discharge increases, involves, step by step, higher frequencies producing, after a first meaningful growth of the lower frequencies, a general rise of the spectrum in the whole frequency field. The behaviour of the amplitude spectrum just downstream the device section is instead very different. In fact, if for $\sigma < \sigma_c$, the spectral character of the pressure signal between the upstream and downstream section is similar, except for negligibile differencies due to the hydrodynamic characteristics of the jet flowing from the device, when the discharge increases, the observed differencies increase reaching high values through a meaningful growth of the spectrum, mainly for lower frequency values. This behaviour can be explained as an effect due to the character of the flow, thanks to the large development of turbulent eddies and to the pulses on the fluid mass generated by the increased number of vapor bubbles implosions. The graph in *fig. 5*[4], referred to the valve, explain this behaviour; infact, it is possible to see that, for light cavitation, there is a slight increase of the spectrum for frequencies lower than 10 Hz and higher than 16 Hz, effect on which the cavitation phenomenon plays an important role, because of the reoccurrence of the involved frequencies varying either Re or the valve opening degree. When the discharge increases, the phenomenon raises in the above mentioned ranges, producing, as a result, the raising of the closer frequencies, whereas it fades totally in the other ones. So, because of the increased fluid compressibility due to the formation of the vapor bubbles, the pressure pulses, coming either from the flow dynamics or from the implosions of the bubbles, cause fluctuations of low frequency; in substance the smaller stiffness of the fluid mass provokes a shift downwards of the involved frequencies. The behaviour of the chocking spectrum improves the previous remarks, infact, for this

---

[4] The σ values are equal to 16,59 for no cavitation condition, to 2,56 for light cavitation, to 1,86 for heavy cavitation and to 1,30 for chocking cavitation.

cavitation condition, it is possible to observe a further shift down of the involved frequency.

In this case, the pulses are returned from the fluid mass by means of a limited number of low frequency harmonics, to point out an increased absorbing capacity of the stresses. The hydraulic behaviour of the orifices confirms the above-mentioned conclusions, even if, in different frequency ranges, because of the different geometry.

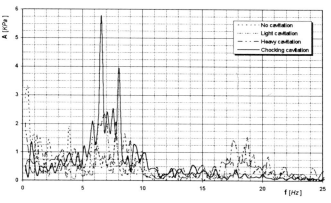

Fig. 5 - Pressure spectra downstream the 100 mm valve for several σ

In view of the previous considerations, it is possible, through a detailed study of the spectral behaviour of the pressure fluctuations downstream, to make a careful analysis of the temporal evolution of the cavitation. Infact, limiting the study to the 20 Hz maximum frequency, a variable representative, by means of a trapezoidal integration, of the spectral behaviour is defined. This variable, called $\Sigma_{[a;b]}$, represents the spectrum area of the observed frequency range $[a;b]$ and has the following expression:

$$\Sigma_{[f_1;f_2]} = \int_{f_1}^{f_2} a(f)\, df \qquad (7)$$

in which $a(f)$ is the amplitude value at the f frequency and f_1 and f_2 are the frequencies that define integration extreme terms.

The chosen integration intervals are [0-20 Hz], [15-20 Hz] e [5-10 Hz] and, through graphs like in *fig. 6*[5], it is possible to set out their peculiarities. The figure shows that the curves have a bell-shaped trend, typical of the cavitation measurements. The curve of the integral values in the [0-20 Hz] range spectrum has obsviously higher values than the other two and defines the critical cavitation limit where the curve has the abrupt change of slope, whereas the other two curves have substantially the same trend, even if the one defined in lower interval is, on the average, higher.

This behaviour confirms some of the above mentioned considerations, showing, in the defined frequency range that, besides an initial constant amplitude values, there is an increase involving first the higher frequencies and then the lower ones. In short, it is possible to see that, for σ value less than σ_{ch}, the observed variables decrease quickly to underline a sudden drop of the cavitation intensity, that evolves in a hydrodynamic condition of fully chocked flow where the flowing jet is surrounded by a vapor cavity instead of water.

---

[5] In the same figure the incipient, critical and chocking cavitation limits are represented, determined previously by accelerometer and by sound level meter.

The last mention is for the analysis made for defining the way in which the pressure fluctuations evolve along the pipeline varying the hydrodynamic conditions. The spectra comparison shows that they have similar shape when the flow conditions define light phenomena of cavitation, unless slight differencies due to the way in which the mean stress transmission occurs. Infact, it is possible to find a

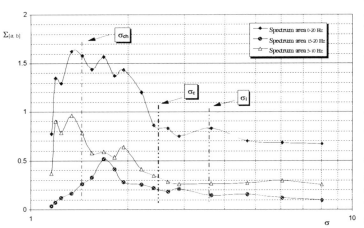

Fig.6 - Plot of $\Sigma_{[a,b]}$ versus σ for the 100 mm valve (C=20%)

complete agreement of the involved frequencies, while the amplitudes have a decreasing behaviour when the X/D ratio increases.

For that purpose, , for the valve, the $\Sigma_{[5-10\ Hz]}$ value is plotted versus the pressure taps position in different cavitation conditions *(fig. 7)*. From the figure, the behaviour of this variable varying the pressure tap location is clear. It, for pressure tap fixed, increases in a monotone way, while decreases, step by step, moving downstream. Thus, for σ fixed, the cavitation phenomena show a maximum intensity value just downstream the device section and have a lower one when the X/D ratio increases, because of expected longitudinal variability of the cavitation intensity.

This remark gives further proof to the theory that states the existence of a close relation between the cavitation intensity

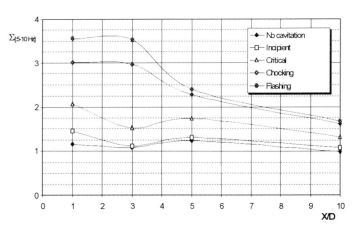

Fig. 7 - Plot of $\Sigma_{[5-10\ Hz]}$ versus X/D at several σ for the 100 mm valve (C=30%)

and the pressure fluctuations spectra, confirming that the pressure measurements, by means of the spectral analysis, are an important source for defining the cavitation development.

5. Conclusions

The described method for detecting the cavitation intensity appears of great interest and, in the future, of possible large use. In fact, the evaluation of the pressure signal allows, through a global analysis, the physical definition of the critical cavitation index. At last, the pressure fluctuation spectra show that the pressure fluctuations intensity increases when the cavitation intensity raises and that the pressure fluctuations, although with a random trend in magnitude and frequency, have a clear drop when the opening degree of the valve increases and, at the same opening degree, going from upstream to downstream.

Notation

C	Stroke, in per cent, of the closing element of the valve	
C_p	specific heat at constant pressure	(J kg^{-1} K^{-1})
D	diameter	(m)
f	frequency	(Hz)
$\bar{P}$	expected value of the pressure	(N m^{-2})
P_i	instantaneous pressure values	(N m^{-2})
P_m	absolute upstream pressure	(N m^{-2})
P_w	vapor pressure of the water	(N m^{-2})
$\bar{r}$	root mean square of pressure fluctuations	(N m^{-2})
T_r	bubble permanency time in low pressure region	(s)
V	mean pipe velocity	(m s^{-1})
$\Delta\gamma$	difference in specific weight across an interface of different densities	(N m^{-2})
ΔP	pressure drop	(N m^{-2})
ε	bulk modulus of water	(N m^{-3})
μ	viscosity of water	(N m^{-2} s)
ρ	density of water	(N m^{-2})
σ	cavitation index	
τ	surface tension	(N m^{-1})
Θ	temperature	(K)
ψ	opening degree or area ratio	

$$Eu = \frac{\Delta P}{\rho V^2} \qquad Ne = \frac{P_m}{\rho V^2} \qquad Ne_r = \frac{\bar{r}}{\rho V^2} \qquad Re = \frac{\rho VD}{\mu} \qquad We = V\sqrt{\frac{\rho D}{\tau}}$$

$$Fr_d = \frac{V}{\sqrt{\frac{\Delta\gamma}{\rho}D}} \qquad Ma = V\sqrt{\frac{\rho}{\varepsilon}} \qquad Ne_w = \frac{P_w}{\rho V^2} \qquad Ec = \frac{V^2}{C_p\Theta} \qquad St = \frac{D}{VT_r}$$

References

1. **ANSI-ISA**: "*Control valve Capacity Test Procedure*" S75.02, Research Triangle Park, NC, 1988.
2. **Anton L.**: "*The effects of turbulence on cavitation inception*" La Houille Blanche n°5, 1993.
3. **Castorani A., De Martino G., Fratino U.**: "*Metodologie di valutazione delle caratteristiche cavitanti delle valvole di regolazione*" XXV Convegno di Idraulica e Costruzioni Idrauliche, Torino, 1996.
4. **Di Santo A., Fratino U., Piccinni A.F.**: "*Criteri di scelta delle valvole di regolazione al servizio dei grandi adduttori*" Convegno A.I.I. Macchine ed Apparecchiature Idrauliche, Baveno (NO), 1993.
5. **Fratino U.**: "*Effetti cavitanti nelle valvole di regolazione*" PhD Thesis, Università di Napoli, 1996.
6. **Li S.C., Zhang Y., Hammitt F.G.**: "*Characteristics of cavitation bubble collapse pulses, associated pressure fluctuations and flow noise*" Journal of Hydraulic Research n°2, IAHR, 1986.
7. **Tullis J.P.**: "*Hydraulics of Pipelines*" Wyles and Sons, New York, 1989.
8. **Tullis J.P.**: "*Cavitation Guide for Control Valves*" prepared for U.S. Nuclear Commission, 1993.
9. **Vigander S.**: "*Wall pressure fluctuations in a caviting turbulent shear flow*" Cavitation in Fluid Machinery, ASME, New York, 1965.

STABILITY OF AIR CAVITIES IN TIP VORTICES

A. CRESPO, F. CASTRO*, F. MANUEL AND D. H. FRUMAN **

E.T.S.I. Industriales, Universidad Politécnica de Madrid
José Gutiérrez Abascal, 2, 28006 Madrid, Spain
* Universidad de Valladolid, Spain
** ENSTA, Groupe Phénomènes d'Interface, Palaiseau, France.

Abstract: A model is presented to interpret the behavior of air cavities formed in the tip vortex issued from hydrofoils in a water tunnel, when air is injected. These cavities have the ability to move upstream and reach a stable position near the hydrofoil. From detailed measurements of the velocity components the conditions for the cavities to appear are examined, and compared to the model results. The proposed model, that is based on the equilibrium balance of the cavity, and takes into account that it is closed downstream by a reentrant jet developing in a Lamb vortex, can explain aspects of the behavior not contemplated by another models.

1. Introduction

If air is fed into a water tunnel containing a finite span wing, it may be captured by the low pressure region appearing in the vortex generated at its tip (Figure 1). Cavities in tip vortices have been observed by Meijer (1981) in an open-channel, and Manuel et al. (1987) in a closed-channel of circular cross-section; Meijer (1981) termed them swimming cavities because of their ability to move upstream. Escudier and Keller (1983) associated these cavities to the vortex-breakdown phenomenon, and obtained analogous shapes by pumping air in the region downstream of the vortex-breakdown; however, for the experiments reported here the vortex core remains smooth and no irregularities resembling vortex-breakdown are observed prior to air injection; nevertheless, the analytical predictions made by Keller et al. (1985) based on the two-stage breakdown hypothesis of Escudier and Keller (1983) coincide with those presented here in the appropriate limiting situations. Vapor cavities can also occur in tip vortices depending on the value of the minimum pressure reached on the vortex axis and the air contents in the water (Arndt et al. (1991)). For large air content values (termed weak water), the inception cavitation numbers at a given Reynolds number become weakly dependant of the lift coefficient (Arndt and Keller (1992)). It appears that, for water containing large amounts of free air, capture of nuclei is the dominant effect and that ventilated cavities are formed prior to vapor cavities occurrence. It has been noticed by Arndt and Keller (1992) that in weak water the inception process is highly intermittent, as it also happens in some cases for the phenomenon presented here.

The investigation of the behavior of ventilated cavities shape and intermittency, is

E. Cabrera et al. (eds.), Hydraulic Machinery and Cavitation, 554–563.

interesting from three points of view : first, there is no need to operate under reduced pressure conditions, second, interesting information can be gathered to understand the behavior of large natural (vapor) cavities and, third, ventilated cavities behavior can be associated with some aspects of the vortex breakdown process. Besides, ventilated cavities occur very often in blades and hub vortices of propellers operating near the free surface, surface-piercing propellers, turbines torches, etc.

As it is discussed by Manuel (1981), Manuel et al. (1987) and Escudier and Keller (1983), two characteristic transition regions can be identified in the cavities of Figure 1. The first region contains the stagnation point at the front, and can be considered as a transition from the unperturbed flow to a uniform cylindrical cavity. The second region is a sudden constriction of the first cavity, followed by a wavy pattern. The second transition region has been studied by Manuel et al. (1987) using an analogy with water wave theory in which hydraulic jumps, progressive waves, and solitary waves move along the interface; similar results for progressive waves were obtained by Ackerett (1930), Uberoi et al. (1972), and Keller and Escudier (1980).

The model presented here to study the first transition region is a continuation of a previous one proposed by Manuel et al. (1992). It is assumed that there is conservation of mass, momentum and Bernoulli's constant. In the closure region there is a reentrant jet that contributes to the overall momentum balance. According to observations, the cross section of the viscous core of the vortex and the area of the reentrant jet are assumed to be small compared with the cross section of the cavity, and this area is small compared to the cross section of the tunnel. Keller et al. (1985) assume that in the interface the velocity is zero, and that there is a thin rotational layer near the interface where the velocity changes from zero to its uniform value outside; Escudier and Keller (1983) suggest that this layer may grow by viscous diffusion or become unstable, thus explaining the various spiral forms of vortex breakdown. In our model it is assumed that in the vicinity of the stagnation point the fluid at the interface is at rest and is dragged along by viscous diffusion of the rotational layer, until it reaches a velocity of the order of that of the external irrotational flow. Experimental evidence of the existence of a reentrant jet at the back of the cavity has been accumulated. The appearance of the jet may be a consequence of the instability of the shear layer as a result of the inflexion point required to satisfy the zero normal velocity gradient condition at the interface.

The model predicts a criterion for the cavity to be steady in the flow as a function of the flow conditions upstream of the cavity and the section of the reentrant jet. In contradiction to Escudier and Keller`s (1983) model, in which the cavity will only occur for a given combination of vortex core diameter and maximum azimuthal velocity, our model predicts that there is a range of equilibrium conditions for the cavity depending on the section of the reentrant jet. This is in agreement with the experimental observation showing that stable positions all along the downstream diffusing vortex are possible. The model retains radial variations of the axial velocity, because an overshoot near the axis has been observed experimentally, and uses the Lamb vortex for swirl velocity, because it fits the experimental results much better than the Rankine vortex used by Keller et al. (1985) and Manuel et al . (1992).

Experiments have been conducted in two water tunnels, ENSTA and ETSII, with rectangular (ENSTA) and circular (ETSII) test sections respectively, using hydrofoils of rectangular planform (ENSTA and ETSII), and elliptic planforms (ENSTA), that have been described by Manuel et al. (1992); here, only the results needed for comparison and interpretation of the model will be presented.

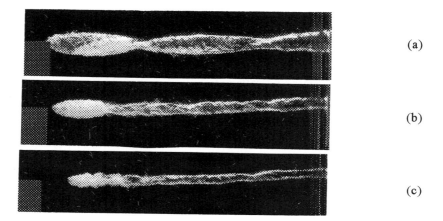

Figure 1. Cavity at three different times (a: 1'06", b: 2' 30" and c: 6') after stopping the air supply. Rectangular planform, 10° incidence angle, W_1=5 m/s. ENSTA tunnel.

2. Experimental results

2.1 CAVITY OBSERVATIONS

Injected air is trapped by the vortex and a cavity, whose leading edge nearly coincides with the transition between the test section (of constant area) and the diffuser (of variable area and adverse pressure gradient) develops. In the case of elliptical planform the bubble remains in the diffuser, whatever the flow conditions. However, if the incidence angle of the foil is above a critical value the cavity is able, occasionally and in a fraction of a second, to move back and forth from the diffuser to the wing. For rectangular planforms and large enough values of water velocity, air flow rate, and incidence angle, the bubble is able to move upstream in the main test section, and, under some circumstances, to get in contact with the foil. By modifying the test conditions, the leading edge of the cavity can be stabilized directly in contact with the foil or at a short distance downstream; however, the range of flow conditions necessary to maintain the cavity at a certain distance from the wing is very narrow: changes in the incidence angle of the order of 1° can send the cavity back to the diffuser or attach it to the wing. Under some circumstances the successive waves pinch off and another reentrant jet occurs.

The cavity in the tip vortex of the rectangular foil can be maintained within the test section even if the air supply is stopped; for a test performed in the ENSTA water tunnel this process lasted about ten minutes. In the absence of injected air the size of the leading bubble and the satellite cavity decrease, and the leading edge moves downstream in a quasi-steady manner, as shown by the three photographs of Fig. 1. With continuous air injection rate an analogous behavior is observed: for larger air injection rates the cavity becomes larger and gets closer to the wing. The reduction of the size of the cavities is related to the continuous draining of the air inside the cavity, while the downstream displacement of the bubble is associated to the force balance over the bubble.

It should also be pointed out that the leading edge of the cavity oscillates around its stable position, with increasing amplitudes and decreasing frequencies as the cavity becomes smaller and moves downstream.

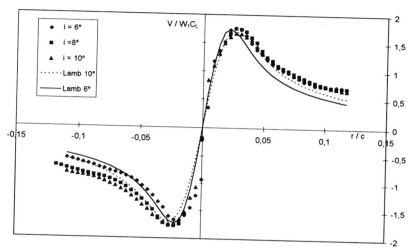

Figure 2. Radial distribution of azimuthal velocities normalized with $C_L W_1$ at three different incidence angles. Rectangular planform, 3 chords downstream. Comparison with Lamb vortex for $i = 10°$. ENSTA tunnel.

2.2 VELOCITY FIELD

Figure 2 shows, for tests performed in the ENSTA tunnel, a typical radial distribution of the azimuthal velocities, V_1, normalized with the free stream velocity, W_1, and the lift coefficient, C_L, for the vortex generated by the rectangular planform wing operating at three different incidence angles, 6°, 8° and 10°, at a location situated three chords downstream of the trailing edge. It is noteworthy that all the curves fall almost in the same one showing that the normalization is appropriate. The profiles follow the usual behavior: solid body rotation within the vortex core and potential velocity distribution outside it. The swirl velocity profile is quite well adjusted by a Lamb vortex (see equations (2) and (3)), as also shown in Figure 2.

The radial distribution of axial velocity has a typical overshoot around the vortex axis, whose diameter is slightly larger than of the viscous core (Manuel et al. (1992)).

The value of the ratio of the maximum azimuthal velocity, V_{MAX} to the axial velocity at the axis, W_0, including the effect of the overshoot, is an important parameter of our model to determine cavity stability,

$$k = \frac{V_{MAX}}{W_0} .$$ (1)

The values of k as a function of the downstream distance, are displayed in Figure 3, for conditions under which the cavity, appears either in a stable position for the rectangular planform or moving back and forth for the elliptic planform. The parameter k in general decreases with distance, except for the rectangular foil of ENSTA.

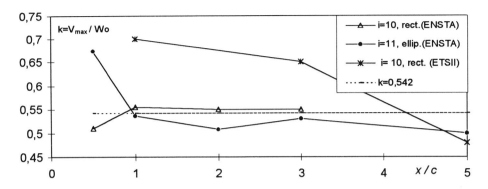

Figure 3. Variation of k with downstream distance. Comparison with analytical prediction. Measurements for both elliptical and rectangular wings at several downstream distances, i=11° and 10° respectively. ENSTA and ETSII tunnels.

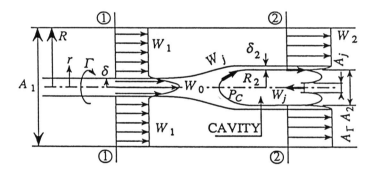

Figure 4. Schematic showing the geometry of the domain under consideration.

3. Proposed model

The geometry of the portion of the flow for which the model will be developed is shown schematically in Figure 4. In section 1 the approaching flow has a uniform axial velocity, W_1, outside a vortex of radius δ and intensity Γ; inside the vortex the approach velocity has an overshoot. Azimuthal velocities are supposed to behave as in a Lamb vortex,

$$V = \frac{\Gamma}{2\pi r}(1 - e^{r^2/\delta^2}) .$$ (2)

The maximum azimuthal velocity and the corresponding radial position are,

$$V_{MAX} = 0.638 \frac{\Gamma}{2\pi\delta} \quad at \quad r = 1.12\delta \quad .\tag{3}$$

In the following δ and Γ will be estimated from experiments using equations (3). Figure 2 shows that the experimental values of the swirl velocity are fitted quite well by equation (2) when equations (3) are satisfied.

The pressure distribution is obtained by integration of the following equation,

$$\frac{dP}{dr} = \frac{\rho V^2}{r} \quad ,\tag{4}$$

that, using equation (2), gives

$$P_1 = P_0 + \frac{\rho\Gamma^2}{4\pi^2\delta^2}\left(\log 2 - \frac{1}{2X^2} + \frac{e^{-X^2}}{X^2} - \frac{e^{-2X^2}}{2X^2} - E_1(X^2) + E_1(2X^2)\right),\tag{5}$$

where, $X = r/\delta$, P_0 is the pressure at the axis, and $E_1(X)$ is the exponential integral. For r (or X) going to infinity a constant value of the pressure is reached,

$$P_{1\infty} = P_0 + \frac{\rho\Gamma^2}{4\pi^2\delta^2}\log 2 = P_0 + 1.703\rho V_{MAX}^2 \quad ,\tag{6}$$

where equation (3) has been used. By integration of the pressure field, equation (5), across the upstream section 1 we obtain the force,

$$\frac{F_1}{\pi R^2} = P_0 + 1.703\rho V_{MAX}^2\left[1 - \frac{1}{\log 2}\left(\frac{\delta}{R}\right)^2 \log\left(\frac{R}{\delta}\right)\right].\tag{7}$$

Where it has been assumed that $\delta \ll R$ and only a first order correction term has been retained.

In the downstream section the flow is supposed to be irrotational, with constant axial velocity, and a circulation equal to that upstream, outside a thin rotational shear layer of thickness δ_2. As the areas of the two rotational regions should be of the same order, because only the fluid particles inside the rotational core of radius δ in region 1 are the ones that can be in the rotational layer of thickness δ_2 in region 2, the value of δ_2 is such that $\delta_2 = O(\delta^2/R_2)$. It will be shown later that the radius of the cavity is much larger than δ, and much smaller than the radius of the test section; the ordering of the different lengths involved is as follows (see also equation (18)):

$$\delta_2 \ll \delta \ll A_j^{1/2} \ll R_2 \ll R \quad , \quad \delta_2 = O(\delta^2/R_2) \quad , \quad A_j = O(\delta^2\log(R_2/\delta)) \quad .\tag{8}$$

For the particular case of Figure 1, typical values are $\delta = 1$ mm, $R_2 = 10$ mm and $R = 100$ mm; the jet area is smaller than that of the cavity, and slightly larger than that of the upstream vortex. As the shear rotational layer is very thin, it can be assumed that over most of region 2 the flow is irrotational and the azimuthal velocity is given by $V = \Gamma/(2\pi r)$ and on using equation (4) the pressure distribution over region 2 is obtained:

$$P_2 = P_C - \frac{\rho \Gamma^2}{4\pi^2} \frac{1}{2} \left(\frac{1}{r^2} - \frac{1}{R_2^2} \right) , \tag{9}$$

where is P_C the cavity pressure, and by integration we obtain the force on region 2,

$$F_2 = P_C \pi R^2 - 2.457 \rho V_{max}^2 \pi \delta^2 \left(\log\frac{R}{R_2} - \frac{A_2}{2\pi R_2^2} \right) , \tag{10}$$

where equation (3) has been used.

It is assumed that the pressure in the cavity is equal to the stagnation pressure along the axis of the upstream region,

$$P_C = P_0 + \frac{1}{2}\rho W_0^2 . \tag{11}$$

Manuel et al., (1992), consistently with the fact that there is a reentrant jet in the back of the cavity, assumed that the constant value of the velocity in the interface is the jet velocity, thus obtaining a lower value of the pressure in the cavity; however, it is not clear how this assumption can hold at the tip stagnation point. It is more reasonable to assume that, if the cavity is long enough for the viscous forces to be significant, the external flow will drag along the stagnant flow in the interface until it reaches a finite velocity, of the order of that of the reentrant jet.

For the control volume of Figure 4, we have conservation of mass,

$$W_1 A_1 + W_j A_j = W_2 A_2 , \tag{12}$$

and the conservation of momentum,

$$F_1 + \rho W_1^2 A_1 = F_2 + \rho W_2^2 A_2 + \rho W_j^2 A_j , \tag{13}$$

where the effect of the overshoot has been neglected because it is of order δ^2. It is also assumed that the jet of area A_j and velocity W_j, that enters in the cavity in the back, produces a force directed upstream, and its mass is incorporated to the liquid again through the interface. It is also assumed that there is conservation of stagnation pressure in the outer irrotational region, and in particular, near the wall, $R \gg R_2 \gg \delta$,

$$P_{1\infty} + \frac{1}{2}\rho W_1^2 = P_{2\infty} + \frac{1}{2}\rho W_2^2, \tag{14}$$

where, from equations (9) and (3),

$$P_{2\infty} = P_C + 1.228 \rho V_{max}^2 \left(\frac{\delta}{R_2}\right)^2 . \tag{15}$$

First, is considered the case in which all higher order terms, according to what is indicated in equation (8), can be neglected. Then, $A_1 = A_2$, from (12) $W_1 = W_2$, and from (14) and (15) $P_{1\infty} = P_{2\infty} = P_C$. Then, from equations (6) and (11), it is obtained that,

$$k = \frac{V_{MAX}}{W_o} = 0.542 . \tag{16}$$

The momentum equation has not been used, and is satisfied identically. The basis of this result is quite simple: the leading point of the cavity is a stagnation point, where the head of the axial velocity just cancels the pressure minimum of the vortex. Manuel et al. (1992) performed a similar analysis using a Rankine vortex and obtained for k a value of $1/\sqrt{2} = 0.707$ instead of 0.542. This value is also obtained using the two-stage transition model for breakdown of Keller et al. (1985), with a Rankine vortex and without considering overshoot (see also Escudier, 1988, the value of k that he defines differs by a factor 2 from the one defined in equation (1)). Spall et al. (1987) give a criteria for breakdown based on a Rossby number, Ro, that is related to k by Ro=0.57/k, without considering overshoot; they find that, for Reynolds numbers based on core radius larger than 100, there is breakdown when $k > k_c$, where $k_c = 0.877$; this result is based on experimental and numerical research both of them and another authors. The comparison of the result of Spall et al. (1987) with equation (16) shows that larger values of vortex intensity are needed for the appearance of vortex breakdown than for the formation of the cavity. This is consistent with the fact that, if there is no air injection, no vortex breakdown appears in the single phase for the value of k given by equation (16).

If higher order terms are retained, equation (16) is modified to give

$$k = \frac{V_{MAX}}{W_o} = 0.542 \left(1 + 3.406 \frac{W_1^2}{W_0^2} \left(\frac{R_2}{R}\right)^2\right) . \tag{17}$$

This means that, as the radius of the cavity increases the maximum azimuthal velocity necessary for the cavity to be stable also increases; equation (16) is then a lower limit for V_{MAX}. From the momentum equation (13), that so far has not been used, the following relation is obtained for the radius of the cavity as a function of the upstream kinematic variables and the jet characteristics:

$$\left(\frac{R_2}{R}\right)^4 = \frac{1}{\log 2}\left(\frac{W_0}{W_1}\frac{\delta}{R}\right)^2 \log\left(\frac{R_2}{\delta}\right) + 2\frac{W_j}{W_1}\left(1 + \frac{W_j}{W_1}\right)\frac{A_j}{A_1} \ . \tag{18}$$

This equation was first obtained by Manuel et al. (1992) without the first term in the right hand side, and a minus sign in the brackets of the last term. The last term represents the effect of the jet; if this term is zero, this equation gives the value of the cavity radius, R_2, as function of the vortex radius, δ, and equation (17) gives the azimuthal velocity, V_{MAX}, necessary for the cavity to be stable; there is then a relationship between δ and V_{MAX} that has to be satisfied; this means that for a vortex whose properties change with distance, stability will only be possible at one position. On the other hand, if we retain the last term of equation (18), the cavity can be at different positions of the vortex: if the cavity gets smaller, equation (17) shows that the bubble should move to another position where V_{MAX} should also be smaller, in this position δ will have a new value, and equation (18) can still be satisfied if the jet characteristics are appropriate. For the two terms in (18) to be of the same order, the area of the jet should be: $A_j=O(\delta^2\log(R_2/\delta))=O(\delta_2 R_2\log(R_2/\delta))$ (as anticipated in equation (8)), slightly smaller than that of the vorticity region close to the interface. For the elliptic wing the radius of the vortex is smaller than the for the rectangular wing, then A_j is also smaller, and the last term of equation (18) may not be large enough to provide the stabilizing effect previously mentioned. Besides, for very small δ all the right hand side of equation (18) becomes very small, the radius of the cavity, R_2, is also very small (this is observed experimentally for elliptic wing), and equation (17) shows that the cavity will only appear for values of k so close to the limiting value given by equation (16), that it is not possible to observe it experimentally.

Equations (17) and (18) indicate that the minimum value of k for the cavity to be stabilized in the test section is 0.542; in figure 3, it is compared with the measurements corresponding to incipient cavity appearance (stable cavity for rectangular wings and oscillating cavity in elliptic wings) in the test section. Since inequalities (8) hold for all the experiments, it is expected that k should be close to this value, whenever the cavity is stable; this is in agreement with the experimental observation that for the cavity to be stable in the test section, the range of variation of the incidence angle is very small. For the rectangular wing in the ENSTA tunnel, k is close to 0.542 all along the vortex. For the elliptic wing at a short distance from it, k is considerably larger than 0.542, and further downstream is very close to 0.542, but as it was already mentioned the cavity is never stabilized and moves back and forth. For the rectangular wing in the ETSII tunnel k gets close to 0.542 about 4 chords downstream, whereas the experimental observations show that the cavity remains stable about 1 to 2 chords downstream of the wing.

4. Conclusions.

Ventilated cavities formed in the tip vortex issued from finite span hydrofoils have different behaviors depending on their planforms. Experiments show that the cavities are generally formed by a leading bubble with a reentrant jet followed by a satellite cavity with longitudinal waves whose amplitude and length decrease with the distance downstream. Moreover, depending on the wing and the flow conditions, the cavity formed initially in the transition between the test section (of constant area) and the diffuser (of increasing area and adverse pressure gradient) can or can not move upstream and, sometimes, get in contact with the foil.

By considering the equilibrium of the leading bubble with a reentering jet, a model is proposed that gives the range over which the ratio of the local maximum tangential velocity to the free stream velocity has to be maintained to allow for cavity stability. Comparison of the model prediction and experimental results show a good agreement and can explain some aspects of the observed behavior.

Further experimental and theoretical work should be conducted to investigate the behavior of the tip vortex and its interaction with the ventilated cavity for a greater range of flow conditions. Further test should also provide information on the cavity characteristics, in particular the length at which instabilities occur in the interface

5. Acknowledgments

The authors wish to express their deep appreciation for the partial support received from the Spanish (grants n°86-B, A.I. and PB 89-0184 of the DGICYT) and French (grants n° 92/205 A.I. and DRET 88/1038) governments.

6. References

Ackerett, J. (1930) *Uber Stationäre Hohlwirbel*. Ing. Archiv. pp 399-402.

Arndt, R.E.A., Arakeri, V.H. and Higuchi, H., (1991) *Some observations of tip vortex cavitation*, Journal of Fluid Mechanics, Volume 229, pp 269-289.

Arndt, R.E.A. and Keller, A.P., 1992, *Water quality effects on cavitation inception in a trailing vortex*, Journal of Fluids Engineering, Volume 114, pp. 430-438.

Escudier, M.P. (1988) *Vortex Breakdown: Observations and Explanations*. Progress in Aerospace Sciences. Volume 25, N° 2, pp 189-229. 1988

Escudier, M.P. and J. J. Keller. (1983) *Vortex Breakdown: a Two-Stage Transition*. AGARD CP N° 342. Aerodynamics of vortical type flows in three dimensions, paper 25.

Keller, J.J., Egli, W. and Exley, J. (1985) *Force- and loss-free transitions between flow states*, J. Appl. Math. Phys. 36(6), 854-889.

Keller, J.J. and Escudier., M.P. (1980) *Theory and Observations on Hollow Core Vortices*. J. Fluid Mech. Vol. 99, part 3, pp 495-511.

Manuel. F. Ph. D thesis. (1981) Universidad Politécnica de Madrid.

Manuel, F., Crespo, A., Castro, F. (1987) *Wave and Cavity Propagation along a Tip Vortex Interface*. Physico Chemical Hydrodynamics, Vol. 9, N° 3/4, pp. 611-621.

Manuel, F., Crespo,A.,Castro,F., Fruman,D.H., Beuzelin,F., Gaudemer, R. (1992) "*Ventilated cavities in tip vortices.*" Proceedings of the Institution of Mechanical Engineers. IMechE 1992-11. International Conference on Cavitation. Cambridge .

Meijer, M.J. (1981) *Swimming Vortex Cavities*. Euromech Colloquium 146. Institut de Mechanique de Grenoble.

Spall, R. E., Gatski, T. B., and Grosh, C. E. (1987) *A criterion for Vortex Breakdown* Phys. Fluids Vol. 30 (11), pp 3434- 3440.

Uberoi, M. (1972) *Stability of a Coaxial Rotating Jet and Vortex of Different Densities*. Phys. Fluids, Vol. 15.

CAVITATION EROSION PREDICTION ON FRANCIS TURBINES-PART 3 METHODOLOGIES OF PREDICTION

J.M. DOREY, E. LAPERROUSAZ
Electricité de France
6, Quai Watier, 78401 CHATOU Cedex, France
F. AVELLAN, P. DUPONT
IMHEF/ EPFL
33 Avenue de Cour, CH 1007 Lausanne, Switzerland
R. SIMONEAU, P. BOURDON
Hydro-Québec
1800 boul. Lionel Boulet, Varennes, Québec, Canada J3X 1S1

Summary: In the frame of a joint research programme between EDF, Hydro-Québec and IMHEF, different methods are investigated to predict cavitation erosion on Francis turbines from model. They are based on measurement of pitting, pressure fluctuations and acceleration. The measurement techniques have been detailed in Part 1 and Part 2. The present article describes essentially the theoretical and practical aspects of the methods and discusses the results obtained until now from the model and prototype tests. The first analysis shows that the methods proposed are suitable to measure cavitation aggressiveness on model and on prototype, and that the level on the model is several orders of magnitude smaller than on the prototype. To adjust transposition laws, a more complete set of data is needed.

1. Introduction

Despite many efforts in the past, cavitation erosion still remains an unsolved problem regarding the acceptance tests on model for hydraulic turbines including Francis turbines. The different methods used today to fix acceptable tailwater level from cavitation model tests (to add a " safety margin" to the efficiency drop tailwater level, to fix a certain acceptable cavity length, to use paint erosion tests or so)do not give full satisfaction since they involve empirical experience and subjective considerations.

As this problem remains essential for sizing and design of hydraulic turbines, EDF, HydroQuébec and EPFL have conducted an important research programme on this subject, including tests on a 260 MWe Francis Turbine and its model. The programme has been previously presented [1]. The measurements of cavitation intensity on the prototype and on the model are described in Part 1 and Part 2.

E. Cabrera et al. (eds.), Hydraulic Machinery and Cavitation, 564–573.
© 1996 *Kluwer Academic Publishers. Printed in the Netherlands.*

2. Cavitation aggressiveness and model-to-prototype similitude

2.1. CAVITATION AGGRESSIVENESS

It must be reminded that cavitation erosion is the result of accumulation of impacts due to the collapse of vapour "structures" near the wall. So cavitation erosion can be split in two different mechanisms :
- first, the hydraulic mechanism: local depression in the flow generates vapour structures, growing and then collapsing, leading to minute impacts on the wall;
- second, the damage mechanism: under cavitation impacts the wall material is damaged, ending with mass removal.

The interface between these two distinct phenomena is called the "cavitation aggressiveness", which is the impact loading applied on the wall by successive collapses. In a first approach, it can be set that cavitation aggressiveness is the pure consequence of the first mechanism and the input of the second.

So the cavitation aggressiveness can be defined as a set of impacts striking the wall, each impact being characterized by its pressure P_i, its duration t_i, and the characteristic size L_i of the surface stroked. Although the mechanisms of collapse are not yet fully elucidated, it is admitted now that these parameters are in the range of gigaPascal for P_i, microsecond for t_i, and 0.1-1 mm for L_i. Up to now, no on-line instrumentation is able to measure cavitation aggressiveness, and the methods proposed hereafter are based on indirect measurements.

2.2. MODEL TO PROTOTYPE SIMILITUDE

Here we remind only the main features of the reasoning leading to scaling laws. More argued statements of these laws can be found in [2]. Between a prototype (reference size: L_{proto} , speed : N_{proto}) and its model (reference size: L_{model} , speed : N_{model}), the hydraulic mechanism can be theoretically ruled by scaling relationships:
- The extend of cavitation area: the well known Sigma similitude between model and prototype theoretically ensures that similar areas are under vapour pressure;
- The size of vapour transient cavities: since cavitation extent is similar, the size of cavities is also similar, i.e. proportional to the scale;
- The rate of production of vapour cavities: if they are produced by a pocket attached on the wall, or by a vortex shedding, their rate of production follows a constant Strouhal number. If the vapour structures are generated by nuclei, this rate will be proportional to the number of nuclei per volume unit;
- The potential power of vapour structures: it is the product of their potential energy (i.e. the volume multiplied by the pressure that forces the collapse minus vapour pressure) by their rate of production. Since size, pressure and rate can be scaled, this also can be scaled.
- The cavitation impacts on the wall: the impact being considered as a shock, the impact pressure P_i due to the collapse of a vapour cavity can be scaled by $\rho c V$

(ρ = water density, c = sound velocity in water, V = reference flow velocity), while the impact size L_i and the time t_i can be scaled by the size of the vapour cavity i.e. by the homology factor.

With these considerations in mind, aggressiveness can be transposed from model to prototype.

On the contrary, the loss of mass under cavitation aggressiveness cannot be ruled by similitude laws. Up to now, too little is known on the processes involved in material damage to lay down a simple modelisation.Experimental data are still needed. These may be provided by devices such as water jet [4] or vortex generator [8], or others.

3. Prediction methods and results

3.1. PREDICTION USING PITTING MEASUREMENTS

3.1.1 *Requirements for pitting tests*
This method is based on the analysis of cavitation pitting obtained after a short duration test (less than an hour) on polished samples made of soft metal (pureCu, pure Al, Ag or so) mounted on the model. The test is run in hydraulic similitude with investigated prototype operating conditions, including cavitation similitude.

To get the highest cavitation aggressiveness, it is important to reach the highest head possible on the model platform. This will be favourable for transposition, to get a sufficient pitting, and to diminish extraneous effects such as damping due to air content, and viscous effects. It is often impossible to reach the same head on the model as on the prototype, but a head of about one half to one quarter the prototype one seems to be a good range. Air content in the water of the loop has to be reduced as low as possible to avoid any damping effects on the collapse.

Preliminary tests are required to fix test duration that must be long enough to get a representative pitting , but not too long to avoid excessive overlapping of pits.

3.1.2. *Techniques and apparatus*
The easiest way to put soft polished metal in the erosion area is to use samples mounted in the blades. If this technique has been successful on the prototype, unfortunately, on the model, preliminary tests with pure aluminium samples shows undesirable cavitation pitting behind small discontinuities (less than 0.2 mm height) between sample and blade. Electrolytic deposition of metal directly on the blades is found safer but it rules out the use of Al (poor quality of deposits) and Cu (variable hardness of deposits). In this case, silver has been found to be suitable.

As for pitting measurement, we used a UBM laser profilometer that provides a mapping of the surface. After levelling, pitting can be extracted by a software named "Adresse" that determines the values of pit depth h_i, pit radius R_i, and pit volume V_i for each pit. This process is detailed in [3]. This software calculates also the pressures and sizes, and perfoms the transposition from model to prototypeaccording to the laws set out hereafter.

3.1.3. Transposition procedure

After a test duration T, pitting obtained is measured on a surface $S_{a\,mod}$ and provides a set of pits $(h_i, R_i)_{mod}$, each one being characterized by its depth h_i and radius R_i. From the pits geometry and the material properties (simple shear stress S_0, elastic Young modulus E, Poisson modulus v, sound velocity C_s), pressure Pi and size Li of the corresponding impact can be deduced using the following relationship obtained by analysis of elastoplastic deformation [5] [6]:

$$\frac{R_i}{L_i} = f_1(\frac{C_1}{L_i} \cdot dt) \qquad and \qquad \frac{h_i}{L_i} = \frac{S_0}{E} \cdot \left(\frac{P_i}{S_0}\right)^m \cdot A^{1-m}$$

where $A = f_2(\frac{C_1}{L_i} \cdot dt, v)$, $m = f_3(\frac{C_s}{L_i} \cdot dt, \frac{C_1}{L_i} \cdot dt)$, C_l is the sound velocity in the water and dt is the time duration of the impact. f_1, f_2 and f_3 are determined by numerical simulation of plastic deformation, $C_s dt / L_i$ and $C_l dt / L_i$ can be assumed to be constant [7]. So the relationship can be written in the following form :

$$(P_i, L_i) = f_{mat}(h_i, R_i)$$

where f_{mat} depends only of the material.
This provides the set of impacts $(P_i, L_i)_{mod}$ obtained on the model.

Given :
- the scale ratio $\lambda = L_{proto} / L_{mod}$, and
- the speed ratio $n = N_{proto} / N_{mod}$

of the model, this set $(P_i, L_i)_{mod}$ is then transposed to prototype $(P_i, L_i)_{proto}$ by means of the following scaling laws:

$$P_{i\,proto} = P_{i\,mod} \cdot \lambda \cdot n \quad and \quad L_{i\,proto} = L_{i\,mod} \cdot \lambda$$

This deduced aggressiveness $(P_i, L_i)_{proto}$ is supposed to be applied on a surface:

$$S_{a\,proto} = S_{a\,mod} \cdot \lambda^2 \quad during\ a\ time\ T_{proto} = T_{mod} / n$$

These relations are deduced from the similitude considerations mentionned above, with the hypothesis of constant Strouhal number for production of cavities.
With the inverse relationship f_{mat}^{-1} between pressure pulse and pit, one can calculate pitting $(h_i, R_i)_{proto}$ that would be produced by this aggressiveness $(P_i, L_i)_{proto}$ on the prototype, on the prototype material. A volume pitting rate can be inferred:

$$V_d = \Sigma V_i /(S_{a\,proto} \cdot T_{proto})$$

where V_i is the volume of pit i ($V_i = f(h_i, R_i)$).
Finally, erosion on the prototype is deduced using correlations between pitting and mass loss obtained in laboratory (high pressure cavitating jet or venturi) for a range of impulse pressures and materials:

$$Er = f(V_d, material)$$

3.1.4. Pitting Results

The pitting obtained on the prototype is described in Part I and in a previous paper[4]. What must be noticed here is the very large size of the larger pits: up to 2 mm diameter despite the hardness of the material (316L stainless steel). This confirms the assumption on scale of pits proportional to the size of the machine.

2 mm ━━━   0.2 mm   ━━━

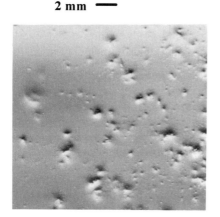   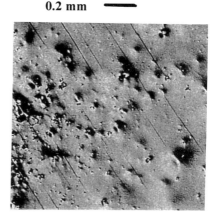

Figure1: typical pitting on the prototype   Figure 2 : pitting obtained on the model
on 316L SS (130 HB)   on pure aluminium (16 HB)

On the model, preliminary tests have been carried out with aluminium samples flush mounted on two blades. After a short duration, a good pitting was obtain, even if on some samples parasite cavitation was observed. This can be seen in figure 2.

Nevertheless, these preliminary tests are not strictly comparable to the prototype tests since, as explained in Part 2, the model is not exactly representative and the cavitation behaviour is different from the prototype, what forbid direct comparison.

3.1.5. Tentative transposition

Despite this fact, an interesting exercise can be carried out using the prototype results : from measurements on stainless steel on sample 2 (one of the more severely damaged, $\sigma = 0.120$, $\phi = 0.27$, vane opening 78%) one can use the inverse procedure to predict what is expected on the model. The test lasted $T_{proto} = 30$ mn. The sample 2 was analysed on $S_{proto} = 100$ mm^2 and gives the following results:
- number of pits : 193
- total volume of pits: $V_{tot} = \Sigma V_i = 4.35\ 10^8\ \mu m^3$

From the pits set (h_i, R_i, V_i) the following values are calculated:
- mean weighted depth: $h_v = \Sigma(h_i.V_i)/\Sigma V_i$
- mean weighted radius: $R_v = \Sigma(R_i.V_i)/\Sigma V_i$
- volume pitting rate: $V_d = V_{tot}/(S_{proto}.T_{proto})$

These values are listed in table 1, as for the successive steps of the transposition.

Then, the mechanical characteristics of 316 L stainless steel beeing $S_0 = 400$ MPa, $E = 200$ GPa, $C_s = 5800$ m/s, the inferred pressures P_i and size L_i of impacts are calculated. They range from 1.9 to 4.5 Gpa on the prototype. Transposition is then applied. Since the model scale is $\lambda = 14.66$ and $n = 0.1125$, pressures are divided by 1.65, while sizes are divided by 14.66. The corresponding surface and duration are:

$$S_{mod} = S_{proto}/\lambda^2 = 0.465 \text{ mm}^2 \quad \text{and} \quad T_{mod} = T_{proto} \cdot n = 3.36 \text{ mn}.$$

Once aggressiveness is transposed, one can calculate the pitting on the model, for a given material with the software "Adresse". On the same material (316L SS), the pitting obtained is described in column 2 of Table 1. What is significant is that, while the radius R_v decreases almost proportionally to the scale, the depth is divided by more than four time the scale. Most of such pits would be hard to measure. The weakness of the aggressiveness on the model justifies the use of a softer material.

The same aggressiveness (set (P_i, L_i)) applied on pure Al ($S_0 = 100$ MPa, $E = 50$ GPa, $C_s = 5000$ m/s) leads to the pitting described in column 3 of Table 1. Here, the pitting is very important since many pits are deeper than large. This seems to be unrealistic. It can be compared with the pitting obtained during one of the preliminary tests on the model ($\sigma = 0.07$, $\phi = 0.27$, vane opening 78%)). After a test duration of 15 mn, a 9 mm^2 surface of one of the most eroded samples has been analysed and the results are given in column 4 of Table 1. The pitting is much lower, pits are less deep and larger than expected but the similitude conditions are not respected (specially σ similitude), so that no conclusion can be drawn up to now. More experimentation is currently carried out, using silver coating of the blades. At the present time, the results show clearly that pitting measurements on model can be obtained, and that transposition has to be adjusted.

Table 1 : Results of transposition from prototype pitting tests

	Prototype tests on 316L SS	Transposed to the model on 316L SS	Transposed to the model on Al 59	Model test on Al 59
Duration (mn)	30	3.36	3.36	15
Surface (mm^2)	100	0.465	0.465	9
hv (μm)	22.2	0.207	36	1.9
Rv (mm)	0.66	0.045	0.045	0.17
V_d (μm/s)	$1.29 \ 10^{-3}$	$1.07 \ 10^{-4}$	$2.05 \ 10^{-2}$	$6.2 \ 10^{-4}$
(mm/hour)	$4.66 \ 10^{-3}$	$3.87 10^{-4}$	0.074	$0.223 \ 10^{-2}$

3.2. PREDICTION USING VIBRATION ANALYSIS

This method aims to measure the amplitude of shock forces applied on the blades by cavitation. Since direct measurement is not possible, a method is proposed to infer these forces from the signal of an accelerometer mounted outside the machine. To avoid extraneous noises from mechanical phenomena or flow, forces are calculated in a high frequency range. The transfer function from the blades where cavitation occurs to the measurement spot can be determined by two different means : spark generator and instrumented impact hammer. From energy considerations, it can be said that inferred forces F_{mod} obtained on the model can then be transposed to the prototype using the following formula:

$$F_{proto} = f(\lambda,n) \cdot F_{mod}$$

where λ is the scale ratio between prototype and model, and **n** is the speed ratio.

That such a relation or another may exist rests on the following considerations. Developed forces for similar flows on model and prototype should scale with the energy of cavities, i.e. their volume. The rate of collapse of these cavities on the runner must also be considered as this parameter may vary between model and prototype and influences the inferred forces values.

With the results of both campaigns in hand, we identify scaling law on the basis of the calibrated values of the mean square inferred high frequency forces. From the knowledge of the damaged areas on the prototype, the inferred forces values can be converted to erosion rates using erosion data obtained on a laboratory jet.

Transmissibility functions (ratio of response acceleration power spectrum to input force spectrum) were measured between the lower guide bearing and the eroded a reason the blades by a reciprocity instrumented hammer impact method. An average response was established by measuring the response of 5 blades on the prototype and 4 on the model with the runner under water in both cases. The 0-25 and 0 -100 kHz frequency ranges were utilized respectively on the prototype and on the model. The useable frequency ranges where adequate coherence (>.8) between the response and the excitation were realized were 0-16kHz on the prototype and 0-80kHz on the model. The need to measure the transmissibility function with the runners in water was confirmed by the significant differences between values measured with the runners in air or in water. On the prototype, smaller values were seen with the runner in water, as anticipated, due to the coupling of vibration energy from the runner blades to the water while the opposite was true on the model where increased coupling from the lower guide bearing to the runner is achieved with the presence of water in the runner crown to bearing gap.

Forces were inferred on the prototype in the .8 to 11.296 kHz as had been done but this time with the transmissibility function measured with the runner underwater. This produces substantially higher force levels than estimated in the previous measurement campaign. On the model, the 20 to 35 kHz range was utilized because of interfering high frequency vibration at the lower guide bearing generated by a 90° 1 to 1 gearbox linking the model shaft to the generator. This underestimates the real force values by a factor of at least 2 or 4. The presence of these undesirable vibrations is inconsistent with reciprocity measurement approach and will be eliminated in the future tests. The results of the prototype inferred force calculations are summarized in figure 3 with maximum values observed on the prototype at 90% guide vane opening with high downstream levels. Figure 4 shows the correlation with volume pitting rate. The highest level on the model is also obtained at 90%

guide vane opening but the inferred forces per unit area on the model are four to five times much weaker than on the prototype.

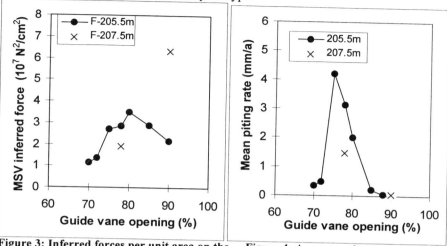

Figure 3: Inferred forces per unit area on the proto (lower guide bearing acceleration)

Figure 4: Average volume pitting rate on 4 disks for same tests on the proto.

Cavitation characteristics on the model however appear to differ slightly from those observed on the prototype. The high frequency acceleration amplitude modulation spectra at the lower guide bearing and in the crown of the runner show that impacts occur at the guide vane passing frequency as on the prototype but modulation components of similar amplitude are also present at the blade passing frequency. This confirms that flow conditions are not fully homologous to those of the prototype as discussed in Part 2.

3.3. PREDICTION USING ELECTROCHEMICAL PROBE DECER

This method based on electrochemical effects during cavitation erosion can give a good representation of cavitation aggressiveness through the activation current delivered by the titanium electrochemical probe which is directly proportional to the actual erosion rate [8]. This localized on-line measurement on both prototype and model is a complement to pitting results to provide the influence of operating parameters on cavitation aggressiveness. On the prototype, this method gives similar results than the others methods, with maximum erosion near optimum opening The maximum current measured is in the range of 3 μA for a 7.5cm^2 of a grade 2 Titanium. (This corresponds to 20 mm/year erosion on Ti grade 2 according to laboratory jet tests [8]). On the model, the maximum current level detected on Ti49 is 0.05μA on a surface of 0.32 cm^2. This is very low and correspond to 0.16 mm/year of Titanium erosion (according to laboratory jet tests [8]). Since the resistance of Ti 49 is ten times lower than Ti grade 2, it means that cavitation erosion found on the model seems to be about 1000 less than on the prototype.

3.4. PREDICTION USING CAVITATION EROSIVE POWER

It has been set previously (Avellan *et al*) that the cavitation erosive power could be evaluated from energy dissipated by collapses of transient cavities and their production rate. This power can be expressed as :

$$P_{er} = \frac{1}{2} K \cdot \rho \cdot \left(c_{p\,max} + \sigma \right) \cdot C_{ref}^3 \cdot St \cdot \frac{V_c}{L_c}$$

where $c_{p\,max}$ is the maximum pressure coefficient in the recovering pressure region downstream the main cavity, σ the cavitation coefficient, C_{ref} a velocity reference, St the Strouhal number corresponding to the production rate of the transient cavities, L_c the main cavity length, V_c the transient cavities volume and K a scaling constant. This parameter can be calculated on the model using visualisation to determine V_c, L_c and St, and transposed from the model to the prototype.
Hydraulic similitude and sigma similitude will ensure that $c_{p\,max}$, σ and St are the same on the model and on the prototype. Then it can be assumed that L_c is proportional to the reference length L, and that $C_{ref} \propto N \cdot L$ (N : rotational speed)
As for V_c, two hypothesis can be proposed:

- if it is assumed that V_c is proportional to L^3, then the ratio between erosive power between model and prototype can be written:

$$\frac{P_{proto}}{P_{model}} = \left(\frac{C_{ref\,proto}}{C_{ref\,model}} \right)^3 \cdot \frac{V_{c\,proto}}{V_{c\,model}} \cdot \frac{L_{c\,mode}}{L_{c\,protol}} = \left(\frac{N_{proto}}{N_{model}} \cdot \frac{L_{proto}}{L_{model}} \right)^3 \cdot \left(\frac{L_{proto}}{L_{model}} \right)^3 \cdot \left(\frac{L_{proto}}{L_{model}} \right)^{-1}$$

$$r = \frac{P_{proto}}{P_{model}} = \left(\frac{N_{proto}}{N_{model}} \right)^3 \cdot \left(\frac{L_{proto}}{L_{model}} \right)^5 = n^3 \cdot \lambda^5$$

(notice: this is equal to the classical ratio on machine powers)
Since $n = 0.1125$ and $\lambda = 14.66$ then in this case: $r = 964$

- if the following law, established in a previous statistical study [10] is applied:

$$V_{cr} = k \cdot L_c^{2.6} \cdot C_{ref}^{-0.4}$$

then the ratio becomes:

$$r = \frac{P_{proto}}{P_{model}} = \left(\frac{N_{proto}}{N_{model}} \right)^{2.6} \cdot \left(\frac{L_{proto}}{L_{model}} \right)^{4.2} = n^{2.6} \cdot \lambda^{4.2}$$

then in this case: $r = 269$

These two results emphasises clearly that, even if some uncertainty remains on the transposition laws, the cavitation intensity is much higher on the prototype than on the model. Then, to predict erosion on the prototype from transposed power, experimental correlations from reference data have to be used.

4. Conclusions

The proposed procedures to predict cavitation erosion from model tests take advantage from a campaign on a large Francis Turbine and its model.
Particular methods are set from each instrumentation type: pitting, vibration, electrochemical probe and erosive power. Results allow to confront them to actual data and to assess the hypothesis on transposition of cavitation aggressiveness. At the present time, since the programme is not completed, only partial conclusions can be drawn. Obviously the four methods proposed are able to quantify cavitation aggressiveness on the model. They all clearly show that cavitation aggressiveness is much smaller on the model than on the prototype, and this is due to less energetic collapses, due to the reduction of both pressures and sizes of impacts.
As for a definitive assessment of transposition and prediction processes, the completion of the programme is needed for a thorough synthesis.

References:

[1] E. Laperrousaz et al, 1994 "Prediction cavitation erosion in Francis turbines on the basis of scale model testing", *Proc. 17th IARH Symposium*, Beijing, China.
[2] Y. Lecoffre, P. Grison;, J.M. Michel , 1986 "Prevision de l'érosion de cavitation pour les turbomachines " *Proc. 14th IARH Symposium*, Montreal, Canada.
[3] J.M. Dorey, R. Simoneau ,P. Bourdon, M. Farhat, F. Avellan 1994 "Quantification of cavitation aggressiveness in three different dvices using accelerometer, DECER, and pitting measurements" , *Proc.Second International Symposium on Cavitation*, Tokyo, Japan.
[4] P. Bourdon, R. Simoneau, J.M. Dorey, 1994 "Accelerometer and pit counting detection of cavitation erosion on a laboratory jet and a large Francis tu rbine" *Proc. 17th IARH Symposium*, Beijing, China.
[5] R. Fortes-Patella, J.L. Reboud, J.M. Dorey, 1991 "Simulation of cavitation impact damageon an elastoplastic solid", *Proc. ASME Cavitation and Multiphase Flow Forum*, Portland, USA.
[6] R. Fortes-Patella ,J.L. Reboud , 1992 "Analysis of cavitation erosion by numerical investigation of solid damage" *Proc. 16th IARH Symposium* , Sao-Paulo, Brasil.
[7] R. Fortes-Patella ,J.L. Reboud, 1995 "A new approach to evaluate cavitation erosion power" *Proc. International Symposium on Cavitation*, Deauville, France.
[8] R. Simoneau ,P. Bourdon, M. Farhat, F. Avellan, J.M. Dorey, 1993 "Cavitation erosion, impact intensity and pit size distribution of jet and vortex cavitation", *Proc. ASME annual winter meeting, Bubble noise and cavitation erosion in fluid systems* , New Orleans, USA.
[9] R. Simoneau, 1995 "Cavitation pit counting and steady state erosion rate" *Proc. International Symposium on Cavitation*, Deauville, France.
[10] F. Pereira, Ph. Dupont, F. Avellan, 1995 "A statistical approach to the study of transient erosive cavities on a 2D profile" *Proc. ASME annual meeting* , South Carolina, USA.

Cavitation Erosion Prediction on Francis Turbines. Part 2 Model Tests and Flow Analysis

PH. DUPONT, J.-F. CARON, F. AVELLAN
EPFL, IMHEF/LMH
33 Av. de Cour, CH-1007 Lausanne, Switzerland

P. BOURDON, P. LAVIGNE, M. FARHAT, R. SIMONEAU
Hydro-Québec
1800 Boul. Lionel Boulet, Varennes, Québec, Canada J3X 1S1

J.-M. DOREY, A. ARCHER
Electricité de France, DER
6, quai Watier, F-78401 Chatou, France

E. LAPERROUSAZ
Electricité de France, CNEH
Savoie-Technolac, B.P. 26, F-73370 Le Bourget-le-Lac, France

M. COUSTON
GEC-Alsthom, Neyrpic
75, rue Général-Mangin, B.P. 75, F-38041 Grenoble, France

ABSTRACT Different measurement techniques have been used to detect cavitation on a Francis turbine model. The results are compared to those obtained on the prototype and presented in the first of this series of articles. The runner model used for that study is build on the basis of a geometrical recovery of one of most eroded blade of the prototype. The results of the different measurements are presented and commented by comparison with prototype measurements. This comparison leads to a proposal of the physics which should be involved in transposition laws for the prediction of prototype erosion from cavitation model tests. The consequences of such scaling laws, as well as their application to the prototype and model results, are part of the third facet of this work.

1. INTRODUCTION

The problem of cavitation erosion on Francis turbines leads IMHEF, Hydro-Québec, Electricité de France (EDF) and GEC Alsthom-Neyrpic to be involved together in a large and ambitious research program on this topic. Preliminary measurements using various techniques have been performed on the prototype of a 266 MW Francis turbine and are fully described in Part I of this work.

The present paper is concerned with the model tests performed at IMHEF. A runner model is manufactured using the data provided by the geometry recovery of one of the prototype blades [1]. The model measurement campaign, similar to the prototype one, is

574

E. Cabrera et al. (eds.), Hydraulic Machinery and Cavitation, 574–583.
© 1996 *Kluwer Academic Publishers. Printed in the Netherlands.*

described. As for the prototype measurements, the model runner is equipped with different transducers in order to follow the cavitation development. The results of hydraulic characteristics, vibratory monitoring, pit counting, pressure pulses and electro-chemical probes measurements are reported. Additionally, a flow analysis is presented using both Euler and Navier-Stokes calculation codes.

One of the goal of the present study is to define, on the one hand, the best way to perform cavitation tests on model to be as close as possible to the cavitation behavior of the prototype, and one the second hand, the transposition laws to apply to predict cavitation erosion on prototype from the model tests and the way to measure it. The measurements show important differences between model and prototype cavitation development. These differences are commented and analyzed. This leads to propose possible explanations about the physics involved in these scale effects.

In the third part of this series of articles, the similitude and the differences between results of both campaigns are used to give recommendations for the prediction of cavitation erosion from model tests.

2. GEOMETRY RECOVERY AND DESIGN OF THE MODEL RUNNER

In order to reproduce the cavitation pattern of the prototype on the model, one of the most eroded blades, namely blade number four, as well as the blade to blade channel corresponding to its suction side have been measured on site. Three techniques were used and compared with each other [1]: the classical template technique, a 3D laser interferometer and a portable 3D digitizer. Because of slight discrepancies between the blade and the blade to blade channel results, it was decided to manufacture the model upon the geometric data of the selected blade. This choice would have lead to bigger outlet openings on the model runner then the measured ones on the prototype. It has been so decided to slightly extend the model blades in order to keep these openings of the runner identical to the prototype. From this result, a model of the runner with a 300 mm diameter has been machined using a five axes CNC machine by GEC Alsthom-Neyrpic. To control the possible differences of the flow between model and prototype, flow calculations on the two geometries have been performed.

3. FLOW CALCULATION

The analysis of the flow in the model and prototype runner is done using both Euler and Navier-Stokes calculation codes. The geometry recovery of blade four is first used to generate the meshing of the prototype runner. The operating condition followed corresponds to the best efficiency point. The comparison of the results of the different calculation codes used with this prototype geometry is given in left part of Figure 1.

This flow analysis shows that the cavitation development seems to be due to a local bulge of the blade, close to the band and at the third of the blade chord length. This local bulge creates a local underpressure located just upstream of the eroded area of the prototype. Moreover, changing the blade to band fillet size of the tested model runner at suction side has shown to change the cavitation inception. It is so suspected that both corner flow along the blade to band fillet together with this local underpressure is responsible for the unexpected erosion observed on the prototype. The influence of the model blade elongation is investigated comparing the pressure distribution calculated for both prototype and model blade geometry. The right part of Figure 1 shows the elongation to slightly increase the pressure level at runner outlet close to the band.

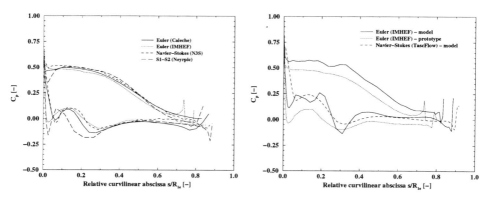

Figure 1:   Pressure coefficient distribution along the blade close to the band.
             On left: Prototype blade #4 geometry - Comparison of different codes
             On right: Comparison between prototype and model geometry results

4. MODEL TESTS SET-UP

The model tests are performed on the high performance research test rig of IMHEF (PF III). This closed loop facility allows to reach flow rates up to 1.5 cubic meters per second, with a maximum net head H_n of 60 meters. A first series of tests are performed using a set of blades without any instrumentation to compare the hydraulic performances of the model with the prototype measurements.

After this first step, 6 of the 13 blades are replaced with instrumented ones. Two blades (2 & 8) are equipped each with five miniature static and dynamic pressure transducers, with a useful 15 to 20 kHz upper frequency determined by a preliminary dynamic calibration. These sensors allow to measure the pressure fluctuations acting on the blade due to cavitation. Two other blades (4 & 10) are instrumented with a total of five DECER electro-chemical cavitation erosion probes, which provide a real time monitoring of the erosion rate. Finally, two blades (6 & 12) are equipped with samples of soft material, namely pure copper, for the pit counting experiment. The locations of the first four transducers (pressure, electro-chemical or samples), are identical to those used on the prototype. Moreover, due to the extra length of the model blades, a fifth location is available downstream of the others. These locations correspond to the erosion area found on the prototype, close to the trailing edge of the blades and near to the band to blade fillet. To prevent any disturbance of the flow due to the mounting of the transducers, it was decided to alternate instrumented and non instrumented blades.

Figure 2:   Conditioning electronics

The hub of the runner is especially designed to allow a sealed volume large enough to install the conditioning electronics connected to the onboard transducers (See Figure 2).

This electronics has an amplifier with a remotely variable gain and a multiplexer that allows the scanning of all the transducers, with two simultaneous outputs. Voltage signals are converted to frequency to avoid electromagnetic disturbances. They are brought out from the rotating part through a slip ring collector and finally reverted to voltage signals which are digitized on high speed waveform recorders. In this same sealed volume is placed an accelerometer, fixed to the runner blade ring. This accelerometer replaces for the model tests the one placed on the blade for the prototype tests.

As for the prototype campaign, in addition to the onboard sensors, four accelerometers are used to monitor the cavitation impacts at the lower guide bearing; another one is fixed on the upstream flange of the draft tube cone. The angular position of the last accelerometer is perpendicular to the meridian plane including a dynamic pressure transducer, fixed in the wall of the draft tube cone. The data acquisition systems are essentially the same as those utilized in the prototype program described in part 1 of this series of papers with the addition of a Metrum RSR 512 16 channel digital data recorder.

5. TEST CONDITIONS

Two operating points are followed during the model tests, corresponding to 78% and 90% of the maximum guide vanes opening. To ensure identical runner inlet flow angles on both model and prototype, model guide vanes opening and discharge coefficient $\varphi_{\overline{1}e}$ are chosen to be equal to the prototype campaign values. Two different model heads, 13 and 51 meters, corresponding respectively to rotational speeds of 800 and 1600 RPM, are investigated to check the influence of the head on the cavitation development. For these two heads, sigma coefficient investigations are done to detect cavitation inception. This inception is determined according to the first pressure peaks detection, corresponding to the first cavity collapses.

Table 1:   Operating conditions for the model tests.

TEST	MA01	MAM1	MAM2	MAM3
Guide vanes opening [%]	78	78	78	90
$\sigma_{\overline{1}e}$ [-]	0.20	0.08	0.07	0.10

To check the influence of the water nuclei content on the cavitation development, tests were done with and without nuclei injection, using the IMHEF nuclei injection system [2] . If overall cavitation performances are affected by this nuclei content, no influence has been observed on the specific cavitation pattern developing along the blade to band fillet. These specific results are presented in an other paper [3]. To avoid any water quality effects, all measurements are performed after a two hours degassing period.

The operating conditions used for the pit counting and vibrations tests, given on Table 1, are similar to the prototype campaign. The sigma values on model are chosen to obtain cavity development in agreement with erosion observed on prototype.

6. MEASUREMENTS RESULTS

6.1 HYDRAULIC CHARACTERISTICS

6.1.1 Efficiency tests
The model results obtained compares very well with those previously obtained with the original model tested by Neyrpic. These hydraulic characteristics, transposed at prototype scale, corresponds quite well to the on site measurements even if two local maximum efficiency are observed on the latest.

6.1.2 Cavitation tests
The σ−η cavitation curves have been performed for the model, and compared to the prototype cavitation behavior. Figure 3 shows the model σ−η cavitation curve for a 90% guide vanes opening, and the two sigma values corresponding to the downstream levels achieved during the prototype campaign. The cavitation development on the model corresponds to the location of the erosion on the prototype and confirms the local underpressure revealed by the calculation. This seems to indicate that, even if the model is built on the basis of only one blade geometry recovery, the cavitation behavior of the prototype is well reproduced. On the other hand, the

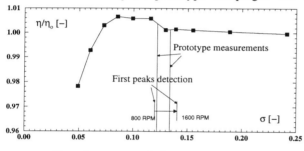

Figure 3:   σ−η cavitation curve at α=90%.

comparison of the cavitation index for a given cavity extent, observed on the model and estimated on the prototype from the pitting tests, shows an important deviation. For an equivalent cavitation extent, the corresponding cavitation index is lower on the model than on the prototype. This last point needs a deeper analysis, in the light of a local cavitation coefficient notion presented here latter. It has also been noticed a influence of the fillet size between suction side blade and band on the cavitation inception. Decreasing the size of these fillets, cavitation inception index is notably increased.

6.2 VIBRATORY MONITORING

6.2.1 Dynamic calibration
To identify an average transmissibility function (ratio of response acceleration to input force autospectra) between the blades and 2 radialy mounted accelerometers at the lower guide bearing, reciprocity instrumented hammer impact techniques [4] are applied between each of these two monitoring points to excite 5 Kistler miniature

accelerometers mounted on runner blades. Each measurement sequence consists of 32 radialy applied hammer impacts at the guide bearing. A frequency range of 100 kHz with 64 Hz of resolution is used. The quality of the measurements is excellent as indicated by the coherence function between the input force spectrum and the underwater blade response acceleration spectrum shown in Figure 4. The transmissibility function is then used in a reciprocity mode to infer forces on the blades from the accelerations measured at the lower guide bearing during the tests.

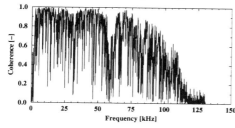

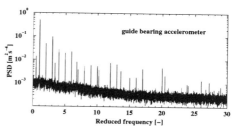

Figure 4:   Coherence function between input force and underwater blade response acceleration spectrums.

Figure 5:   Power spectral density of the 20-35 kHz accelerations signal envelope at the guide bearing.

A 20 to 35 kHz frequency range is utilized to avoid high frequency interference caused by a 90 degree translating gearbox and to minimize artificial inferred force amplitude caused by the lower resonance frequency of the guide bearing accelerometers in comparison to that of those used on the blades.

6.2.2 Measurements

In the test cases, 128 blocs of 4096 points are acquired with Hanning weighting and their autospectra averaged to minimize statistical measurement errors on the random data. So, a 90% confidence interval ranging better than +/- 1 dB is achieved at each frequency. Forces acting on the blades are inferred from these autospectra with the transmissibility functions. Forces inferred from the lower guide bearing monitoring accelerometer in the 20 to 35 kHz frequency range are listed in the following table.

Table 2: Inferred forces

TEST	MA01	MAM1	MAM2	MAM2	MAM3
Inferred forces (N^2)	533.6	2130.7	2101.3	1843.8	5222.9

High frequency acceleration amplitude modulation studies are also performed on data recorded digitally on the Metrum data recorder. The modulation of the 20-35 kHz acceleration at the different monitoring points is studied by filtering the data with a band pass elliptical filter prior to the envelope detection.

Power spectral densities with Hanning weighting are evaluated with a frequency resolution of 0.0625 Hz. The frequency axis is normalized to orders of rotation. The high frequency vibration envelopes are seen to be mainly modulated by two sets of frequencies centered around the guide vane passing frequency 20X and the blade passing frequency 13X. The power at these frequencies +/- 1 and 2 times the rotation frequency are extracted. The power is integrated over a 0.4 Hz bandwidth around each of these frequencies. An example of these PSD is given in Figure 5 showing the forcing effects of these frequencies on the cavitation dynamics.

6.3 PIT COUNTING MEASUREMENTS

Three pitting tests (MAM1-2-3) have been performed at the highest head, to ensure the highest pitting rate. These tests are done using Al 59 samples mounted on two different blades. The pitting results are shown on Figure 6 for MAM3 test and blade #6. The exact flush mounting of the samples is rather difficult due to the blades size. A misalignment has induced side effects corresponding to highest local pitting related to the leading edge of samples, as shown on Figure 6.

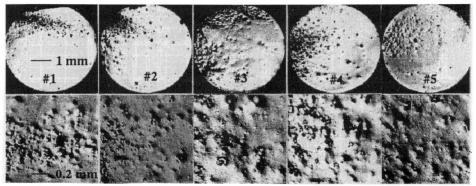

Figure 6:   Pitting results on blade #6 for test MAM3 (Flow direction from left to right)

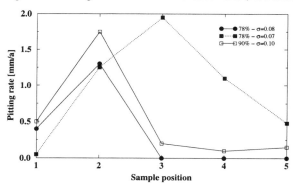

Figure 7:   Volume pitting rate for MAM1-2-3 tests using Al 59 samples.

In spite of that, the mean radius of the biggest pits is 0.2 mm, which corresponds to about the tenth of the pits obtained on prototype. This is coherent with the geometric scale ratio equal to 14.67. The volume pitting rates are compared for the three tests on Figure 7. As for the prototype measurements, the maximum pitting is moving closer to the trailing edge of the blade when decreasing the sigma value. As well, a equivalent pitting rate is obtained at the same location and for a highest sigma value for a guide vane opening of 90% compared with 78%.

6.4 PRESSURE PULSES MEASUREMENT

Pressure fluctuations due to cavitation development are measured on two blades on five locations, indicated by dots on Figure 8. These unsteady pressure transducers are able to detect high frequency pressure variations as shock waves due to cavity collapses. This measurement technique is shown to be very efficient to detect the beginning of cavitation. On Figure 8 the incipient cavitation is detect at an opening of 90% for a σ value of 0.132. Using visual observations, this limit is set for a σ value of 0.120.

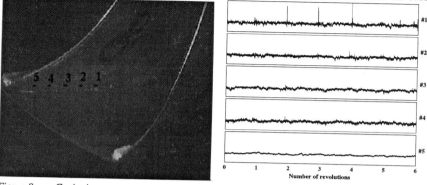

Figure 8:   Cavitation pattern visualization and instationnary pressure time signals at five different locations on blade #8 (Operating conditions: α=90%, σ=0.132, H$_n$=50 m).

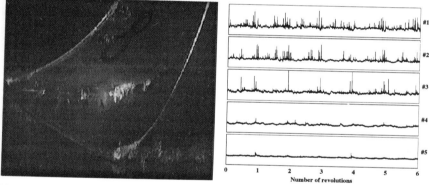

Figure 9:   Cavitation pattern visualization and instationnary pressure time signals at five different locations on blade #8 (Operating conditions: α=90%, σ=0.100, H$_n$=50 m).

For developed cavitation, pressure measurements gives good information about the modulation of the collapses. On Figure 9 for a σ value of 0.100 at 90% of guide vane opening, a displacement of the collapses closer to the trailing edge is observed. A larger number of small collapses occurs close to pressure transducers #1 and #2, while fewer but larger collapses occur close to pressure transducer #3. This shows the biggest vapor cavities to be convected farer downstream to the local minimum pressure area corresponding to there generation point. It can be observed on Figure 9 the strong modulation of the biggest peaks at each revolution. This will lead for a static observer to vibrations frequency modulation at blade passage frequency as shown by the vibratory monitoring. On the contrary, the guide vane passage frequency is more difficult to detect on pressure fluctuations, may be due to the more intermittent characteristics of this modulation.

6.5 ELECTRO-CHEMICAL PROBES MEASUREMENTS

Because of the small size of the electro-chemical probes and the weakness of the cavity collapses on the model, most of the followed operating points are under the detection threshold of this technique. Only the most erosive situations lead to measurable values corresponding to erosion rates more then three orders of magnitude weaker then those observed on the prototype.

7. ANALYSIS

In order to compare the cavitation development on the model and on the prototype, a local cavitation factor χ_E is used instead of the usual Thoma number σ. This local cavitation factor is defined as:

$$\chi_E = \frac{p_{\bar{1}x} - p_v}{\rho E}$$

where ρ is the mass density, E is the energy, p_v corresponds to the vapor pressure and $p_{\bar{1}x}$ is the local static pressure at the runner outlet.

Using the well known Bernoulli equation between the runner outlet and the draft tube outlet, and taking into account the losses in the draft tube, one can written:

$$\chi_E = \sigma + \frac{1}{Fr^2} \frac{Z_{ref} - Z_{\bar{1}x}}{D_{\bar{1}}} - \frac{C_{\bar{1}x}^2}{2E} + \frac{E_{rd}}{E}$$

where σ is the Thoma number, Fr the Froude number and E_{rd}/E the relative energetic losses in the draft tube. Knowing the flow rate coefficient $\varphi_{\bar{1}}$ at the whirl free operating point (φ_o), the energy coefficient at the runner outlet $\psi_{\bar{1}}$, these relative losses can be expressed as a function of the flow coefficients:

$$\frac{E_{rd}}{E} = k_{rd} \frac{\varphi_{\bar{1}}^2}{\psi_{\bar{1}}} + k_{dcin} \left[\frac{\varphi_{\bar{1}}}{\varphi_o} - 1 \right]^2 \frac{1}{\psi_{\bar{1}}}$$

where k_{rd} is the viscous loss coefficient and k_{dcin} the residual kinetic energy loss coefficient. Using a model geometrically homologue to the prototype and investigating corresponding flow conditions, the differences in cavitation behavior between model and prototype are due to both Froude and Reynolds effects. The influence of the head on the incipient cavitation is investigated using the pressure transducers. The incipient cavitation index for a net head of 13 and 51 meters, corresponding to a rotational velocity of respectively 800 and 1600 RPM, is indicated on Figure 3. One can noticed that if, at the smallest model head, the first pressure peaks are detected on the model at a σ value of 0.120, lower then the lowest prototype measurement, the incipient cavitation at the highest model head corresponds to a σ value of 0.140, higher then the highest prototype measurement. As the influence of the Froude term is small, this difference is explained by the modification of the relative losses between the two model heads.

8. DISCUSSION AND CONCLUSIONS

The model, build on the basis of one of the most eroded prototype blade (#4), shows a cavitation development similar to the one suspected on prototype, but occurring for lower σ values. This difference is attributed to two main reasons. The first one is the geometry used to build the model. The cavitation behavior at the suction side of a given blade is not only due to its eigen geometry but also to the blade to blade channel geometry. In that meaning, the model is closer to the cavitation developing on the previous blade (#3), which is one of the less eroded blade on the prototype. The second reason is the scale effects between model and prototype. As shown by the measurements done at 800 and 1600 RPM, Reynolds effects on incipient cavitation are observed. The usage of a local cavitation coefficient leads to much better comparison, but needs to be able to evaluate the different losses on both model and prototype.

The difficulty to compare cavitation erosion on model and prototype is increased by the difficulty to evaluate it on a model. One of the goal of this study is to test different measurement techniques to be used on model to predict erosion on the prototype. The vibratory technique is shown to be well suited to follow developed cavitation, and can be used to quantify erosion rate on model. It is with no doubt the easiest technique to be applied on both model and prototype, because it does not required onboard equipment. It seems on the contrary to be not sensitive enough to detect cavitation inception. The pitting technique on soft material samples gives good information about the localization and the rate of the erosion but it is very sensitive to local geometry defaults. Pressure peaks detection using unsteady pressure transducers is shown to be efficient to detect incipient cavitation as well as the modulation of the collapses. It is on the contrary difficult to be used to quantify the erosion because it is difficult to relate the peaks amplitude to an energy, the location of the collapse being unknown. Finally, electro-chemical probes technique, which is able to give on line erosion rate on a prototype, needs to be more sensitive to be used on model.

In closing, the authors would like to thank IMHEF technical staff for their contributions to the success of this project. They are also indebted to Jean-François Combes from EDF/DER, for the Navier-Stokes (N3S) and Euler (Calèche) flow calculations.

9. REFERENCES

[1]   Dumoulin, Ch., Avellan, F., Henry, P. [et al.], « Geometry recovery of a Francis runner prototype at site », Proceedings of the 17th IAHR Symposium, section on hydraulic machinery and cavitation, 15-19 September, 1994, Beijing, China

[2]   Brand, C., Avellan, F., « The IMHEF system for cavitation nuclei injection », Proceedings of the 16th IAHR Symposium, section on hydraulic machinery and cavitation, 14-18 September, 1992, Sao Paolo, Brasil.

[3]   Arn, Ch., Dupont, Ph., Avellan, F., « Efficiency alteration of Francis turbine by travelling bubble cavitation: Experimental and theoretical study », Proceedings of the 18th IAHR Symposium, section on hydraulic machinery and cavitation, 1996, Valencia, Spain.

[4]   Bourdon, P., Simoneau, R., Avellan, F., « Erosion Vibratory Fingerprint of Leading Edge Cavitation for a NACA Profile and a Francis Model and Prototype Hydroturbine », Proceedings of the ASME Symposium on Bubble Noise and Cavitation Erosion in Fluid Systems, pp. 51-67, New-Orleans, Dec. 1993

IMPACT OF VAPOUR PRODUCTION AND CAVITY DYNAMICS ON THE ESTIMATION OF THERMAL EFFECTS IN CAVITATION

Daniel H. FRUMAN
Ecole Nationale Supérieure de Techniques Avancées
Groupe Phénomènes d'Interface (GPI)
91125 Palaiseau Cedex, France

Jean-Luc REBOUD, Benoît STUTZ
Laboratoire des Ecoulements Géophysiques et Industriels
Institut de Mécanique de Grenoble (LEGI - IMG)
B.P. 53, 38041 Grenoble Cedex 9, France

Abstract

Vapour production through cavitation extracts heat from the fluid surrounding the cavity and creates a temperature difference between liquid and vapour. This thermal effect is particularly significant in cryogenic liquids. Estimates of the temperature difference can be made provided : a) the rate of vapour production required to sustain a given cavity, and b) an appropriate model for the heat exchange at the interface of the cavity are known. The vapour production is usually estimated by assuming that it is equal to the non condensable gas flow rates necessary to sustain ventilated cavities of equal geometry. It has been shown that experimental results previously obtained can be quite satisfactorily predicted if the interface is assimilated to a rough flat plate through which the amount of heat necessary to generate the vapour is being fed. In order to obtain data for cavities developed over walls whose geometry and pressure gradient are analogous to those of turbopump inducers, and to achieve a better precision, tests were conducted in a specially designed cavitation loop operating with Freon 114. It is shown that the experimental results are well predicted by the proposed model.

1. Introduction

Vapour production through cavitation extracts heat from the fluid surrounding the cavity and creates a temperature difference between liquid and vapour. This thermal effect is particularly significant in cryogenic liquids such as those utilised in the turbopumps of the European space launcher Ariane and preferentially occurs in the inducer stage. Estimates of this thermal effect can be made provided : a) the rate of vapour production required to sustain a given cavity, and b) an appropriate model for the heat exchange at the interface of the cavity are known. The vapour production is usually estimated by assuming that it is equal to the non condensable gas flow rates necessary to sustain ventilated cavities of equal geometry. However, this globalised vapour production does not tell much about how it is distributed along the interface and if vaporisation is indeed

584

E. Cabrera et al. (eds.), Hydraulic Machinery and Cavitation, 584–593.
© 1996 *Kluwer Academic Publishers. Printed in the Netherlands.*

restricted to the interface. To predict the heat exchange at the interface an appropriate model has to be selected and its suitability checked against existing experimental results. Previous estimations (Fruman *et al.*, 1991, Fruman and Beuzelin, 1992) have shown that the experimental results obtained by Hord (1973) can be quite satisfactorily predicted if the interface is assimilated to a rough flat plate through which the amount of heat necessary to generate the vapour is being fed. For that purpose, the vapour is assumed to be uniformly distributed along the interface (rough wall) and the relative roughness has to be properly tuned.

In order to obtain data for cavities developed over walls whose geometry and pressure gradient are analogous to those of turbopump inducers, and to achieve a better precision, tests were conducted in a specially designed cavitation loop operating with Freon 114 ($C_2Cl_2F_4$), developed with the support of the Agence Française de l'Espace (CNES) and the Société Européenne de Propulsion (SEP) and operated by the Centre de Recherches et d'Essais de Machines Hydrauliques de Grenoble (CREMHyG). This investigation was coupled with a research program, conducted at GPI and LEGI and sponsored by the Agence Française de l'Espace (CNES), which consisted of determining:

- the morphology and dynamics of vapour cavities,
- the rate of growth of vapour cavities (Larrarte *et al.*, 1995),
- the void fraction and the two phase flow velocities within the vapour cavities (Stutz and Reboud, 1994),
- the air flow rate required to sustain a ventilated cavity of equal length,
- the morphology and dynamics of ventilated cavities (Larrarte *et al.*, 1995).

Nomenclature

B	non dimensional temperature difference	U_∞	free stream velocity for a flat plate or velocity on the cavity interface
B	cavity width	V_{ref}	reference velocity in the Venturi test section
C_Q	flow coefficient		
c_p	specific heat at constant pressure	V_g	mean velocity of the vapour (air) uniformly distributed over the cavity interface
C_f	flat plate friction coefficient		
L	latent heat of vaporisation		
ℓ	cavity length	x	distance along the flat plate (interface)
p_{ref}	reference upstream pressure in the Venturi test section		
P	Prandtl number	ΔT	temperature difference ($T_\infty - T_p$, $T_{ref} - T_p$)
Q	vapour production or air injection flow rate		
		ΔT_{visc}	increase of temperature due to viscous dissipation
$Re_{x,\ell}$	Reynolds number computed with the distance $x(\ell)$	α	void fraction
		ε	roughness of the flat plate (interface)
T_{ref}	reference temperature in the Venturi test section		
T_p	wall (interface) temperature	φ_p	heat flux transferred to the flat plate (interface)
T_∞	free stream temperature	ρ	liquid specific mass
u_b	velocity at the edge of the viscous boundary layer	ρ_v	vapour specific mass
		σ	cavitation number

The results have shown that, for equal length, the vapour and ventilated cavities have a different morphology and dynamic behaviour. In spite of these differences, the estimation of the mean vapour production rate can be made, with a good degree of accuracy, by using the air flow rate necessary to sustain the ventilated cavities. The above mentioned model (Fruman *et al.*, 1991, Fruman and Beuzelin, 1992) was applied to predict the thermal effect for cavities obtained in experiments carried out at CREMHyG in the Freon 114 test loop. The predicted and measured difference between the temperature of the circulating liquid and the temperature in the cavity (at the interface) compare very favourably.

2. Experimental

The Freon 114 experimental facility (called Plateforme d'Essais de Cavitation de Liquides à Effet Thermodynamique (PECLET)) is a closed loop operating with a reference pressure, obtained by pressurising, up to 35 bars, a reservoir with gas Nitrogen. To prevent gas dissolution into the circulating liquid, a flexible membrane separates the liquid and gas phases. The loop is fitted with a test section having the shape of a two dimensional Venturi, designed in such a way that the pressure distribution is analogous to the one existing on the upper surface of an inducer blade (Kueny *et al.*, 1991). The throat section of the Venturi is 44 mm width and 43.7 mm height and the velocity can be varied between 16 and 45 m/s. Temperature can be varied and controlled from 20 to 40°C. Under these experimental conditions, it is possible to achieve cavities having a maximum length of 12 cm before choking. Temperatures on the Venturi wall next to the cavity were measured using five micro-thermocouples situated at 3, 20, 38, 75 and 115 mm from the throat. One reference temperature was measured 20 cm upstream. Mean cavity interfaces were visualised with a video camera and the length determined by image processing (Kueny *et al.*, 1991).

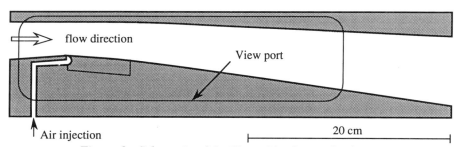

Figure 1 : Schematic of the Venturi in the test facilities.

Tests with air ventilated cavities were conducted in a facility having the same test section and operating with water and pressurised by air, Figure 1. However, maximum throat velocity was only 20 m/s and the pressure was set in order to avoid vapour production at the ventilated cavity interface. Air flow rate was measured using either a turbine flow meter or a rotameter. The same loop was used to perform vapour cavitation tests. Void fraction and phase velocity in the cavity were measured using specially designed optical probes (Stutz and Reboud, 1994, 1996). Vapour and gas flow rates were obtained by integration of the velocity profiles (Stutz, 1996). Independent tests with ventilated and vapour cavities were conducted using an hydrofoil with a sharp leading edge and water as circulating fluid (Larrarte et $al.$, 1995).

3. Proposed model

Since over the interface of the cavity the pressure gradient is nearly zero and the curvature of the interface is small we assume that the heat exchange is analogous to the one occurring at a solid flat plate with zero pressure gradient from which heat is being removed in the amount required for the vapour production given by the flow coefficient C_Q,

$$C_Q = \frac{Q}{U_\infty B \ell} = \frac{V_g}{U_\infty} \qquad (1)$$

where Q is the air flow rate, B an ℓ are the width and length of the two dimensional cavity, U_∞ is the free stream velocity and $V_g = Q/B\ell$ is the mean velocity of the vapour assuming that it is brought to the cavity through the whole interface. We will also assume that the flow is turbulent and that the surface behaves as a rough plate. The validity of the above assumptions can only be verified if the results are favourably compared with available data. This is the purpose of section 5.

It can be shown (Brun et $al.$, 1970) that the wall temperature on a flat plate over which a heat flux, φ_p, is being transferred by the fluid is given, for Prandtl numbers $P \approx 1$, by,

$$T_p = T_\infty + \frac{\varphi_p}{\frac{1}{2}\rho c_p U_\infty C_f}\left(1 - \frac{u_b}{U_\infty}(1-P)\right) \qquad (2)$$

where T_∞ is the temperature of the liquid, ρ its specific mass, c_p is the specific heat at constant pressure, C_f is the local friction coefficient, u_b is the velocity at the edge of the viscous boundary layer and,

$$\varphi_p = -\rho_v L V_g = -\rho_v L C_Q U_\infty \qquad (3)$$

where L is the latent heat of vaporisation and ρ_v the specific mass of the vapour phase. For a rough plate at very high Reynolds numbers, the friction coefficient is given by,

$$\frac{1}{2}C_f = 0.00695\left(\frac{x}{\varepsilon}\right)^{-1/7} \qquad (4)$$

where ε is the equivalent roughness. Finally, the ratio u_b/U_∞ is assumed to be equal to that of a smooth plate and given by,

$$\frac{u_b}{U_\infty} = \frac{2.1}{Re_x^{0.1}} \qquad (5)$$

Combining (3) to (5) and (2), we obtain,

$$T_p = T_\infty - \frac{\rho_v L C_Q}{0.00695\,\rho c_p}\left(\frac{x}{\varepsilon}\right)^{1/7}\left[1 - 2.1\,Re_x^{-0.1}(1-P)\right] \qquad (6)$$

A better data correlation of Hord's results was obtained by using a roughness (in meters), varying with the Reynolds number, of the form (Fruman and Beuzelin 1992),

$$\varepsilon = 2.2 \, Re_x^{-0.5} \tag{7}$$

4. Experimental results

4.1. THERMAL EFFECT RESULTS

The cavity length, ℓ, as a function of the cavitation number,

$$\sigma = \frac{p_{ref} - p_v(T_{ref})}{\frac{1}{2} \rho V_{ref}^2} \tag{8}$$

where p_{ref}, V_{ref} and T_{ref} are the reference pressure, velocity and temperature measured upstream of the throat, p_v is the vapour pressure, is plotted in Figure 2 for a variety of test conditions in the PECLET facility. For comparison purposes, results obtained in the water loop have been also indicated. They show that the cavity length increases very sharply for σ values of less than 0.60. The difference between the Freon and the water results can be explained because of the thermal effect. Figure 3 shows the temperature difference, ΔT, between the free stream temperature and the temperature measured on the Venturi wall, as a function of the distance to the throat, non dimensionalized with the cavity length (50 mm), for three reference temperatures and flow velocities. Temperatures decrease linearly with distance whatever the experimental conditions. Maximum differences between the reference temperature and the wall temperature, ΔT, are presented as a non dimensional coefficient $\boldsymbol{B}$,

$$\boldsymbol{B} = \frac{\Delta T \, \rho \, c_p}{\rho_v \, L} \tag{9}$$

for a variety of test conditions as a function of σ, Figure 4. In all these tests, ΔT is comprised between 1 and 5°C.

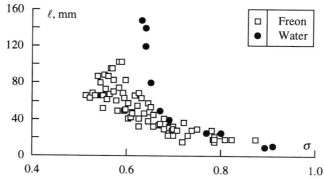

Figure 2 : Cavity length as a function of cavitation number in the PECLET and water facilities.

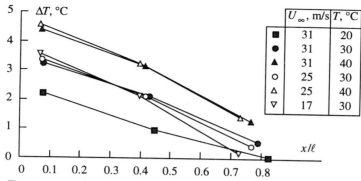

Figure 3 : Temperature difference as a function of distance to the throat for ℓ = 50 mm.

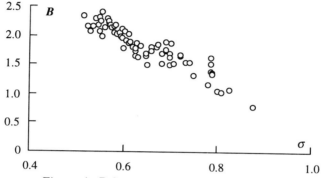

Figure 4 : B factor versus cavitation number.

4.2. FLOW COEFFICIENT FOR VENTILATED CAVITIES

The flow coefficients and the Reynolds numbers were computed using as the free stream velocity the one on the cavity interface given by the Bernoulli's equation, $V_\infty = V_{ref}(1 + \sigma)^{1/2}$. The mean flow coefficient is independent of Reynolds numbers in the range tested, much smaller than those usually encountered in practical applications and in the PECLET loop, and has a value of $C_Q = 3.85 \times 10^{-3}$, with a standard deviation of 0.45×10^{-3}. The latter is consistent with the precision of the measurement of the cavity length, estimated to ±10 percent and due essentially to the fluctuations of the cavity closure. Stutz (1996), has conducted velocity and void fraction measurements within a 8 cm long ventilated cavity. They show, Figure 5a, that near the leading edge of the cavity the void fraction peaks at the interface while becoming nearly constant over the cavity height downstream. The velocities are in the direction of the main flow on the upper half of the cavity thickness and show a backflow in the lower half, Figure 5b. This backflow justifies the homogenisation of the void fraction by the intense mixing so created. The air (vapour) flow rate was computed by integrating, over the cavity thickness at different stations, the product of the velocity and the void fraction. In spite of the difficulties associated with this type of measurements and the precision of the integration, conducted with a limited number of data points, the comparison with the injected flow rate is very much satisfactory, Figure 6. It has to be pointed out, however, that the experimental C_Q is in the lower range as compared to earlier experiments. This may be due to the estimate of the cavity length made by the experimenter.

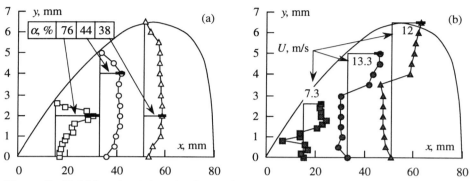

Figure 5 : Void fraction (a) and velocity (b) profiles at three stations for $V_\infty = 14$ m/s and a ventilated cavity length of 80 mm.

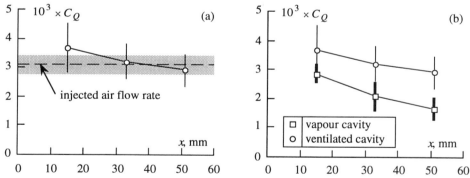

Figure 6 : Local flow coefficients obtained by integration of the product of the void fraction and the velocity from Figure 5 and 7.

4.3. FLOW COEFFICIENT FOR VAPOUR CAVITIES

For the geometry of the Venturi used in the ventilation tests, Stutz (1996) performed velocity and void fraction measurements for a 8 cm long natural stable cavity, Figure 7. As compared to the ventilated cavity situation, the void fraction do not show a significant augmentation near the leading edge in the vicinity of the interface. Further downstream, the void fraction is larger near the wall than near the interface. The velocity profiles are surprisingly close to the ones of the ventilated cavity. In terms of vapour flow coefficients, Figure 6b shows that they are slightly below the non condensable gas flow coefficients but close enough to justify "a posteriori" the hypothesis of the entrainment theory (Holl *et al.*, 1975).

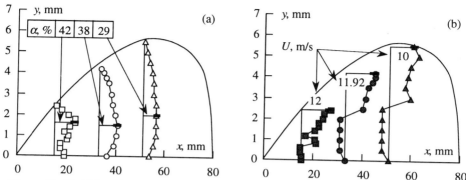

Figure 7 : Void fraction (a) and velocity (b) profiles at three stations for $V_\infty = 14$ m/s and a vapour cavity length of 80 mm.

For unsteady nearly periodic cavities, Larrarte *et al.* (1995) have shown that the vapour production can be assimilated to the cavity growth rate during the initial phase. Assuming that there is no vapour production during the phase of cavity detachment and cloud convection, the mean vapour flow rate during the total duration time of a period gives a flow coefficient 4 to 6 times larger than the one of the stable cavity. However, Stutz and Reboud (1996) have shown that the mean void fraction of these cavities during the growth phase is around 20 percent. If this void fraction is taken into account, the vapour flow coefficient is reduced (by a factor of about five) and the flow coefficient estimates are then close to the ones for stable cavities.

4.4. PRELIMINARY CONCLUSIONS

The results presented above show that the hypothesis of the entrainment theory, assuming that the vapour flow rate can be estimated by the non condensable gas flow rate required to sustain a ventilated cavity, are quite well justified by the experiments if due account is taken of the difficulty of conducting very precise measurements in vapour cavities. The methodology developed by Fruman and co-workers to estimate the thermal effect will be now applied to the flow conditions of the PECLET loop tests.

5. Comparison of experimental and predicted temperature differences

By using expressions (6) and (7) and making $x = \ell$, the temperature differences were computed with $C_Q = 3.85 \times 10^{-3}$ The results are shown in Figure 8a as a function of the experimental ΔT. It can be seen that the predicted values of ΔT are much larger than the experimental ones. Among the many factors which can justify these differences, a screening investigation has firstly shown that the vapour phase characteristic parameters in expression (6), calculated for the reference temperatures instead of the cavity temperatures, are the most significant. Secondly, it has been observed that because of the high velocities of the liquid flow (20 m/s < V_∞ < 50 m/s), the heat produced by viscous effects might not be neglected. According to Eckert and Drake (1972), in the case of turbulent boundary layers the increase of temperature induced by viscous effects can be estimated (from experiments) by,

$$\Delta T_{visc} = \frac{U_\infty}{2\,c_p}\,P^{1/3} \tag{10}$$

A computation was then performed using the vapour phase parameters at the cavity temperature and taking into account the heat provided by viscous effects by subtracting ΔT_{visc} given by equation (10) to the right term of equation (6). The results are plotted in Figure 8b. Here the agreement is very satisfactory and gives an "a posteriori" justification to the assumption sustaining the model. However, it must be stressed that the assumption that the vapour production is uniformly distributed over the whole cavity interface is not justified by the results presented in Figure 6 showing that the flow coefficient has reached its maximum value at the first measuring station downstream the cavity leading edge. Moreover, the size of the adopted roughness can not be justified either on physical grounds and has to be considered as an adjustable parameter whose physical meaning is yet to be found.

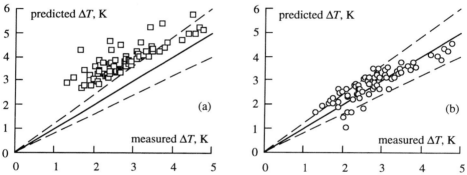

Figure 8 : Predicted versus measured temperature difference for (a) values of the parameters at the free stream temperature and (b) corrected values of the parameters for the vapour phase.

6. Conclusion

Results of tests conducted in two cavitation loops having the same geometric Venturi type test section are presented. In one of the test loops the working fluid is Freon 114 and allows to determine the temperature reduction due to the vapour produced to feed a natural cavity. In the other loop, operating with water, tests were conducted with ventilated and natural stable cavities to determine the non condensable gas and vapour production rates by performing void fraction and velocity measurements within the cavities.

A model that allows the computation of the temperature reduction by assuming that : i) the vapour production is equal to the air injection necessary to sustain a ventilated cavity of equal mean length, ii) the vapour production is uniformly distributed on the whole surface of the cavity interface, and iii) the interface behaves as a rough flat

commenced (zone without aeration), is underway but a balance does not yet exist between the air entering and the water leaving (gradually aerated zone), or whether this balance has been reached (uniform aeration zone). Four areas can exist transversely, depending on whether the air or the water is predominant: surface, mixture, intermediate or without air. Both structures can be seen in figure 2.

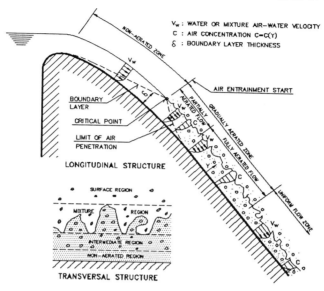

Fig. 2.- Natural aeration. Longitudinal and transversal flow structure.

Aeration theoretically starts at the point where the Boundary Layer reaches the curent surface (Critical Point), but in practice, the irregularities of both surfaces -nappe and Boundary Layer- determine the existence of a strip in which zones without air co-exist with others where entrainment is already underway. It is necessary to know the profile of both surfaces and to use the Boundary Layer theory, in order to locate the Critical Point, and this has been studied by several authors. However, the results obtained reveal considerable differences, so it is advisable weigh up the results.

The "gradual aeration zone" is characterised by a longitudinal variation in the air concentration. According to Wood (1991), the average concentration and the depth throughout the chute [$C(x)$; $h_w(x)$], can be determined by integrating the energy equations for the water-air mixture by numerical methods[4] (finite

---

[4] When the chute width and its slope are constant, the equations are simplified. The air continuity equation (1) can be directly integrated, thereby obtaining $C(x)$; that for the energy (2), with which $h_w = h_w(x)$ is calculated, the finite differences method is still used for solving (Gutiérrez Serret, 1994).

differences) . Once C(x) and $h_w(x)$[5] are known, the rest of the parameters that characterise the flow can be obtained - concentration at the bottom (C_o), velocity, emulsified flow (h_a) - by applying the expressions that are indicated later on for the "uniform aeration zone". These equations are expressed as follows:

-    Air continuity.

$$\frac{d\overline{C}}{dx} = (1-\overline{C}) \; [\frac{(V_T)_u \cos\alpha}{q_w} \; (k_e\overline{C}_u - K_T\overline{C}) \; (1-\overline{C}) \; + \frac{\overline{C}}{b}\frac{db}{dx}] \tag{1}$$

-    Energy for the water-air mixture.

$$\frac{dh_w}{dx} = \frac{\sin\alpha \cdot (1+h_w\frac{d\alpha}{dx}) \; + E\frac{q_w^2}{gh_w^2}\frac{1}{b}\frac{db}{dx} - I}{\cos\alpha \; - \; \frac{E}{g}\frac{q_w^2}{h_w^3}} \tag{2}$$

in which:
x : Distance above the chute axis from the Critical Point.
b : Chute width, generally variable; b = b (x)
α : Angle of chute with horizontal, generally variable; $\alpha = \alpha(x)$
$q_w; q_a$: Specific water/air discharge (m^3/s.m).
h_w: Equivalent depth.
C : Average concentration in each section; $\overline{C} = q_a/q_a + q_w$.
V_e: Inflow velocity of air into the current.
V_T: Limit velocity of the bubbles in turbulent flows,
   [0.17 m/s (Wood 1991) < V_T < 0.40 m/s (Chanson 1992)]. $K_e = V_e/(V_e)_u$;
   $K_T = (V_T)/(V_T)_u$[6]. Subindex u: Values $\overline{C}$, V_e and V_T in the uniform aerated zone.
E : Corrective coefficient of the kinetic energy[5]; g : Apparent gravity.
I : Energy line slope:$I \simeq q_w^2 \cdot f_w/4gh_w^3$ (Darcy-Weisbach)
   f_w and f: Roughness Coefficient "equivalent flow" and flow without aeration

---

[5] Equivalent flow: theoretical situation in the hypothesis in which the water and the air flow separately, but at the same velocity that the water-air mixture would flow at (enclosed figure).

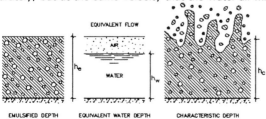

h_c is usually adopted up to heights with an air concentration of 90% (h_{90}).

EQUIVALENT FLOW

AIR

WATER

EMULSIFIED DEPTH   EQUIVALENT WATER DEPTH   CHARACTERISTIC DEPTH

[6]   The following is usually considered when integrating these equations: $K_e = K_t = 1$ y E = 1

$$\frac{f_w}{f} = \frac{1}{1+10\overline{C}^4} \tag{3}$$

In the "uniform aerated zone" it is of interest to know the average concentration ($\overline{C}$), the bottom concentration (C_0), the emulsified depth (h_c) and the mixture velocity.

Numerous formulae have been proposed for calculating the average concentration ($\overline{C}$) and they are not all consistent. Those of Gangadharaiah et al. (1970) and Hager (1991) have been selected from these, and their expressions are:

- Gangadharaiah et al. (1970):

$$\overline{C} = 1 - \frac{1}{1 + \Omega\ F_w^{3/4}} \tag{4}$$

in which:

F_w:"Equivalent Fronde number"; $F_w = V/\sqrt{g\ h_w}$

Ω: Constant depending on the roughness and the shape of the channel. $\Omega = 1.3$-5·n   rectangular and $\Omega = 2.16 \cdot n$ trapezoidal (n: n° Manning).

- Hager (1990):

$$\overline{C} = 0,75\ (\text{sen}\ \alpha)^{0,75} \tag{5}$$

The bottom concentration (C_0) can be calculated, using the formulae of Rao and Gangadharaiah (1971) or Hager (1991), amongst others. Hager's expressions are:

$$C_0 = 1.25\ (\frac{\pi}{18D}\alpha)^3;\qquad 0^\text{o} \leq \alpha \leq = 40^\text{o} \tag{6}$$

$$C_0 = 0.65 \cdot \sin\ \alpha;\qquad 40^\text{o} \leq \alpha \leq 80^\text{o} \tag{7}$$

The emulsified flow (h_e), or to be more precise, the characteristic depth (h_c), is:

$$h_c = h_w/(1 - \overline{C}) \tag{8}$$

3.- Artificial aeration. Aeration devices (Aerators).

As we have already said, when there is a risk of damage due to cavitation and the natural air entrainment is not sufficient ($\sigma > 0.2$-0.25 and $C_0 < 8\%$), artificial aeration has, since the 60's and 70's, been one of the most widely used solutions for dealing with the cavitation problem. Aeration devices (one or more[7]) are usually constructed down the length of the chute, and these aerators suck the air into the flow.

Its typologies are varied (Fig.3), but they are all combination of four basic elements: ramp, offset, groove and aeration duct, so that the advantages these elements can be enhanced and the operation can be made more satisfactory. They are usually located 30 to 100 m. apart, and the chute's slope changes are suitables locations.

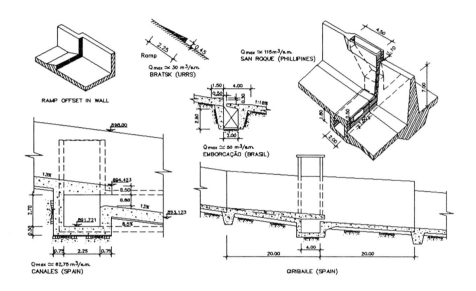

Fig.3.- Aerators. Usual designs.

Ramps provide significant quantities of air for slight water discharges, but they are by no menas ideal when the variability of such discharges is high, and they also cause disturbance to the flow; their height ranges from 0.1 to 1 m. and the angles are from 5° to 15°. **Offsets** cause less disturbance and work well when the discharges are high, heights of 0.5 to 2 m. or even less, are common. **Grooves** distribute the air evenly at the bottom, but a risk of flooding exists when the discharges are slight; heights generally range from 0.2 to 2 m. and widths from 1 to 2 m. **Aeration ducts** are normally rectangular and are designed for air velocities

---

[7] Spillways with slopes gentler than 20°, usually require more than one aerator to be placed down the length of the chute, but when the slope is steeper than 30°, one single aerator will suffice at the head of the chute. No generalisations can be made about intermediate slopes.

ranging from 30 to 45 m/s and, in exceptional circmstances, from 90 to 100 m/s, the maximum depressions being from 0.5 to 2 w.c.m. (water colunm metres).

The basic operation of aerators (See Fig. 4), is due to the pressure difference between that of the atmosphere and the cavity formed in the aerator, which sucks air in from outisde (air demand). Furthermore, the flow turbulence is greater, bringing about an increase in air entrainment at the upper surface of the nappe. according to the mechanism described, the flow structure in an aerator is made up of five zones (Fig 4.): "approach", "transition", "aeration", "impact" and "bottom aerated flow".

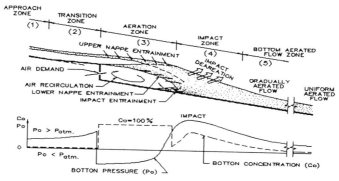

Fig.4.- Artificial aeration. Longitidinal structure of the flow.

The flow conditions in the "approach zone" determine to a large extend the way the aerator operates. A pressure increase, with respect to the hydrostatic, greater friction and an increase in turbulence, all occur in the "transition zone". Air suction takes place in the "aeration zone", and it rejoins the flow via the lower nappe of the cavity, and there is also an increase in entrainment at the upper nappe.

Furthermore, in addition to these aeration processes in the "impact zone", caused by the existing turbulence and pressure gradients, the following occur: air entrained by impact, recirculation of this element in the cavity and later de-aeration. Downstream, in the "bottom aerated zone", the mixture flows in the same way as for natural aeration, a "gradual aerated zone" appearing first, and this is followed by a "uniformly aerated zone", if the chute is long enough.

Quantification of all these entrainments, has been carried out by several authors (see summary by Gutiérrez Serret 1994, 1995), but information is still not complete, especially where air entrainment through the upper surface of the nappe and by impact are concerned, and as regards the de-aeration that takes place in the zone.

4.- Computer program for positioning of aerators. Example.

In accordance with what has been indicated, and on the basis of the gradually aerated flow equations (Eq. 1 and 2.), a computer program has been developed which makes it possible to determine the aerator positioning requirements.

It ought to be pointed out, that the results, should not be considered as more than an approach to the problem, and in the light of this, they must be compared with experience gained to date, as well as with laboratory tests and even prototype.

The program is structured into three stages. At the First stage (A), an analysis is conducted into the way the spillway operates without aerators (natural aeration). Stage (B) determines the position of the first aerator, if necessary, where $\sigma > 0.2$-0.25 and $C_o < 8\%$, and the aerator must be designed. At Stage (C), the depth (h_i) and the average concentration C_i) downstream from the impact zone are taken as initial flow conditions (Chanson, 1992; Falvey 1990) and, equations (1) and (2) are used again to obtain $\sigma = \sigma(x)$ and $C_o = C_o(x)$; a second aerator is then placed if $\sigma <$ 0.2-0.25 and $C_o < 8\%$, and the process is repeated, downstream from this second aerator.

By way of example, figure 5 shows the application of the Giribaile dam (Spain).

BIBLIOGRAPHY.

- ARNDT, R.E.A. (1977). "Recent Advances in Cavitation Research". Advances in Hydroscience, Ven Te Chow ed., Academic Press, Vol. 1.
- CHANSON, H. (1993). "Self-Aerated Flows on Chutes and Spillways". ASCE, Jour. Hydr. Div. vol.119, HY2, pp. 220-243.
- FALVEY, h. (1980). "Air-Water Flow in Hydraulic Structures". Engineering Monography N⁰ 41, Bureau of Reclamation. U.S.A.
- FALVEY, H. (1990). "Cavitation in Chutes and Spillways". Eng. monogr.42. Bureau of Reclamation. EEUU.
- GANGADHARAIAH, T., LAKSHMANA RAO, N.S. & SEETHARAMIAH, K. (1970). "Inception and Entrainment in Self-Aerated Flows". ASCE, Jour. Hydr. Div. vol.96, HY7, pp. 1549-1565.
- GUTIÉRREZ SERRET, R. (1994). "Aireación en las Estructuras Hidráulicas de las Presas: Aplicación a los Aliviaderos". Tesis doctoral. Esc. Téc. Sup. de Ing. de Caminos, Canales y Puertos. Madrid. España
- GUTIÉRREZ SERRET, R. & PALMA, A. (1995). "Aireación en las Estructuras Hidráulicas de las Presas: Aliviaderos y Desagües Profundos". Premio José Torán, Comité Español de Grandes Presas Madrid.
- HAGER, W. (1991). "Uniform Aerated Chute Flow". ASCE, Jour. Hydr. Div. vol. 117. HY4, pp. 528-533.
- IAHR. (1991). "Air Entrainment in Free-Surface Flows". Design Manual N⁰ 4, Balkema Ed. Netherlands.
- ICOLD. (1992). "Spillways. Shockwaves and Air Entrainment". Boletín N⁰ 87.
- LECOFFRE, Y (1994). "La Cavitation". Ed. Hermès. Paris.
- MATEOS, C.(1987). "Aireación y Cavitación en Desagües". Curso sobre Comportamiento Hidráulico de los Desagües de las Presas. CEDEX, Ministerio de Obras Públicas y Transportes. Madrid, España.
- MAY, R. (1987). "Cavitation in Hydraulic Structures: Ocurrence and Prevention". Report Research SR79, Hydraulic Research Wallingford. Oxfordshire UK.
- PINTO, N.L. de S. & NEIDERT, S.N. (1983). "Evaluating Entrained Air Flow through Aerators", Water Power and Dam Construction, vol. 35, pp. 40-42.
- WOOD, I. (1985). "Air Water Flow. Keynote Address". XXI IAHR Congress, Melbourne, Vol.6 pp. 18-21.

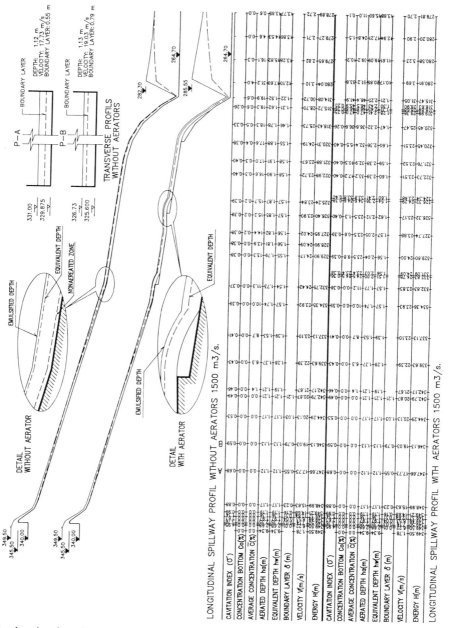

Fig 5.—Aerator Location. Design Process. Application to the Giribaile Dam Spillway (Spain).

LEADING EDGE CAVITATION IN A CENTRIFUGAL PUMP: NUMERICAL PREDICTIONS COMPARED WITH MODEL TEST!

R. HIRSCHI, PH. DUPONT AND F. AVELLAN
IMHEF-EPFL
33, av. de Cour, CH-1007 Lausanne, Switzerland
E-mail: roland.hirschi@imhef.dgm.epfl.ch

AND

J.-N. FAVRE, J.-F. GUELICH AND W. HANDLOSER
Sulzer Pumps P.O. Box 65
8404 Winterthur, Switzerland

Abstract. The aim of this paper is to present the results obtained with a 3-D numerical method allowing the prediction of the cavitation behaviour of a centrifugal pump and to compare this prediction to model tests.
The proposed method consists in assuming the cavity interface as a free surface boundary of the computation domain and in computing the single phase flow. The unknown shape of the interface is predicted by an iterative procedure matching the cavity surface to a constant pressure boundary (p_v). The originality of the presented method is that the adaptation process is done apart from the flow calculation, allowing to use any available code.

1. Introduction

Cavitation behaviour prediction of hydraulic machines, such as inception ψ_{c_i} and standard ψ_{c_s} cavitation coefficients, partial cavity length and its associated performance drop is of high interest for the manufacturers.
When upgrading existing hydraulic installations or designing new geometries, the cavitation guarantees are often the main limiting features. A precise prediction of this phenomenon by numerical simulation is then essential. In recent years, models to predict cavitation development have been refined and applied with success on isolated profiles, for steady [3] and unsteady flows[2][7]. Some 3-D models, based on S1/S2 and Euler flow computation were also developed, but without taking into account the 3-D turbulent and viscous effects of the cavity on the flow [8].
To present the accuracy of the new proposed method, numerical calculations are performed on a centrifugal pump with a commercial Navier-Stokes code (*TASCflow*). A first analysis enables an intrinsic verification of the numerical results given by the solver and make sure they are physically meaningful. Once this analysis is done, particular operating points are chosen to predict the impeller cavitation behaviour.

E. Cabrera et al. (eds.), Hydraulic Machinery and Cavitation, 604–613.
© *1996 Kluwer Academic Publishers. Printed in the Netherlands.*

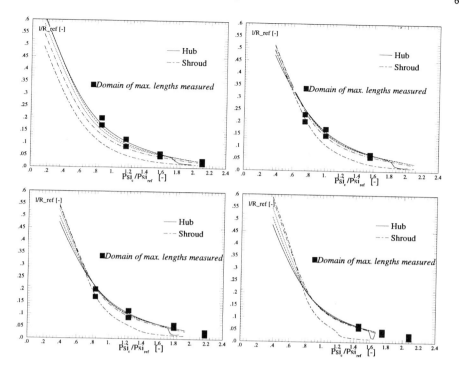

Figure 7. Predicted cavity lengths compared with measurements

6.3. CAVITY DEFORMATION

Once the initial cavity determined, its shape is iteratively modified until the calculated pressure at its interface is constant and equal to the vapor pressure. The pressure coefficient distributions (grid lines 2=hub, 9, 13 & 16=shroud) obtained after the flow computation without cavitation and for a ψ_c/ψ_{ref} value of 0.265 are compared in Figure 8. The cavity effect on the pressure distribution corresponds to the constant pressure region on the suction side. In this example, the pressure loss at the leading edge is partially compensated by the pressure increase near the cavity closure region.

6.4. HEAD DROP COMPUTATION

The comparison of the torque, obtained by the pressure and the viscous forces integration on the blades and impeller side walls, to the moment of momentum in the relative frame of reference allows to quantify the energy loss due to the cavity. In term of torque this is expressed by:

$$\vec{T}_t = \vec{T}_f + \vec{T}_r \qquad (6)$$

Where $\vec{T}_t$ represents the transferred torque, $\vec{T}_f$ the provided torque and $\vec{T}_r$ the loosed torque.

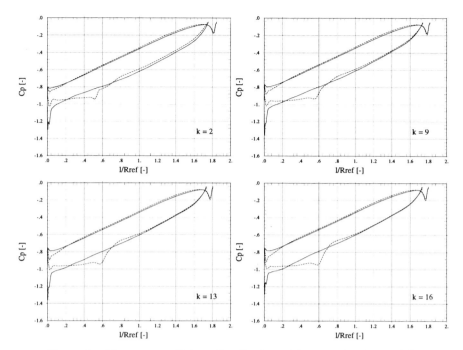

Figure 8. Pressure coefficient distribution without cavitation and for $\psi_c/\psi_{ref} = 0.265 : \phi_1/\phi_{ref} = 1.21$

Transferred power The transferred torque given by the impeller is:

$$\vec{T}_t = \int_{S_{imp}} \vec{r} \wedge (p\vec{n})ds + \int_{S_{imp}} \vec{r} \wedge (\overline{\overline{\tau}}\,\vec{n})ds \qquad (7)$$

where S_{imp} represents the blades and impeller side walls, p the calculated pressure, $\vec{r}$ the position vector and $\overline{\overline{\tau}}$ the constraints tensor, which correspond to the molecular and turbulent viscosity. The transferred energy can be expressed under a non dimensional number ψ_t through the transferred power:

$$P_t = \vec{T}_t \cdot \vec{\Omega} = \rho Q E_t \quad \rightarrow \quad \psi_t = \frac{2E_t}{\omega^2 R_1^2} \qquad (8)$$

Provided hydraulic power The expression of the provided power is given by the moment of momentum equation in the relative frame of reference. Let $\vec{T}_f$ to be the torque representative of the provided moment of momentum:

$$\vec{T}_f = -\int_{S_{tot}} \vec{r} \wedge (\rho \vec{w})\ \vec{w}\vec{n}ds + \int_V \vec{r} \wedge \left(\rho\vec{S}\right) dv + \int_{S_{tot}} \vec{r} \wedge (\overline{\overline{\tau}}\,\vec{n})\, ds \qquad (9)$$

where $\vec{S}$ corresponds to the centrifugal and Coriolis effects, and where S_{tot} represents all the domain surfaces, delimited by the inlet and outlet of the impeller. It is then possible to express the provided energy in the form:

$$P_f = \vec{T_f} \cdot \vec{\Omega} = \rho Q E_f \quad \rightarrow \quad \psi_f = \frac{2E_f}{\omega^2 R_1^2} \tag{10}$$

In the Table 1 are reported, for three relative ψ_c values, the energy coefficient representative of the provided energy (ψ_f) and the transferred energy (ψ_t). For a quite constant transferred energy value, one remarks that the provided energy is decreasing with the pressure level ψ_c at the impeller inlet section. If one plots the calculated relative head evolution, which corresponds to the provided energy, versus the ψ_c value (Figure 9), it can be noticed that the ψ_{c_s}, which corresponds to the head drop beginning, coincides very well with the measured one. The results exposed in Figure 9 show that the calculated impeller head is over-estimated in comparison with the measurements. However, it has to be taken into account the fact that the presented measurements are representative of the entire pump, with the fixed parts (diffuser, inlet pipe,..). It is then reasonable to think that this over-estimation is mainly due to the losses in the fixed parts of the machine, which are not taking into account by the computation.

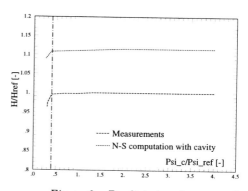

ψ_c/ψ_{ref}	$\psi_f[-]$	$\psi_t[-]$
cav. free.	1.129	1.28
0.383	1.12	1.285
0.265	1.1	1.28

TABLE 1. Impeller provided and transferred energy as a function of ψ_c ($\phi_1/\phi_{ref} = 1.21$)

Figure 9. Predicted and measured head drop

7. Conclusion

Considering an impeller geometry, its behaviour and its main cavitation characteristics can be described with the results of a Navier-Stokes 3-D computation, using models of cavitation developments. The predicted behaviour of the cavitation flow in a centrifugal pump, using the proposed method, compares very well with the measurements.

It was brought to the for that the knowledge of the impeller inlet velocity field is essential for the cavitation behaviour prediction. Therefore, a flow computation of the inlet pipe was first performed.

The pressure level corresponding to cavitation inception is determined for all calculated operating points and compared with measurements. The mean standard deviation between the measured and calculated points is lower than 4%.

The cavity lengths computation using the *Rayleigh-Plesset* equation provides very good results. Indeed, the measured cavity lengths behaviour is faithfully reproduced by the calculation, and this for all measured operating points.

The prediction of the head drop beginning is done by a direct flow computation taking into account the cavitation sheet. The very good results obtained with this method proves its efficiency to predict head drop due to cavitation. As a conclusion, the results analysis of this study demonstrates that the presented method provides an interesting tool for the 3-D cavitation development prediction in hydraulic machines.

Figure 10.   3D view of the cavitation sheet

8. Acknowledgments

The authors are particularly grateful to the members of the IMHEF Cavitation research group. We wish to thank *SULZER Pumps* which made possible the publication of this paper and the staff of the test-rig who performed the measurements. This work is financially supported by the Swiss Federal "Commission d'Encouragement à la Recherche Scientifique", *SULZER Pumps, Hydro-Vevey S.A.* and *Rolla S.P. Propeller*.

9. Nomenclature

e	:	thickness	(m)
k_{rd}	:	coefficient of viscous losses	
p	:	static pressure	(N/m^2)
p_v	:	vapor pressure	(N/m^2)
Cp	:	pressure coefficient	$Cp = \frac{p - p_{\bar{1}}}{\frac{1}{2}\rho U_{\bar{1}}^2}$
E	:	energy	(J/kg)
H	:	head	(m)
P	:	power	(W)

3. Analysis of Stability of Viscous Sublayer

Based upon linear instability theory, in this section, flow stability of viscous sublayer in turbulent boundary layer is analyzed. Suppose two dimensional viscous sublayer is disturbed by some external perturbation which can be represented as stream function with a form of Tollimen -Schlichting travelling waves :

$$\phi = \phi(\bar{y})\exp[i\alpha(\bar{x} - \bar{c}t)]$$

The linear perturbations meet the Orr-Sommerfeld equation :

$$(\bar{u} - \bar{c})(\phi''' - \alpha^2 \phi) - \bar{u}''\phi = (\frac{i}{\alpha R})(\phi^{(4)} - 2\alpha^2 \phi'' + \alpha^4 \phi) \quad , \tag{8}$$

where δ is the thickness of viscous sublayer, λ is wave length of perturbation, $\alpha = 2\pi\delta/\lambda$ is wave number, $\bar{u} = u/U$ is non-dimension phase velocity and U is external flow velocity, $R = U\delta/\nu$ is Reynolds number of viscous sublayer, while all derivatives are with respect to $\bar{y} = y/\delta$.

On boundary there are non-slip definition:

$$\phi(0) = \phi'(0) = 0 \tag{9}$$

For convenience of analysis, assume that flow regions outside viscous sublayer are completely turbulent flow with characteristic Reynolds number being very large, the flow can be regarded as non-viscosity flow. So Orr-Somerfield equation reduces to non-viscosity differential equation. At the outside boundary of viscous sublayer(y= δ), there is,

$$\phi'' - \alpha^2 \phi = 0, \qquad (\bar{y} \geq 1). \tag{10}$$

General solution of (10) can be write as:

$$\phi = C\exp(\pm\alpha\bar{y}). \tag{11}$$

where a solution corresponding to signal "+" is not consistent with physical meanings and will be omitted .Differentiate (11)and we have

$$\phi' + \alpha\phi = 0 \qquad (\bar{y} \geq 1). \tag{12}$$

Hence there have already been four boundary conditions, that is, (9), (10) and(12) for O-S equation. On the numerical method to get the solution of the equation (8), one can refer to literature[3]. Here are the numerical results quoted from literature [4] (see Table 1).

Table 1 Relations between Reynolds number (R) and
the parameters of neutral stable perturbation waves

$\bar{c}$	0	0.10	0.20	0.30	0.40	0.50	0.60	0.70
α	0	0.05	0.118	0.213	0.342	0.537	0.809	1.066
R	∞	2.415×10^5	1.292×10^4	2124.9	558.2	191.9	70.0	33.4

Due to the character that in viscous sublayer Reynolds number is approximately constant, let $R=136$, then the parameters of Tollimen-Schlichting travelling wave in neutral stability can be interpolated from Table 1:

$$\bar{c} = c/U \approx 0.52, \qquad \alpha = \frac{2\pi\delta}{\lambda} \approx 0.625,$$

Namely $\lambda \approx 10\delta$.

If amplitude Y of perturbation is the same order of magnitude as the thickness δ of viscous sublayer, approximately there will be $\lambda \approx 10Y$. Although this result is very close to the experimental values of literature [6] in which the ratios of wavenumber of erosion ripples to its amplitudes are changed from 9.2 to 10.5, the mechanism of the ripple formation are not simply attributed to solid particle motion in viscous sublayer, because the thickness of viscous sublayer is not only very thin (10^{-2} cm order of magnitude), but also the flow velocity in it is very slow. In fact, suspended particles' diameters in rivers are greater than or equal to the above magnitude, so in viscous sublayer there are not enough energy is to control particle's motion.

On the other hands, the effects of boundary's roughness on the instability of viscous sublayer should be taken into account. Assume that there is not flow separation in viscous sublayer, then the micro-irregularity of boundary is bound to be reflected into flows of viscous sublayer. So in viscous sublayer, streamline patterns should be similar to the shape of micro-irregularities of rigid boundary. We express boundary roughness function Δ as Fourier series:

$$\Delta = \sum_{k=1}^{\infty} \alpha_k \exp(ikx)$$

(13)

To sum up, it is reasonable to consider that in the viscous sublayer turbulent boundary layer, small and instable wavelike disturbance exists and can be written as Fourier series following:

$$\phi = \sum_{k=1}^{\infty} \phi_k(\bar{y}) \exp[ik(\bar{x} - \bar{c}t)]$$

(14)

4. Turbulent Coherent Structures near the Wall and Dilute Particle's Motion in it

The analysis in paragraph 2 indicate that viscous sublayer has no direct effect on particle's motion in turbulent boundary. So we can imagine that as a trigger of turbulence near the wall, it is possible for turbulent coherent structure to control the solid particle's motion. Although coherent structure is very complex, in order to investigate particle's motion in the boundary layer, Here is a simplified model of coherent structure.

As shown in Figure 3, turbulent boundary layer flow is roughly divided into two section, i.e. , viscous sublayer and logarithm law's region. In turbulent boundary layer, a section I--I having been chosen arbitrarily, it is obvious that distribution of average velocity within viscous sublayer, as mentioned in paragraph 1, is parabolic while that in the other section, i.e., turbulent core region observes logarithm law. Under limit conditions, the velocity at the interface between viscous sublayer and "logarithm law " region is discontinued and corresponding, the velocity gradient normal to streamwise direction is infinity. Practically although velocity should be continuous due to viscosity and there is a intermediate layer between viscous layer and logarithm law region, the velocity gradient is very great

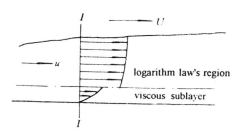

Figure 3   flow sketch in

turbulent boundary

in this intermediate layer, so it is possible for large scale spanwise vortex to generate in the intermediate layer. We now describe it with spanwise K-H vortex layer which are nearly distributed continuously along stream wise direction. It is well known that K-H vortex is extremely instable. Hence the small disturbance which always lies in viscous sublayer will have effects on the K-H vortex layer, Providing that K-H vortex layer waver at limited amplitude under act of small disturbance, and the waves are similar to (14) but its amplitude is amplified. For convenience of analysis, the coordinates of position for vortex center are accepted in followings instead of stream function, that is,

$$f(x,t) = \sum_{k=1}^{\infty} A_k \exp[ik(x-ct)]$$

(15)

When the vortex layer wavers into high velocity region, it will be accelerated. otherwise it decelerated. For any vortex line, there are high velocity section and slow velocity section alternatively along vortex line in the spanwise direction, and due to action of acceleration and deceleration, the vortex line is twisted and stretched with "horseshoe" shape or "hairpin" pattern structure being formed. At last vortex structure breaks off into random turbulence. After that, the above quasi-periodic process begin again. It is necessary to point out that the above process doesn't last very long and is often covered by the random characteristics of turbulence ,So, for a long time coherent structure is not understood well. We think that K-H vortex layer and its wave like movement is the main feature of coherent structure near wall ,with other characters of coherent structure being derived from it. What follows are application of above simplified model of coherent structure to analysis of particle's motion in the turbulent boundary. Under the condition of dilute particles phase the existence of particles hasn't great effects on the flow structure in liquid phase boundary layer. Further more, if collision between particles is neglected, dilute particles motion will be similar to that of a single particle. For a particle the diameter of which is d, when moving near the simplified coherent structure as shown in Figure 4, it will be drawn into the vortex layer by pressure gradient. Suppose the particle moves at the same streamwise velocity as the vortex layer, but at the direction normal to wall the particle is forced to vibrate under the action of coherent structure. If we take the effects of particle's inertia into account, the harmonic wave which governs the particle's vibration will be only one which wave number is the same as or near to particle's harmonic frequency. So particle vibration at balance point can be expressed as:

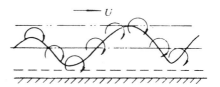

*Figure 4   simplified model of
coherent structure*

$$f_d(x,t) = \sum_{k=1}^{\infty} A_{kd} \exp[iK_d(x-ct) - \delta_d]$$
(16)

Where K_d is the wave number corresponding to particle's harmonic frequency, A_{kd} is amplitude corresponding to K_d, δ_d is phase lag due to velocity difference between fluid and particle.

Given that particle is in balance point when $t=0$, that is,

$$x_0 = 0, \qquad y_0 = 0 \qquad ,$$

then after the time interval of t the displacement of the particle becomes

$$x=ut \quad , \tag{17}$$

$$y = \text{Real}\{f(x,t)\}$$

$$= A_{kd} \cos[K_d(x-ct) - \delta_d]$$
(18)

Substituting (17)into (18)to eliminate parameter variable t, one obtains

$$y = A_{Kd} \cos[K_d(1-c/u)x - \delta_d] \quad . \tag{19}$$

From equation (19), it is indicated that the trajectory of particle motion in turbulent boundary layer is a cosine wavelike curve. Since solid wall is always flat plate or regular curved surface, particle's motion result in impingement on the wall, and the effect of impact is attempted to have boundary shape suitable to particle's motion. At last surface ripples have arisen.

4. Conclusion

Through analysis of the similarities between turbulent boundary layer and laminar transitional boundary layer, in this part, it is revealed that the erosion ripple formation on wetted surface of hydraulic turbine is association with the instability wavelike motion of fluid near wall region of turbulent boundary layer. Based on the relation between viscous sublayer and coherent structure, a simplified turbulent coherent structure model that an unstable wavelike motion of the Kelvin-Helmholtz vortex layer with approximately continuous distribution along flow direction under the action of perturbation is proposed. This model is applied to analyze dilute particles' motion in

In Time Domain. For steady pressure,collapsing time of a bubble is about 75 persent of that of growing. For non-steady pressure, with dynamic pressure increasing, the time of callopsing and growing and the size of the bubble are increased.

In amplitude domain (sonic pressure). During a bubble growing and collapsing, there are a hegetive low frequency vibrating component and a positive sharp peak of sonic pressure. It is indicated by analyzing the sonic pressure equation that the positive peak occurs at the time of the bubble compressed to its minmum size. If the gasous pressure rate within the bubble was larger than 0.2, the positive peak value would be less than that of the negetive.If the rate was less than 0.2, the peak would hundreds times of that of the negetive.

In frequency domain. During a bubble growing, the sonic pressure is negetive with low frequency. During a bubble collapsing, the pressure appears pluses with high frequency components increasing significantly. The feature of pulses determines, the attenuation character of the high frequency componenents. It can be proved theoretically that, for a sharp pulse, a right angle pulse, and a triangle pulse a trapezoid pulse, the spectra attenuate in a rate of 0 dB/oct, 6 dB/oct, 8~10 dB/oct, 10~12 dB/oct, respectively, after vibration modulation of bubbles with dominant distributed radius. This is one of the basis of judging with or without, which type of, and damage potanlial of cavitation.

For small gas content, the sonic pressure appears shock wave. Its spectrum has low attenuate rate,which indicates vapour cavitation with high cavitation erosion potontial. For large gas content (in case of gas was cavitation or vortex cavitation), the sonic pressure spectrum has high attenuate rate, which means little cavitaion erosion but may cause sonic vibration.

In sonic energy. The sonic radiant energy is almost less than 1 percent of the potontial energy of bubble in growing. However, the energy ratio of radiant and potential of the bubble is far larger in collapsing than in growing.For air content rate of a bubble being less than 0.5 persent, The ratio is about 67 percent. For the rate being 0.5 to 10 persent, the ratio is 66 to 15 percent. For the rate larger than 10 persent, the ratio is less than 15 persent.

In acturally flow field, the underwater noise is determined by the character of averaged dominant bubbles,which is basically similar to that of a single bubble but with wider frequency band.Therefore, the acoustic charactiristics of a single bubble is the basis of the acoustic method of judging cavitation.

2.3 MEASURING AND ANALYZING SYSTEM

The measuring and analyzing system consists of hydrophones, signal amplifiers, a tape recorder, a FFT dynamic singnal analyzer, a PC and output equipment. To measure reliablly underwater noises and to judge cavitations, the basical requirements of the system are multichannels, high sensitivity, wide frequency band (200HZ to 250KHZ), good directivity (to receive singnal from a given direction but the whole flow field), convenient usability, and high data process ability.

2.4 JUDGING METHODS BY CAVITATION NOISE

By using underwater noises, some methods have been used in our cavitation studies both in models and prototypes such as methods of sonic wave pattern, correlation function, spectrum and its attenuation, sonic energy, shock wave, air supply, and 3D display, ect. They gave satisfactory results and good accordance with field cavitation inspections, felled vibrations and other simultaneous measured data as pressures, vibrations and swings, air entretment and load variations. The four methods used in this paper are as fellow:

Spectra level difference. Because of the difference of bubble air content rate,the sonic energy caused by bubble collapsing may be times or hundreds times of that bubble growing. The sonic energy level difference may be up to 10dB to 20dB between both cases.Therefore, this method can be used for comparing different sonic processes such as testing condition with "backgrowd" in a vaccum tank, different operation processes, and testing processes with different vaccum. In practice, a Fourier spectrum is used for comparing sonic energy, which called sonic pressure spectrum level method. The spectrum is got from one third octave band analyzing and frequency band modifing. In the spectrum level comparison, the larger the different in high frequency components of the spectra, the more intense the cavitation.
Spectrum Attenuation. According to a single bubble character and our practice, within 2 to 3 octaves, if the spectrum attenuation rate is larger than 8 dB/oct, there is no cavitation, if the rate is about 6 to 7dB/oct, there is vortex type or gasous cavitation,which would not cause erosion but might produce sonic vibration;if the rate is less than 6 dB/oct, there is developed cavitation. The less the rate, the more intense the cavitation.
High Frequency Shode Ware. This method is used to analyze high frequency characters of sonic preaaure. (frequency higher than 10KHZ). If the spectrum attenuation was very small near to zero in the frequence band, There would occure sharp pulses in sonic pattern, which would mean shock wave type cavitation developed.
Sonic Enercy. Sonic energy can be obtained by three ways. One is to get tatol sonic enercy level directly from the one third octave band analyzing. Another is to sum up the componants of different frequence from the spectrum. The third is to take the value of correlation function at $\tau = 0$. The sonic enercy method can be used to comparing different conditions.

Above methods can be used individually or combinedly.The shock wave method may used to judge vapour cavitation collapsing and rebounding with high frequency. The spectrum attenuation method is suitble to reveal gasous cavitation or cavatation incipience with low frequency.To ensure a reliable judgement,different methods may be used as comparing to analyze whole process of cavitation development.

3. Applications In Hydraulic Turbine Study

3.1 APPLICATION IN A FRANCIS TURBINE MODEL

The diameter of turbine runner is 250mm.14 operating cases were tested with water head being Hmax=198m, Hp=162m, Hmin=135m in prototype. The opening ratios of the guide blade were full open, 60 and 30 persents,. and that in accordance with optimum efficiency and 5 percent output restraint curve of the turbine.

In each case, the test was conducted with different σ_y by changing vaccum in the model. Four hydrophones were installed at the top cover Z_{13}, the base ring Z_0, the entrance and exit of the vertical cone of the tailrace Z_{21} and Z_2 . High speed photographing, video recording, and pressure fluctuation measuring were simultan- eously taken for comparision.

Four types of cavitation were revealed as vane cavitation of blade, vetical axis vortex cavitation, eccentric vortex cavitation and cavity cavitation. For better under- standing, the cavitation of blade is graded into four developing stages as   ,   , and   , which in turns describe the cavitation bubbles looking like a thin foil, sheets, a cloud and a intermittent vortex band.

Figure 2 gives the relationships of the measured sonic energy $Z(0)+ \Sigma Z(f_i)$ with cavitation index σ_y for four cases. The following table gives the test conditions and results of critical cavitation indexes.

The critical indexes were determined by sonic energy method. For gasous or vortex cavitation, the sonic energy would increase with σ_y to a maximum value and then would decrease sharpdly. the reaso is that with this type cavitation develo- ping, the air content of the bubble is increased.   At a certain air concentration the sonic energy would be absorbed,   that caused sonic signals becoming " weak ". Therefore,the inflection point before the maxmum value will accordete to the critical index. For vapour type cavitation, the sonic energy would increase with σ_y to a inflection point and then it would increase rapidly with a steeper gradient, because of sonic energy character of this cavitation development. Hence, the inflection point is accorded to the critical index.

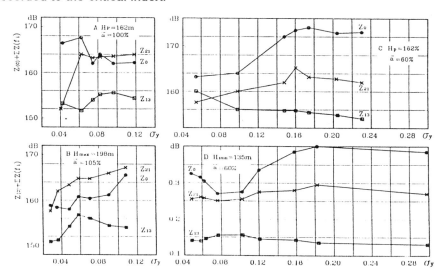

Figure 2.   Relation ship of sonic energy with vaccum

For cases in above table, the cavitation type and development can be described as follow.

Tabie. Criticai Caoiiation Index of onic Energy Method anei eotranai characteristics Method										
Test No	Test Condition			onic Energy Method						ebtetnai characieristis method	
	Hp(m)	a (%)	6r	6k			6i			6k'	6i'
				Z_{21}	Z_0	Z_{18}	Z_{21}	Z_0	Z_{18}		
1	162	100	0.041~0.119	0.062	0.083	0.090	>0.119	0.094	>0.119	0.064	0.119
2	198	105	0.048~0.127	>0.127	>0.127	0.078	>0.127	>0.127	>0.127	0.068	0.102
3	162	60	0.037~0.212	0.140	0.154	<0.08			>0.212	0.053	
4	135	60	0.049~0.299	0.183	0.183	0.102			0.183	0.062	0.092

Test No.1   For σy = 0.119, vane cavitation in grade   occured at the upper middle part and the lower part of the blade. With σy decreasing, the cavitation at the upper part would tend to develop into middle of the blade, and the cavitation at the lower part would develop much intensely. For σy = 0.062, the cavitation at the lower part developed into grade   , that there was full of cavitation cloud.

Test No.2   For σ = 0.127, vane cavitation in grade   occured at the middle and the lower parts of the blade entrance side; vortex cavitation appeared at the vertical cone, which sized like a "rugby" ball. With σy decreasing, the vane cavitation moved up wards and the vortex cavitation developed looking like a " sylender " . For σy = 0.048,the vane cavitattion at the upper middle part of the blade back moved into the head and the middle of blade; and those at the lower part developed into grade.The vortex cavitation also developed with larger size, looking like a "gyro".

Test No.3   For σy = 0.212, vane cavitation in grade   occured at the exit of the blade back, and in grade   at the head of the blade.Some small vortex band cavitation appeared at the exit of blade. With σy decreasing, the cavitation at the exit remained in grade   , but at entrance of the blade back, the cavitation developed into grade   and grade   . The vortex band appeared in the cone. For σy = 0.087, the flow field was fully covered will vortex band, the vane cavitation could not be recognized.

Test No.4   For σy = 0.29, No cavitation occured at the blade, some small vortex bands could be seen occasionally in the cone. For σy = 0.183, vane cavitation in grade   occured at the lower part of the entrance of blade; and vane cavitation in grade   appeared at the upper and the middle parts of the blade, the vortex band developed into a larger size. With σy decreasing, the cavitation at the blade entrance side developed into grade   .The eccentric vortex bands became more intense and finally there were full of cavitation clouds the field.The vane cavitation could not be recognized.

With the cavitation description and the critical indexes given in the table, the followings were indicated.

The critical cavitation indexes determined by the sonic energy method were good accordance with the cavitation bubble development and characterstics.

The cavitation at the tailrace is gasous type vortex cavitation, the cavitation at the blade is varpour type cavitation, and the cavitation near the base ring is combined of the above 2 types on the test conditions. The results of cavitation judgement by the method are basically accordance with the principle and practice of hydranulic turbine cavitation.

The sonic energy method could measure and judge the type and location of critical cavitation index, but the external method has only one critical index, which does not accordant with the prinicple and practice of turbine cavitation.

The critical index in external method is less than that in the sonic method, which is not safe for turbine design. This has been proved by many turbines in operation.For example,if vane cavitations in grade appeared at the blade back,which was caused by uniform negitive pressure, the left force would increase, that means higher energy efficiency. However, when the energy efficiency decreased, the vane cavitations would have much intensly developed. Therefore, if the critical index was determined according to one persent of energy efficiency decreasing, it would be too small to be safety.

3.2 APPLICATION IN A FRANCIS TURBINE

There are 4 Francis turbines installed in Geheyian hydropower plant, Hubei, with capacity 300MW for each unit. The water heads are designed as H_{max}=125m, H_p =103m and H_{min}=87m. The turbines 1# and 2# are made in Canada, 3# and 4# in china. A prototype observation was conducted at water head H=115m on the turbine 1#. The purposes of the observation were to study the hydraulic steability of the turbine, the transient character of the intaking system, the efficiency of the turbine,and safety problems as cavitation and vibration, ect.The observations included pressure and pressure fluctuation, underwater noise, air suppliment and air noise, energy output and discharge (by water hammer method),vibration and swing, ect. Before the observation, the turbine system was dewatered for inspecting cavitation erosion. This paper will focus the cavitation problem only.

Three hydrophones were installed outside the steel plant, in which Z_1 was at the scroll case as the sonic " background ", Z_2 was at the top cover for measuring sthe cavitation noise from the runner,Z_3 was at the entrance of the tailrace to measuring the vortex cavitation noise.

3.2.1 Sonic Energy Method

Figure 3 gives the sonic energy variations with the turbine output. It indicated:

The everage sonic energy at the tailrace was 10dB larger than that at the scroll case. The maximum sonic energy occured at 60 persent output (177MW). This is a gasous type vortex cavitation, that might cause sonic vibration. Other measurement conformed the pressure fluctuation was 0.6HZ in frequency, which accorded to 1/3 frequency of the runner being the same of the vortex band frequency.The fluctuation value was upto 4.34 persent of the total head, which was slightly larger than the value guaranted by the maker. Meanwhile, the swing amplitude of the main shaft reached maximum value 0.283mm, which exceeded the swing allowance 0.25mm by the maker. The maximum air noise was also occured in the tailrace and the runner chamber. At the condition of full output, retalive large sonic energy was also measured, that was caused by the runner cavitation as mentioned below.

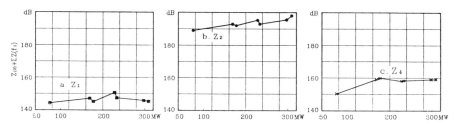

Figure 3. Sonic energy vs output

The sonic energy measured by Z_2 at the top cover was 30dB to 40dB larger than that at the scroll case Z_1. The higher the output load, the larger the energy. It was indicated that there would be vapour type cavitation occured with bubbles collapsing and rebeunding,which might cause cavitation erosion on the runner.Cavitation inspection by dewatering the turbine proved there were cavitation erosions at beginning stage on the blade of the runner.

3.2.2 High Frequency Shock Wave Method

By 1/3 octave band analyzing at 30 frequecies from 125HZ to 100KHZ, the noise spectra for frequency higher than 10KHZ were shown in Figure 4, it revealed:

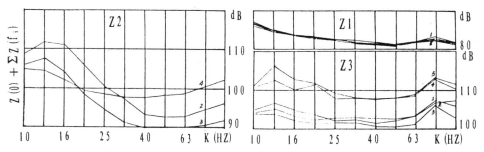

Figure 4.   Spectra of underwater noise

At the scroll case (Z_1), the components in high frequency of the spectra were small and did not vary with output, which means no cavitation.

At the tailrace (Z_3), for outputs 177MW and 228MW, the spectrum attenuation was larger than 10 dB/oct at frequency band of 2KHZ to 40KHZ, which belong to flow turbulence noise. The sonic pressure for 177MW was more intense than for 228MW, which indicated that vortex cavitation occured at the tailrace. This judgement is in accordance to that by sonic energy method. For output 297MW, the sonic pressure was the most intense in the conditions and its spectrum attenuation was small, which indicated cavitation with character of high frequency shock wave.

Comparing the results in sonic energy method, the cavitation noise came from the exit side of the blade where high frequency cavitation occured.

At the top cover (Z2). The spectrum components in high frequency band, were 20dB higher than that at the scroll case (Z1), except output 75MW, the spectrum attunuation was very small, which revealed cavitation developed much intensively.

4. Conclusion

It is reliable therotically and practively to measure and judge turbine cavitation by accoustic method. Futhermore, the method is more reasonable and practive than the external characteristics method, which could reveal the different types,locations and danage potentials of cavitation. However, the method is still preliminary studied. Further study and practice are underway.

5. References

[1] CHENG LIANGUN (1981) HYDRAULIC TURBINE, China Mechanical Industry Publishing Company

[2] Н.И.Пылаев, Ю.У.Эделъ (1974) Кавитация в Гицротурбинах

[3] LIU KEHUANG (1993) Brief Intruduction Of Hydraulic Prototype Observation Development In YRSR1 In Past 10 Year, National Symposium On Safely Monitoring Technique on Large Dams, Hunan, China

[4] LIU KEHUANG (1988) Measuring and Judging Cavitation by Underwater Noise, National Symposium on cavitation and Its Erosion, Xinjiang, China

[5] LIU KEHUANG (1991) Approches to Discernment of Cavitation by Acoustic Characteristics, International Symposium on Modelling Cavitation Phenomena, Wuhan, China

[6] DONADL ROSS (1976) Methanics Of Under Water Noise

[7] HE ZOUYONG, ZHAO YUFANG (1981) Acoustic Principle, National Fefence Industry Publishing Company

[8] SUN QINGLONG (1982) Spectrum Analyse and Measurement, People's Telecommunication Publishing Company

[9] LIU KEHUANG (1988) Prototype Study On Sonic Vibration and Its Elimination of No.2 Ship Lock in Gezhouba Project, National Symposium on Hydraulics, Cheng Du, China

[10] HUANG BOMIN, LIU KEHUANG (1991) Operation Experiment Sumary on the 170Mw Huge Turbine of Gezhou ba Hydropower Plant, International Symposium On Hydraaulic Research In Nature And Laboratury

NUMERICAL SIMULATION FOR DILUTE
SANDY WATER FLOW IN PLANE CASCADE

X. B. LIU
Sichuan Institute of Technology
Chengdu, Sichuan, P.R. China

Q.C. ZENG
University of Queenland
Brisbane QLD 4072, Australia

L. D. ZHANG
Sichuan Institute of Technology
Chengdu, Sichuan, P.R. China

The flow of dilute sandy water (sand carrying capacity less than 1.0kg/m^3) in plane cascade is numerically solved. The collision efficiency is defined. The influence of the Reynolds number based on the characteristic length of the cascade, the angle of attack and the sand diameter on the collision efficiency are discussed. The trajectories and impinging action of sand particles moving in plane cascade flow are calculated by including the effects of the wall, the boundary layer, the wake, the sand particle size and the collision of the sand particle-wall.

1 INTRODUCTION

There are many hydraulic turbines operated in silt laden rivers in the world. The presence of solid particles in the flow often cause abrasive degradation of certain parts of the hydraulic turbine which are especially exposed to the flow. The flows in hydraulic turbine gate vane and runner are cascade flows, therefor, the study of the cascade flow is very important. In this paper, the flow of dilute sandy water in plane cascade is numerically solved. The

632

E. Cabrera et al. (eds.), Hydraulic Machinery and Cavitation, 632–640.
© 1996 Kluwer Academic Publishers. Printed in the Netherlands.

collision efficiency is determined. The influence of the Reynolds number based on the characteristic length of the cascade, the angle of attack and the sand diameter on the collision efficiency are discussed. In this solid-liquid flows, Eulerian equations are used for the description of the liquid phase dynamics, whist the solid particles are treated within the Lagrangian framework. The trajectories and impinging action of sand particles moving in plane cascade flow are calculated by including the effects of the wall, the boundary layer, the wake, the sand particle diameter and the collision of the sand particle-wall. The following assumptions are implied in the work:

(a) The particles are spherical. The physical properties of each phase are constant.

(b) The particle-particle interactions are neglected.

(c) The flow is steady.

2   BASIC MOTION EQUATIONS

2.1 Particle Motion Equations

Tchen (1947) extended the equation for the slow motion of a spherical particle in a stagnant fluid derived by Basset, Bousinesq and Oseen to the case of a particle suspended in an unsteady velocity field. Here, the modified Tchen's equation of motion for a solid particle in arbitrary flow field by Liu (1994, 1995) in the i direction is given as:

$$\frac{dV_{pi}}{dt} = \frac{1}{K_m + S}\left[\frac{3}{4d}C_D\left|\bar{V}_f - \bar{V}_p\right|(V_{fi} - V_{pi}) + \frac{3}{2d}K_B\sqrt{\frac{v_f}{\pi}}\int_{-\infty}^{t}\frac{\frac{dV_{fi}}{dt} - \frac{dV_{pi}}{dt}}{\sqrt{t-\tau}}d\tau\right.$$

$$+ \frac{3}{4}C_M\Omega_i \times (V_{fi} - V_{pi}) + \frac{6}{\pi d}K_S\left|v_f\frac{\partial V_{fj}}{\partial x_i}\right|^{1/2}(V_{fj} - V_{pj})\text{sgn}\left(\frac{\partial V_{fj}}{\partial x_i}\right)$$

$$\left. + (1 + K_m)\frac{dV_{fi}}{dt} - v_f\nabla^2 V_{fi} - (1-S)g_i\right]$$

$$(1)$$

where, V denotes the velocity; $\Omega_i = \omega_{pi} - 0.5\nabla \times V_i$ (ω_p is the angular speed of particle freely rotating); d is the particle diameter; $t,\ \tau$ denote the time; S is the particle-fluid density ratio (ρ_p/ρ_f); g is the gravitational acceleration; x_i are the components of coordinate; v is the kinematics viscosity; sgn is the sgnum function; $K_m(\approx 0.5)$, $K_B(\approx 6.0)$, $K_S(\approx 1.615)$ and $C_M(\approx 1.0)$ are the coefficients of the virtual mass, the Basset, the Saffman lift and the Magus lift forces (In main streams, the Saffman lift and the Magus lift are negligible), respectively. For dilute liquid-particle flows the particle motion drag coefficient C_D can be expressed as:

$$C_D = \begin{cases} 24/\mathrm{Re}_p, & \mathrm{Re}_p = \left|\vec{V}_f - \vec{V}_p\right| d/\nu_f < 1 \\ (24/\mathrm{Re}_p)(1+0.15\mathrm{Re}_p^{0.687}) & 1 \le \mathrm{Re}_p \le 1000 \\ 0.44 & \mathrm{Re}_p > 1000 \end{cases}$$

the subscripts i and j are the coordinate tensors. f and p refer to the fluid and the particle, respectively.

In fact, the solution of equation (1) is very difficult. The calculation flow area is generally divided into two regions: a. the main flow region (① region in Figure.1), b. the boundary layer flow and the wake flow region (② region in Figure.1). While it is a well-known fact that a spining sphere in a moving fluid experiences a lateral lift force due to its rotation (Magus effect), it is perhaps not so generally appreciated that a similar effect is present when a sphere moves through a viscous fluid

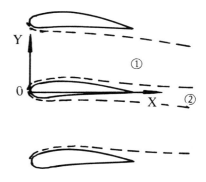

Figure.1 The coordinate of particle motion

in shear flow. The formar mechanism is known as "slip-spin" motion while the latter is known as "slip-shear" motion. At low Reynolds number, Saffman [1965] showed that, for a freely rotating particle, unless the rotating velocity is very much greater than the rate of shear, the lift force caused by the particle rotation is less by an order of magnitude than that caused by the shear. Recently, using the laser photography technique, the result of the particle motion research shows that the particle do not rotate at most areas in the flow field.

Hence, the Saffman lift and the Magus lift as well as $-\nu_f \nabla^2 V_{fi}$ terms in equation (1) can be neglected for ① region, and the particle reduced equation may be expressed as

$$\frac{dV_{pi}}{dt} = \frac{1}{\tau_p}(V_{fi} - V_{pi}) + \frac{6\beta}{d}\sqrt{\frac{\nu_f}{\pi}}\int_{-\infty}^{t}\left(\frac{dV_{fi}}{d\tau} - \frac{dV_{pi}}{d\tau}\right)\frac{d\tau}{\sqrt{t-\tau}} + \beta\frac{dV_{fi}}{dt} \qquad (2)$$

where, τ_p is the particle response time, $\beta = 3/(2S+1)$.

For ② region, we consider the coordinate system which is that the $x0$-coordinate is the surface distance measured from the O point and the $y0$-coordinate is measured normal to the surface. Because of effects of viscous and the particle rotating velocity is very low in the boundary layer. Thus, In this region, the effect of the Magus lift force and the component of Saffman lift force in the $x0$-direction are neglected. Hence, in ② region, the Lagrangian equations of particle motion in the $x0$ and $y0$ directions can now be expressed as, respectively,

$$\frac{dV_{px0}}{dt} = \frac{C_{x0}}{\tau_p}(V_{fx0} - V_{px0}) + \frac{6\beta}{d}\sqrt{\frac{v_f}{\pi}}\int_{-\infty}^{t}\left(\frac{dV_{fx0}}{d\tau} - \frac{dV_{px0}}{d\tau}\right)\frac{d\tau}{\sqrt{t-\tau}} + \beta\frac{dV_{fx0}}{dt} - \frac{2}{3}\beta v_f \nabla^2 V_{fx0} \qquad (3)$$

$$\frac{dV_{py0}}{dt} = \frac{C_{y0}}{\tau_p}(V_{fy0} - V_{py0}) + \frac{6\beta}{d}\sqrt{\frac{v_f}{\pi}}\int_{-\infty}^{t}\left(\frac{dV_{fy0}}{d\tau} - \frac{dV_{py0}}{d\tau}\right)\frac{d\tau}{\sqrt{t-\tau}} + \beta\frac{dV_{fy0}}{dt} - \frac{2}{3}\beta v_f \nabla^2 V_{fy0}$$

$$+ \frac{6.46}{\pi d}\beta\left|v_f\frac{\partial V_{fy0}}{\partial x0}\right|^{1/2}(V_{fy0} - V_{py0}) \qquad (4)$$

where, C_i is the component of correction factors which account for the effects of the wall on the drag. We adopt the following expressions for C_{x0} and C_{y0}:

$$C_{x0} = (1 - \frac{9}{16}\delta + \frac{1}{8}\delta^2 - \frac{45}{246}\delta^4 - \frac{1}{16}\delta^5)^{-1}, C_{y0} = 1 + \frac{9}{8}\delta + \frac{81}{256}\delta^2 \qquad (5)$$

where $\delta = d/(2y0)$, $y0$ denotes the coordinate normal to the wall.

2.2 Fluid Motion Equation

Before the particle equation (1) of motion is solved, the velocity and the pressure distributions of fluid fields must be obtained according to the particle equation (1) of motion. For this cascade flow, the velocity and the pressure distributions of fluid fields are calculated with the aid of a 2-D numerical approximation solving the N-S equations.

3   PARTICLE REBOUND MODE

After impacting a solid boundary, the magnitude and direction of the particle rebounding velocity is dependent on the particle material, the surface material, and on the impact conditions. The experimental studies were carried out in special erosion tunnels, which were designed to include the aerodynamic effects in the rebound characteristics. The following empirical relations (Tabakoff and Hamed 1977) for the rebound-to-impact restitution rations were used in the trajectory calculates:

$$\begin{cases} V_{py02}/V_{py01} = 1.0 - 0.4159\beta_1 - 0.4994\beta_1{}^2 + 0.292\beta_1{}^3, \\ V_{px02}/V_{px01} = 1.0 - 2.12\beta_1 + 3.0775\beta_1{}^2 - 1.1\beta_1{}^3 \end{cases} \tag{6}$$

The rebounding velocity magnitude V_2 and the rebounding angle β_2 are calculated using the following expressions:

$$V_{p2} = \sqrt{V_{py02}{}^2 + V_{px02}{}^2}, \quad \beta_2 = \mathrm{tg}^{-1}(V_{py02}/V_{px02}) \tag{7}$$

where, V_{py0} and V_{px0} represent the particle velocity components normal and tangent to the solid surface, and the subscripts 1 and 2 refer to the conditions before and after impact, respectively. In the above equations, β is the angle between the velocity and the tangent to the surface.

4 DEFINITION OF COLLISION EFFICIENCY

To a great extent, the collision efficiency η indicate the erosive wear rate of the duct surface. Hence, the collision efficiency η is an important parameter. The definition of the collision efficiency η is that at unit time the total number N_I of the particle impacting the duct surface is divided by the total number N_V of the particle in the flow field. That is

$$\eta = \frac{N_I}{N_V} \tag{8}$$

5 NUMERICAL SIMULATION

5.1 Boundary and Initial Conditions

(a) Boundary of fluid phase:
The coming flow velocity $V_{f\infty}$ of the fluid at infinity is given. The velocity of the fluid at the wall can be written as:

$$V_{fi} = 0 \tag{9}$$

(b) Initial conditions of particulate phase:

The relations of the particle and the fluid initial velocity components can be expressed as:

$$V_{pi0} = e_0 V_{fi0} \tag{10}$$

where e_0 denotes a initial tracking coefficient of particle to fluid. The larger particle size and the weightier particle, the smaller e_0. If the particle impact the solid wall, the reflection of the particle must be considered in terms of particle rebound model.

5.2 Nnmerical Simulation Method of Particle Motion

To simulation the motion of suspended particle, the particle initial position and velocity need to be given. A fourth-order Runge-Kutta scheme is used to integrate numerically the Lagrangian particle equations of motion. The Basset history term in the equation is singular at $\tau = t$ and its discretization needs particular attention, to include its effect on the particle motion accurately, it is discreatized in the following manner,

$$\int_0^t \left(\frac{dV_{fi}}{d\tau} - \frac{dV_{pi}}{d\tau} \right) \frac{d\tau}{\sqrt{t-\tau}} = 2 \sum_{k=0}^{K-1} \left\{ \frac{1}{\sqrt{\Delta t}} \left[V_{fi}(k+1) - V_{fi}(k) - V_{pi}(k+1) + V_{pi}(k) \right] \right.$$
$$\left. \cdot (\sqrt{K-k} - \sqrt{K-k-1}) \right\} \tag{11}$$

where, Δt is the time step, $V(k)$ is the velocity at time $k \Delta t$, $K \Delta t$ is the time duration t.

Also using the Runge-Kutta integration technique, the trajectories of aparticle can be found from its velocity components,

$$\frac{dx_i}{dt} = V_{pi} \tag{12}$$

Knowledge of particle velocity and trajectories make possible extrapolation of incidence velocities and angles at the impact locations. Then erosive wear rates may be determined with the erosive wear model.

6 NUMERICAL RESULT AND DISCUSSION

In this study, the cascade spacing T of the plane cascade is 0.5m. The wing chord length is 0.4m. The coming flow impact angles α_0 are given as -10°, -5°, 0°, 5°, 10°. The particle sizes d are 50μm, 100μm, 500μm, 800μm,

1mm. The sandy particle density ρ_p is $2650 \text{kg}/\text{m}^3$. The water density ρ_f is $1000 \text{kg}/\text{m}^3$. The water kinematics viscosity of the water v_f is $1.006 \times 10^{-6} \text{m}^2/\text{s}$. The particle initial position X_0 is -0.5m. The initial tracking coefficient e_0 of particle to fluid is 0.9. The coming flow Reynolds Re_L are 7.0×10^5, 1.4×10^6 and Re_L is defined as

$$\text{Re}_L = \frac{V_{f\infty} L}{v_f} \tag{13}$$

where L is the characteristic length of the body (in this paper, L is the largest thickness of the profile, that is 0.07m)

Figure 2 shows the motion of particles in the cascade flow.. It can be clearly seen from the figure that the impact region mainly occurs in the profile head and the suction pressure side of the profile tail for the larger positive impact angle, and in the profile head and the pressure side of the profile tail for the larger negative impact angle. The vortex flow near the profile tail leads to increasing collision of particles.

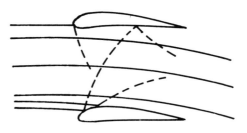

(a) the coming flow impact angle $\alpha_0 = 0°$

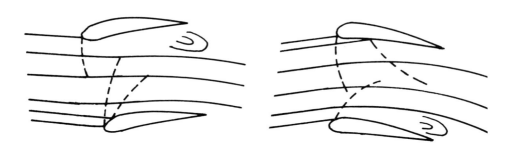

(b) the coming flow impact angle $\alpha_0 = -10°$ (c) the coming flow impact angle $\alpha_0 = 10°$

Figure. 2   The motion of particles in the cascade flow for
the coming flow impact angle $\alpha_0 = 0°$, $-10°$, $10°$
$\text{Re}_L = 1.4 \times 10^6$, d=500μm,   – – – the particle rebounding trajectories